Advances in Atom and Single Molecule Machines

For further volumes:
http://www.springer.com/series/10425

Fabio Benfenati · Enzo Di Fabrizio
Vincent Torre
Editors

Novel Approaches for Single Molecule Activation and Detection

Springer

Editors
Fabio Benfenati
Department of Neuroscience and Neurotechnology
University of Genoa, The Italian Institute of Technology
Genoa
Italy

Vincent Torre
Neuroscience Area
Scuola Internazionale Superiore di Studi
Trieste
Italy

Enzo Di Fabrizio
Department of Nanostructures
The Italian Institute of Technology
Genoa
Italy

ISSN 2193-9691 ISSN 2193-9705 (electronic)
ISBN 978-3-662-43366-9 ISBN 978-3-662-43367-6 (eBook)
DOI 10.1007/978-3-662-43367-6
Springer Heidelberg New York Dordrecht London

Library of Congress Control Number: 2014943416

Printed on acid-free paper

Springer is part of Springer Science+Business Media (www.springer.com)

Preface

This book provides a selection of single molecule investigations, presented at the International meeting held at Palazzo Franchetti in Venice on "Single Molecule Activation and Detection in Molecular Devices" on October 13–14, 2012. This meeting gathered scientists active in developing tools and methodologies for single molecule analysis and was organized within the FOCUS project, sponsored by the EU FP7 FET Proactive 7 Programme: "Molecular Scale Devices and Systems [ICT-2009.8.7]" (grant agreement n, 270483).

The book captures selected contributions from members of the European "Focus" consortium as well as from other distinguished scientists operating in the field of single molecule analysis. The contributions collected here provide not only a good view of the activities of the FOCUS project, but also offer a wide spectrum of novel investigations in an emerging and important field at the interface between material science and biology. Andolfi and Lazzarino have modified the methods of SMFS to develop Single Cell Force Spectroscopy for analysing adhesion processes in cells and neurons. Bauer, Kendziora, Ahmed, Hung and Fruk describe the use of DNA stretches used to tether the computing molecules in a protoptype of the molecular device developed within the FOCUS project. Bianco, Bongini, Melli, Falorsi, Salvi, Cojoc and Lombardi have introduced the force clamp methodology in the optical tweezers apparatus so to improve the detection and resolution of chemical reactions. Dal Maschio describes the development of holographic beams of light able to reveal and analyse molecular properties at a very fine scale and to characterize also circuit properties. Diaspro, Bianchini, Cella Zanacchi and Vicidomini describe how it is possible to detect single molecule using conventional far field illumination by using stochastic reading methods. Endeman, Feyen, Ghezzi, Antognazza, Martino, Colombo, Lanzani and Benfenati show how, by using light sensitive organic semiconductors, it is possible to modulate and control the firing of neurons and neuronal networks. Izquierdo-Serra and Gorostiza describe the recent progress in optogenetics in which endogenous ion channels can be rendered light sensitive by the introduction of a light switchable gate. Petrini and Barberis have attached quantum dots to synaptic receptors and in this way they can monitor and follow the lateral diffusion of these membrane proteins in their physiological environment. Limongi, Ferrara, Das, Moretti, Marini, Miele, Accardo, Raimondo,

Gentile and Di Fabrizio make use of superhydrophobicity to develop a variety of novel devices with which they can detect single or very few molecules. Finally, Naumenko, Cassese, Lazzarino and Bek illustrate how it is possible to detect and activate single molecules by using the special properties of tips with nanometer dimensions.

The Venice workshop was very stimulating and inspiring for all the participants, and we hope that this book will be of interest and inspire its readers.

Acknowledgement

This book would not have been possible without the invaluable help and dedication of Dr. Tania Limongi, who gave a fundamental contribution in coordinating and assembling the above-listed contributions.

Fabio Benfenati
Enzo Di Fabrizio
Vincent Torre

Contents

Contributors

Angelo Accardo Nanostructures, Istituto Italiano di Tecnologia, Genoa, Italy

Ishtiaq Ahmed Center for Functional Nanostructures, Karlsruhe Institute of Technology, Karlsruhe, Germany

Laura Andolfi CBM S.c.r.l. Basovizza, Trieste, Italy; CNR-IOM Basovizza, Trieste, Italy

Maria Rosa Antognazza Center for Nanoscience and Technology @PoliMi, Istituto Italiano di Tecnologia, Milan, Italy

Andrea Barberis Department of Neuroscience and Brain Technologies, Italian Institute of Technology (IIT), Genoa, Italy

Dennis M. Bauer Center for Functional Nanostructures, Karlsruhe Institute of Technology, Karlsruhe, Germany

Antoni Bautista-Barrufet Institut de Bioenginyeria de Catalunya (IBEC), Barcelona, Spain; Institut Català d'Investigació Química (ICIQ), Tarragona, Spain

Alpan Bek Middle East Technical University, Cankaya Ankara, Turkey

Fabio Benfenati Department of Neuroscience and Brain Technologies, Istituto Italiano di Tecnologia, Genoa, Italy; Department of Experimental Medicine, University of Genova, Genoa, Italy

Paolo Bianchini NanoBioPhotonics—LAMBS, Nanophysics, Istituto Italiano di Tecnologia, Genoa, Italy

Pasquale Bianco Laboratory of Physiology, Department of Biology, University of Florence, Florence, Italy

Lorenzo Bongini Laboratory of Physiology, Department of Biology, University of Florence, Florence, Italy

Damiano Cassese IOM-CNR-Laboratorio TASC, Area Science Park—Basovizza, Trieste, Italy

Dan Cojoc IOM—National Research Council, Trieste, Italy

Elisabetta Colombo Department of Neuroscience and Brain Technologies, Istituto Italiano di Tecnologia, Genoa, Italy

Marco Dal Maschio Department of Neuroscience and Brain Technologies, Fondazione Istituto Italiano di Tecnologia, Genoa, Italy; Department of Genes, Circuits and Behavior, Max Planck Institute for Neurobiology, Martinsried, Germany

Gobind Das Physical Science and Engineering Division, King Abdullah University of Science and Technology, Thuwal, Kingdom of Saudi Arabia

Alberto Diaspro NanoBioPhotonics—LAMBS, Nanophysics, Istituto Italiano di Tecnologia, Genoa, Italy

Duco Endeman Department of Neuroscience and Brain Technologies, Istituto Italiano di Tecnologia, Genoa, Italy

Enzo Di Fabrizio Physical Science and Engineering Division, King Abdullah University of Science and Technology, Thuwal, Kingdom of Saudi Arabia

Giulia Falorsi Laboratory of Physiology, Department of Biology, University of Florence, Florence, Italy

Lorenzo Ferrara Nanostructures, Istituto Italiano di Tecnologia, Genoa, Italy

Paul Feyen Department of Neuroscience and Brain Technologies, Istituto Italiano di Tecnologia, Genoa, Italy

Ljiljana Fruk Center for Functional Nanostructures, Karlsruhe Institute of Technology, Karlsruhe, Germany

Francesco Gentile Nanostructures, Istituto Italiano di Tecnologia, Genoa, Italy; Laboratory BIONEM, Dipartimento di Medicina Sperimentale e Clinica, Università degli Studi Magna Graecia di Catanzaro, Catanzaro, Italy

Diego Ghezzi Department of Neuroscience and Brain Technologies, Istituto Italiano di Tecnologia, Genoa, Italy

Pau Gorostiza Institut de Bioenginyeria de Catalunya (IBEC), Barcelona, Spain; Centro de Investigación Biomédica en Red en Bioingeniería Biomateriales y Nanomedicina (CIBER-BBN), Zaragoza, Spain; Institució Catalana de Recerca i Estudis Avançats (ICREA), Barcelona, Spain

Yu-Chueh Hung Institute of Photonics Technologies, National Tsing Hua University, Hsinchu, Taiwan

Mercè Izquierdo-Serra Institut de Bioenginyeria de Catalunya (IBEC), Barcelona, Spain

Dania M. Kendziora Center for Functional Nanostructures, Karlsruhe Institute of Technology, Karlsruhe, Germany

Guglielmo Lanzani Center for Nano Science and Technology @PoliMi, Istituto Italiano di Tecnologia, Milan, Italy; Department of Physics, Politecnico di Milano, Milan, Italy

Marco Lazzarino CBM scrl Area Science Park—Basovizza, Trieste, Italy; IOM-CNR-Laboratorio TASC, Area Science Park—Basovizza, Trieste, Italy

Tania Limongi Nanostructures, Istituto Italiano di Tecnologia, Genoa, Italy; Neuroscience and Brain Technology, Istituto Italiano di Tecnologia, Genoa, Italy; Physical Science and Engineering Division, King Abdullah University of Science and Technology, Thuwal, Kingdom of Saudi Arabia

Vincenzo Lombardi Laboratory of Physiology, Department of Biology, University of Florence, Florence, Italy

Monica Marini Physical Science and Engineering Division, King Abdullah University of Science and Technology, Thuwal, Kingdom of Saudi Arabia

Nicola Martino Center for Nano Science and Technology @PoliMi, Istituto Italiano di Tecnologia, Milan, Italy; Department of Experimental Medicine, University of Genova, Genoa, Italy

Monica Mazzolini Neuroscience Area, Scuola Internazionale Superiore di Studi Avanzati (SISSA), Trieste, Italy; CBM S.c.r.l. Basovizza, Area Science Park, Trieste, Italy

Luca Melli Laboratory of Physiology, Department of Biology, University of Florence, Florence, Italy

Ermanno Miele Nanostructures, Istituto Italiano di Tecnologia, Genoa, Italy

Manola Moretti Nanostructures, Istituto Italiano di Tecnologia, Genoa, Italy; Computer, Electrical and Mathematical Sciences and Engineering (CEMSE) Division, King Abdullah University of Science and Technology, Thuwal, Kingdom of Saudi Arabia

Denys Naumenko CBM scrl Area Science Park—Basovizza, Trieste, Italy; IOM-CNR-Laboratorio TASC, Area Science Park—Basovizza, Trieste, Italy

Enrica Maria Petrini Department of Neuroscience and Brain Technologies, Italian Institute of Technology (IIT), Genoa, Italy

Raffaella Raimondo Laboratory BIONEM, Dipartimento di Medicina Sperimentale e Clinica, Università degli Studi Magna Graecia di Catanzaro, Catanzaro, Italy

Luca Salvi Laboratory of Physiology, Department of Biology, University of Florence, Florence, Italy

Vincent Torre Neuroscience Area, Scuola Internazionale Superiore di Studi Avanzati (SISSA), Trieste, Italy

Giuseppe Vicidomini NanoBioPhotonics—LAMBS, Nanophysics, Istituto Italiano di Tecnologia, Genoa, Italy

Francesca Cella Zanacchi NanoBioPhotonics—LAMBS, Nanophysics, Istituto Italiano di Tecnologia, Genoa, Italy

Introduction to Single-Molecule Analysis and Computation: The Focus Project

Monica Mazzolini and Vincent Torre

The FP7 project *FOCUS* "Single-molecule activation and detection" belongs to the FET Proactive 7 Programme: "Molecular Scale Devices and Systems" and is aimed at investigating and developing molecular devices (MD) where single molecules become computing elements. X-ray diffraction and NMR spectroscopy provides information on the structure of molecules with an atomic resolution, but the generated data are averaged over millions of molecules. Single-molecule analysis, in contrast, investigates properties of individual molecules not averaged over a large ensemble of similar molecules. The great majority of these properties, and of the related measurements, refers to nonequilibrium conditions. In order to obtain classical thermodynamic quantities such as changes of the free energy ΔF, special methods of statistical mechanics, such as those introduced by Jarzinski [5, 6] can be used. Single molecule experiments provide a direct view of molecular events in action and offer the possibility to verify basic notions of chemical reactions such as Kramers theory and Eyring's transition state theory. Indeed, it is possible to characterize very precisely Kramers diffusion coefficient and free-energy barrier by measuring the temperature and viscosity dependence of the transition path time for protein folding [2].

The first single-molecule analyses were those obtained with patch-clamp recordings [3] providing a direct measurement of the current flowing through a single ion channel. These recordings completely changed electrophysiology and our way to understand the operation of ion channels. Subsequently, single-molecule analysis was extended to probe the folding and unfolding dynamics of single

M. Mazzolini · V. Torre (✉)
Neuroscience Area, Scuola Internazionale Superiore di Studi Avanzati (SISSA), via Bonomea, 265, 34136 Trieste, Italy
e-mail: torre@sissa.it

M. Mazzolini
CBM S.c.r.l. Basovizza, Area Science Park, 34139 Trieste, Italy

F. Benfenati et al. (eds.), *Novel Approaches for Single Molecule Activation and Detection*, Advances in Atom and Single Molecule Machines,
DOI: 10.1007/978-3-662-43367-6_1,

molecules by using force spectroscopy [4, 8, 11] either by optical tweezers (OT) [9] or by atomic force microscopy (AFM) [1]. In force spectroscopy, a molecule is mechanically stretched and its elastic response is measured in real time.

Single molecule force spectroscopy (SMFS) is an application of AFM and consists in the application of the force (F) required to unbind a ligand or to unfold a polymer while the distance (D) between the AFM cantilever tip and the sample is measured as the coordinate reaction, with pN and nm resolution. Force-distance (F-D) curves characterizing the stretching of a protein allow the identification of folded and unfolded regions giving insight in the secondary structure [4].

An OT is composed of a highly focused laser beam creating an optical trap in which samples and biological specimen [9] are analysed. This optical trap is equivalent to a restoring force i.e. a simple spring, which follows the well-known Hooke's law. Therefore by knowing the stiffness of the spring, the OT can be used to measure forces in the pN range. An OT is capable of manipulating nanometer and micron-sized dielectric particles with a very high precision in the nm range and is used to manipulate and study single molecules by interacting with a bead that has been attached to that molecule.

Fluorescence can be used to observe one molecule at a time by using photomultiplier tubes, or avalanche photodiodes or special-purpose CCD cameras. Single molecule spectroscopy usually requires the attachment of a fluorescent probe to the molecule under investigation, which can be quantum dots (QD), gold nanoparticles or organic fluophores [7, 13] Single molecule fluorescence resonance energy transfer (smFRET) [12] is used to measure distances at the 1–10 nm scale and is based on a single donor and acceptor FRET pair. Data collection with cameras will then produce movies of the specimen that must be processed to derive the single-molecule properties.

Single fluorophore can be chemically attached to proteins or DNA, and the dynamics of individual molecules can be tracked by monitoring the fluorescent probe. For instance, single molecule labelling has yielded a vast quantity of information on how molecular motors move along in cells and over microtubules. A typical fluorophore used in many biological applications is a QD, which is a semiconductor nanocrystal with excellent optical properties. Antibodies, streptavidin and DNA can be used to target QDs to specific proteins in cells and neurons.

The core of the FOCUS project is to address a long standing issue of programmable assembly, which Nature masters on a fine molecular level, but scientists have not yet achieved: Nature has developed highly sophisticated systems able to detect single photons and molecules and to perform reliable computations in a noisy environment dominated by thermal motion. Can we design and assemble molecular devices with similar properties?

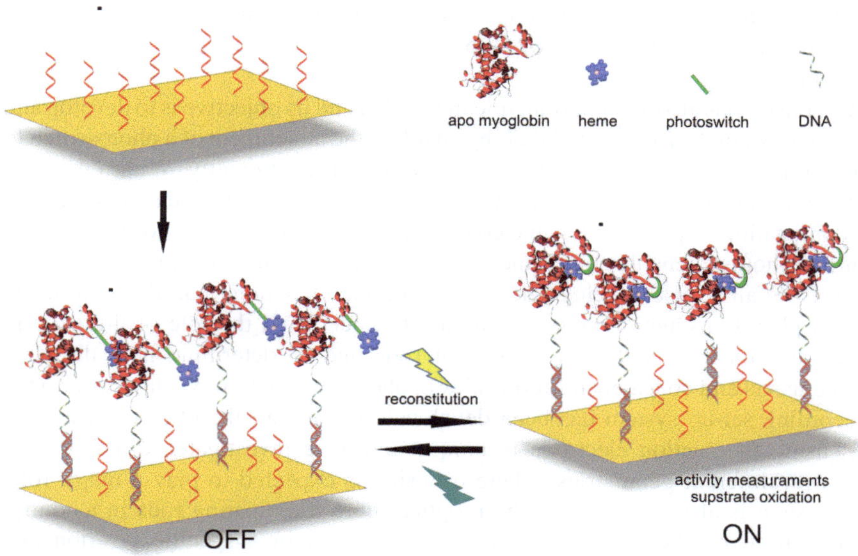

The diagram of the MD developed within the FOCUS project. (Fruk L., FET-NEWS, 2013)

This project wants to build a novel generation of biologically inspired MDs based on the developments of new photonic devices. These devices will use plasmon polariton and two-photon technology, enabling focused light spots with a diameter around 10 nm, i.e. *the size of a single protein.* FOCUS is also developing new light-sensitive molecules that are selectively activated by our new photonic devices. These new technological innovations will provide a way to control activation of single light-sensitive molecules and will allow the investigation of molecular computation in a biological environment and with an unprecedented resolution. A guiding idea beyond the FOCUS project is diffusion, which is seen as a basic component of biological computation.

FOCUS has formed a highly interdisciplinary consortium composed of nano-technologists able to fabricate the new photonic devices—i.e. Enzo Di Fabrizio (IIT), Alpan Bek (METU) and Marco Lazzarino (CBM), chemists able to develop the photoswitches and assemble the MDs—i.e. Pau Gorostiza (IBEC) and Ljiljana Fruk (KIT) and biologists able to understand molecular mechanisms—i.e. Vincent Torre (SISSA) and Fabio Benfenati (IIT). The two companies RappOptoElectronic and NT-MDT Europe BV will transform the new tools and devices into marketable products.

The nanotechnological component of FOCUS has the objective to design and fabricate new photonic plasmonic devices able to provide a well-focused beam of light with a diameter of 10 nm, close to the size of a single molecule. In order to identify the best possible geometrical configuration capable of strong light focusing with low background signal, simulations on three different systems were performed. In particular, we have chosen (1) single nanowire antenna, (2) sub-wavelength holes in metallic slab (3) hollow conical structure. The three devices were also fabricated by means of a combination of focus ion beam and electron beam lithography.

Chemistry has also an important role in FOCUS and its objective is to develop and validate procedures for affinity labelling enabling the attachment of a photoswitch to different proteins, and to evaluate their toxicity and functionality. Within the consortium, chemical and biological assays were implemented and validated to evaluate the conjugation of photoswitchable compounds to the kainate receptor GluK2, and four full photoswitchable compounds were designed, synthesized and assayed.

FOCUS aims also to elucidate how molecular reactions based on diffusion control phototransduction and in particular to understand the role of diffusion in signal amplification and gain control in rods and to determine how diffusion interconnects and concatenates molecules in phototransduction. Within the project, a recording set-up system has been developed to measure the photocurrent from rods using the suction electrode technique and have investigated differences in rods with different geometries, where diffusion is expected to play a major role. This system is also combined with an optical fibre illuminating rods in a configuration allowing the stimulation of few rhodopsins in a specific position and timing. These experiments are expected to solve long standing issues in phototransduction.

These novel probes provide also the opportunity to analyse the biological effects of the activation of a reduced number of ion channels, receptors and proteins possibly at a single molecule level. In this case, it is mandatory to have light beams with well defined wave-front profiles. In order to use light-sensitive molecules for single molecule analysis, it is important and mandatory to be able to have very restricted beams of light, exciting a single molecule or a confined and selected region. Usual light beams obtained with conventional optical components like optical fibres, have Gaussian profiles with characteristic side lobes, which do not allow the controlled excitation of a single light-sensitive molecule. Another result obtained within the FOCUS project is the observation that peculiar light exits from "apertureless" tapered optical fibres (TOFs), commercially available. What is more important is that by using TOFs fed by a laser beam it is possible to have confined spots of light with almost no side lobes, and we demonstrated that their use reveal new fundamental features of phototransduction in vertebrate rods that contain the rhodopsin, a molecule intrinsically light-sensitive at the basis of vision in the animal kingdom. A rod cell is subdivided in a synaptic region, an inner segment (IS) and an outer segment (OS) that contains a stack of lipid discs surrounded by a plasma membrane. In dark condition, cyclic nucleotide-gated (CNG) channels that are localized in the plasma membrane are open and sodium and calcium ions enter inside the cell but this current is abolished in presence of light; in fact when photons reach rhodopsin molecules that are densely packed on disc membranes—phototransduction starts [10]. Using an apertureless TOF, we deliver restricted illumination at various positions of the OS at the base, middle and tip. Our results show that by using apertureless TOFs novel properties of phototransduction are obtained demonstrating that there are differences along the

OS. These apertureless TOFs can also be used in optogenetics for the activation of light-sensitive ionic channels in specific regions of a neuron, such as distal dendrites and/or single spines. These TOFs allow the activation of a very limited number of light-sensitive receptors and proteins and by improving their performances a single light-sensitive molecule could be excited. These TOFs could become useful tools for single-molecule investigations.

During the first period of the project's life, the consortium also identified and validated experimental configurations aiming to the activation of subsets of receptors in the perspective to track the triggered diffusion processes and to reveal basic computational mechanisms involving the neuronal input processes. In the same period of time, the input-output relationships in molecular computation have been investigated. A theoretical study of the input–output (stimulus-response) relationships using both analytical methods from statistical physics and applied mathematics and numerical simulations was performed.

Within the consortium, scientists and engineers have optimized the integration of the AFM cantilevers with plasmonic antennas. We have produced PC–PA (photonic crystal–plasmonic antenna) directly on AFM cantilever by combining focused ion beam milling and induced electron growth. Partners have examined, tested and investigated the properties of PC–PA cantilevers using their modified commercial AFM.

The core of the FOCUS project is where the new photonic devices, the photoswitchable gates and the biological investigation of diffusion are integrated in the prototype of the new molecular computing device, schematically represented in the figure above. DNA-proteins that conjugate for attachment onto the surface and where the length of attachment controls the diffusion of anchored proteins have been developed. The activation of these proteins is controlled by light through use of azobenzene photoswitch, and the new photonic devices developed within the project. A set of DNA oligonucleotides was designed so that the length of the DNA can easily be readjusted if needed for future studies of diffusion and cofactor reconstitution. Regarding protein modification, several mutant myoglobin (Mb) proteins have been expressed and purification methods optimized. Mutants were designed as the native Mb did not have a suitable anchoring group for attachment of two different classes of molecules. In addition, new chemical procedures were developed for attachment of modified DNA strands onto tyrosine residues of the protein as the crystallographic data analysis has shown that the tyrosine residues are a suitable target for DNA attachment.

Another objective is the development of the capability to concentrate light to very small spots of light in the 10 nm range using two photon plasmonic engineering. In accordance with the two tasks of building a nanoparticle-based molecular switch and addressing the individual molecular switch, we have the objectives to fabricate a substrate with a dense array of metal nanoparticles and to set-up an instrument capable of performing nano-manipulation. At the same time, the project also has the objective to realize an optical characterization tool for testing plasmonic properties of the nanoparticle decorated substrates.

The methodologies developed within FOCUS will certainly have wide implications not only to study biological processes but also in the fields of photonics, single molecule manipulation and detection, chemical modification and programmable assembly.

References

1. Binning G, Quate CF, Gerber Ch (1986) Phys Rev Lett 56(9):930–935
2. Chung & Eaton (2013) Single-molecule fluorescence probes dynamics of barrier crossing. Nature 502:685–689
3. Hamill OP, Marty A, Neher E, Sakmann B, Sigworth FJ (1981) Improved patch-clamp techniques for high-resolution current recording from cells and cell-free membrane patches. Pflugers Arch 391(2):85–100
4. Hoffmann T, Dougan L (2012) Single molecule force spectroscopy using polyproteins. Chem Soc Rev 41(14):4781–4796
5. Hummer G, Szabo A (2001) Free energy reconstruction from nonequilibrium single-molecule pulling experiments. Proc Natl Acad Sci 98(7):3658–3661
6. Hummer G, Szabo A (2010) Free energy profiles from single-molecule pulling experiments. PNAS 107(50):21441–22144
7. Michalet X, Pinaud FF, Bentolila LA, Tsay JM, Doose S, Li JJ, Sundaresan G, Wu AM, Gambhir SS, Weiss S (2005) Quantum dots for live cells, in vivo imaging ad diagnostics. Science 307:538–544
8. Müller DJ, Wu N, Palczewski K (2008) Vertebrate membrane proteins: structure, function, and insights from biophysical approaches. Pharmacol Rev 60(1):43–78
9. Neuman KC, Block SM (2004) Optical trapping. Rev Sci Instrum 75(9):2787–2809
10. Pugh EN Jr, Lamb TD (2002) Phototransduction in vertebrate rods and cones: molecular mechanisms of amplification, recovery and light adaptation. In: Handbook of biological physics. Elsevier Amsterdam, pp 183–255
11. Rief M, Gautel M, Oesterhelt F, Fernandez JM, Gaub HE (1997) Reversible unfolding of individual titin immunoglobulin domains by AFM. Science 276(5315):1109–1112
12. Roy R, Hohng S, Ha T (2008) A practical guide to single-molecule FRET. Nat Methods 5(6):507–512
13. Zhang J, Fu Y, Conroy CV, Tang Z, Li G, Zhao RY, Wang G (2012) Fluorescence intensity and lifetime cell imaging with luminescent gold nanoclusters. J Phys Chem C Nanomater Interfaces 116(50):26561–26569

Computer-Generated Holographic Beams for the Investigation of the Molecular and Circuit Function

Marco Dal Maschio

Abstract One of the goals of the modern research is to understand the signal processing and information propagation in biological systems clarifying the peculiar contribution of the different components in order to explore novel paradigms of computation and circuit wiring. For several years, the research in the field of neuroscience took advantage of a large set of fluorescence-based sensors to study the activity patterns ongoing in the neuronal circuits within the brain and the molecular mechanism underlying them. However, only recently, the advent of light-triggered molecular actuators has allowed a change in the research perspective moving the focus toward the investigation of the causal relations involved in the dynamics of the neuronal networks and in their function-optimized wiring. In parallel to the development of light-triggered molecules, scientists are looking for effective optical methods to increase the accuracy they could achieve by means of the light in the control of these tools and in the conditioning of the brain circuits. Moreover, the optical approaches, with respect to other investigation methods, are unique in allowing for a minimal-invasive investigation of the molecular mechanisms of the signal transduction underlying the cellular activity and the neuronal circuit function. In this perspective, computer-generated holographic beams, where the phase of the light wave front is controlled to obtain a specific light intensity distribution within the sample space, represent a parallel, dynamic, and powerful tool to investigate the molecular mechanism and the function of neuronal circuits.

M. Dal Maschio (✉)
Department of Neuroscience and Brain Technologies, Fondazione Istituto Italiano di Tecnologia, Genoa, Italy
e-mail: dalmaschio@neuro.mpg.de

M. Dal Maschio
Department of Genes, Circuits and Behavior, Max Planck Institute for Neurobiology, Martinsried, Germany

F. Benfenati et al. (eds.), *Novel Approaches for Single Molecule Activation and Detection*, Advances in Atom and Single Molecule Machines,
DOI: 10.1007/978-3-662-43367-6_2, © Springer-Verlag Berlin Heidelberg 2014

1 Introduction

Many optical techniques share the common property of being based on different methods for engineering the electric field distributions in the three-dimensional sample space to achieve specific illumination profiles. The generation of the doughnut-shaped intensity profile in stimulated emission depletion (STED) microscopy [1], the induction at the cell-substrate interface of an evanescent electric field in total internal reflection fluorescence (TIRF) [2, 3] or the enhancement of the near-field components in scanning near-field microscopy [4], the grating illumination patterns used for structured-illumination microscopy (SIM) [5], and the use of light sheets in low-scattering samples [6] are a few of the more clear examples. The optimization of the light distribution is usually leading per se or in combination with other cofactors to a significant enhancement of important figures in the signal detection such as the achievable optical resolution, the signal-to-noise ratio, the acquisition speed, the amount of accessible volume, or the number of accessible features of interest. The signals acquired from biological systems with these techniques come in most of the cases from specific labeling of molecules of interest, from synthetic or genetically encoded sensors of the cellular activity or of the molecular dynamics, and from reporters of the change in the cellular and extracellular environment, whose detection is based frequently on the excitation-emission properties of fluorescent moieties. Such fluorescent molecular tools allow for an extremely precise understanding of the cellular function and are continuously opening new windows in the characterization of the mechanisms underlying the brain circuits [7].

Optical methods in neuroscience, with the exception of the applications of the optical tweezers [8, 9], traditionally have been focused on the imaging aspects, in other words on monitoring the status of the sample in a short time interval or over a longer temporal range to characterize the changes induced by external stimuli. Until recently, optical methods for controlling molecular mechanisms and neuronal activity lagged behind methods for monitoring [10]. But, in the past few years, an impressive set of photochemical tools has become available to the purpose, including caged neurotransmitters, natural photosensitive proteins and small-molecule photoswitches introducing light sensitivity in ion channels and receptors [10]. Caged molecules are based on a photolabile protecting group, which when is removed upon light illumination liberate a bioactive compound. MNI-caged glutamate is now the most popular form of caged glutamate [11], with good 2P cross section and very low rate of spontaneous event in the dark. Caged versions for many other neurotransmitters have also been published for GABA [12] and anandamide [13]. On the group of natural photosensitive protein, ChR2 channel rhodopsin 2 [14], a light-sensitive cation channel from algae Chlamydomonas, is the most common example. In this kind of molecules, a trans to cis structural change in the chromophore induced by visible light absorption leads to a transient conformational rearrangement of the transmembrane protein, making it permeable to free ions and very suitable for a fine temporal control of the neuronal activation [15]. Many other different types have been discovered either for excitation or for inhibition of neuronal firing [16]. More traditional examples of this class are the retinal photoreceptors, where rhodopsin and their associated proteins regulate ion channel activity through G protein biochemical cascade. The third class, the one named small-molecule photoswitches, includes tools rationally designed and synthetically manufactured to couple photosensitivity to non-intrinsically photosensitive ion channels or receptors. The photosensitivity is introduced by coupling a photoisomerizable molecule into ordinary channel in such a way that the light absorption induces the delivery or the removal of the ligand for the receptor binding site, thereby regulating its activity.

The most representative molecules of this group is the gated ionotropic glutamate receptor (LiGLuR) [17], based on glutamate derivative attached to an engineered kainite receptor via a tethered azobenzene photoisomerizable molecule.

Computer-generated holography (CGH) is one versatile method for the engineering of the light illumination pattern based on the modulation of the wave front phase. In combination with the different classes of optical actuators, this approach can allow for a spatial and temporal precise control of the molecule activity and brain circuit function. In this chapter will be first presented the basic optical principles behind the phase modulation scheme from the point of view of the light propagation theory. Then, the basic properties of the hardware aspects will be introduced starting from the core of the approach, the spatial light modulator (SLM), to the explanation of the different optical configurations achievable. Finally, an overview of the state of the art of the application of the method will be discussed.

2 Background and Computational Aspects in CGH

2.1 Computation of the Phase Maps

The propagation of a light from a source in the free space as within an optical system results in a light intensity distribution at the observation plane dictated by the interactions experienced by the light wave front with the different elements constituting the propagation medium. Such interactions are involving the properties of the electric field such as its amplitude and its phase. In CGH, the phase profile of the light wave front is carefully controlled in order to obtain at the observation plane a specific intensity distribution. From an optical point of view, the propagation of an arbitrary wave front can be considered as an envelope of spherical waves emitted from a point source (ξ, ζ) and moving in a certain direction. At the observation plane (x, y), the contribution of all the different waves is integrated to give an electrical field distribution $D(x, y)$ as expressed by

$$D(x,y) = \frac{1}{j\lambda} \iint\limits_{\Sigma} W(\xi,\zeta) \frac{e^{jkr}}{r} cos\beta \mathrm{d}s$$

where $W(\xi, \zeta)\ e^{jkr}$ is the single wave front component propagating from the source to the point (x, y) in the observation plane located to a distance r, β is the angle between the propagation direction z and the vector r, and k is the wave vector. From this equation, it is possible to derive, in the approximation at the second order for the vector r, a Fresnel diffraction integral in the following form

$$D(x,y) = \frac{e^{jkz}}{j\lambda} e^{j\frac{k}{2z}(x^2+y^2)} \iint\limits_{-\infty}^{+\infty} W(\xi,\zeta) \left(e^{\frac{jk}{2z}(\xi^2+\zeta^2)}\right) \left(e^{\frac{-j2\pi}{\lambda z}(x\xi+y\zeta)}\right) \mathrm{d}\xi \mathrm{d}\zeta$$

This form can be further reduced, considering a far-field approximation with the first term into the brackets substituted by 1. The field distribution in this case can be approximated by the Fraunhofer diffraction integral

$$D(x,y) = \frac{e^{jkz}}{j\lambda} e^{j\frac{k}{2z}(x^2+y^2)} \iint\limits_{-\infty}^{+\infty} W(\xi,\zeta) \left(e^{\frac{-j2\pi}{\lambda z}(x\xi+y\zeta)}\right) \mathrm{d}\xi \mathrm{d}\zeta$$

This expression clearly shows that the electrical field distribution $D(x, y)$ obtained at the observation plane is, beside a quadratic phase factor, the two-dimensional Fourier transform of the complex field distribution $W(\xi, \zeta)$ at the source plane, linking amplitude and phase of the field at the observation plane to the component of the complex field at frequencies $(x/z, y/z)$ in the source plane. In other words, an input wave front can be modified in the phase of its spatial components in order to obtain upon propagation a specific light intensity distribution at the observation plane (Fig. 1).

In the application of the phase modulation in the light propagation scheme for CGH, the light intensity distribution at the observation plane $D(x, y)$ is an input parameter and finding the optimal field configuration at the source plane $W(\xi, \zeta)$ requires the use of mathematical algorithms and computation to retrieve the proper phase correction or diffractive optical element (DOE). Considering the last equation, in the condition with $W(\xi, \zeta)$ is a spatial uniform planar wave front, the propagation gives the definition of the Dirac delta, corresponding to a spot illuminated at the center of the observation plane. Similarly, introducing a correction of the wave front phase with a linear function of the position along one of the axis, i.e., $\varphi(\zeta) = \varepsilon\zeta \times 2\pi/\lambda z$, it leads to a lateral offset of the illumination spot of an amount proportional to ε and depending on the optical characteristics. These considerations are the basic blocks for a phase computation approach, known as gratings and lenses. This method [18], combines the optical properties of prisms to steer the light beam with those of the lenses to offset the illumination spot along

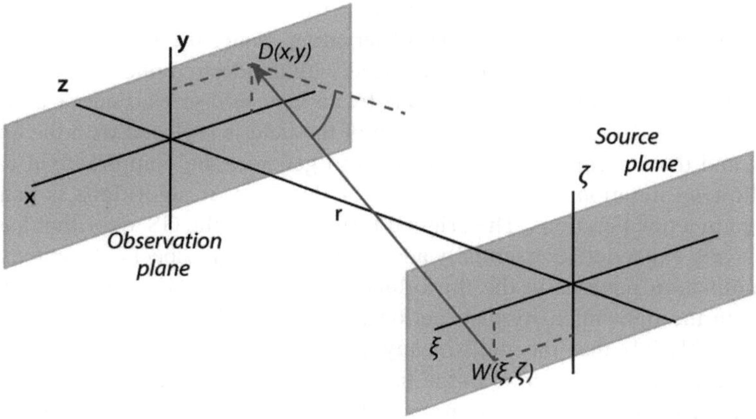

Fig. 1 Geometric representation of the optical conjugation between the source plane of a light beam and the observation or reconstruction plane where all the spatial components of the wave front are constructively or destructively contributing to the intensity distribution of the electric field and to the light intensity pattern

the optical axis. It is the simplest and computationally the fastest, because the required phase maps or diffractive optical elements (DOEs) can be obtained from the following analytical expression:

$$\varphi_{\text{grat}}(\xi,\zeta) + \varphi_{\text{lens}}(\xi,\zeta) = 2\pi(\xi\Delta x + \zeta\Delta y) + \frac{2\pi\Delta f}{\lambda f^2}(\xi^2 + \zeta^2)$$

where $\Delta x, \Delta y$, are the lateral offset values in the two directions of the observation plane, Δf is the desired step along the longitudinal direction z, and **f** is the effective focal distance of the system. This method can be easily extended to allow multiple beams to be generated in a volume and independently controlled by adding together the single field components at the source plane. For many applications, this approach is an effective method for DOE computation, but shaping capability is limited to ensembles of points only and the degree of longitudinal control is reduced by the far-field approximation [19]. A robust method to produce complex illumination patterns is the Gerchberg–Saxton [20] algorithm and the derived formulations. Compared to the gratings and lenses approach, where the phase are assigned and fixed for every spot in the pattern, this class of algorithm introduces phase freedom with an iterative optimization of the phase distribution at the DOE plane. The phase optimization procedure is based on the use of discrete Fourier transforms (DFT) to propagate the source field to the observation plane and to back-propagate the resulting complex field to the source plain again, imposing opportune constraints in the amplitude of the complex field.

The different implementations of these iterative Fourier transform algorithm (IFTA) share few common steps. Initially, from the DOE plane, a complex field with uniform amplitude and random phase distributions is propagated by means of DFT. From the resulting field at the observation plane, the phase distribution is kept and the amplitude information replaced with the desired one. The new field is then back-propagated to the source plane by an inverse discrete Fourier transform. After the back-propagation, the amplitude of the field is replaced with the uniform pattern and the information phase retained to generate the complex field used in the following iteration. The convergence of this kind of algorithms is generally achieved in a few iterations. The original formulation of the GS algorithm has been successively expanded to essentially address two different aspects: the extension of the illumination patterns in the third dimension and the optimization of the uniformity in the light intensity distributions. Regarding the first point, it was demonstrated [21, 22] that the field in any plane of the observation space can be computed introducing a kernel in the field at the DOE and calculating the Fourier transform. This additional kernel represents the phase profile required to shift the plane of interest to the lens focal plane. This development allows for the generation in principle of extended illumination regions located at different plane in the image volume [23–25]. Regarding the research for higher uniformity, this has been motivated by the peculiar characteristic of the 2D-generated patterns to present a certain degree of intensity fluctuations. Specific algorithms have been developed, where, with respect to the original GS loop, more constraints are applied at the level of the observation plane to maximize the SNR of the generated pattern or a correction feedback accounting for the difference between the desired distribution and the one obtained is introduced to minimize the deviations.

3 Hardware for CGH: LCoS-SLM

3.1 LCoS-SLM: Working Principle

The common configuration to control the phase profile of a light wave front is based on nematic liquid-crystal-on-silicon spatial light modulators (LCoS-SLM). These devices are fabricated in the form of large bidimensional arrays typically covering an area of $16 \times 12\ \mathrm{mm}^2$ with 800×600 cells. The working principle of the SLM is based on the combination of the optical and dielectric properties of the liquid crystals (LC). From the structural point of view, these molecules are characterized by a uniaxial anisotropy, molecules with one main axis of symmetry called director and with two equal minor axes in the perpendicular directions giving them the particular physical properties. Depending on the relative angle θ of the light propagation direction with respect to the molecule orientation, the material presents an effective refraction index n_{eff} to the propagating radiation given by the following formula

$$n_{\text{eff}} = \frac{n_o n_e}{\sqrt{n_o^2 \cos^2\theta + n_e^2 \sin^2\theta}}$$

where n_o and n_e are respectively the ordinary and the extraordinary refractive indices experienced by an electrical field with polarization orientation perpendicular or along the director axis. The uniaxial anisotropy is reflected also in the dielectric properties of these materials that cause the LC to interact with an external electric field. The average direction of the molecules will orient as to align with the direction of the applied electric field, but because the rotation varies with the relative position within the thickness d of the LC layer, the effective refractive index experienced by the light traveling all the way in the material is given by the integral computed over the thickness interval. Because of the optical path change induced by the difference in the refraction index upon application of an electrical field, the portion of the wave front propagating along the direction z through this part of the material will experience a phase delay expressed in the following formula

$$\Gamma = \frac{2\pi}{\lambda}\int_0^d (n_{\text{eff}}(\theta(z)) - n_o)\mathrm{d}z$$

In general, the maximal phase stroke of the final device, in other words the amount of single-cell phase modulation in terms of π fractions, is dictated by the ratio between the cell thickness and the wavelength and is dependent on the material birefringence $\Delta n = n_e - n_o$ (Fig. 2).

Being the angular director distribution depending not exclusively on the external electric field applied to the cell but also on the elastic and viscous interactions affecting the LC molecules, the performance of these devices are generally a compromise between the LC responsiveness, dictating the actual refresh rate and reliability and precision in the phase modulation. At this purpose, theoretical frameworks like those described by Eriksen–Leslie theory [26, 27] are normally considered to characterize and to evaluate the dynamics of an LC compound when an electrical field is applied, according to the dynamic equation:

$$K_2\frac{\partial^2\vartheta}{\partial z^2} + \frac{\Delta\varepsilon E_z^2}{4\pi}\sin\vartheta\cos\vartheta = \gamma_1\frac{\partial\vartheta}{\partial t}$$

where K_2 is the twist elastic coefficient, γ_1 is the rotational viscosity, $\Delta\varepsilon$ the dielectric anisotropy, and E_z is the amplitude of the applied electrical field. The approximated solutions to the linearized form of the equation are generally assumed to be a superposition of many spatial modes factorized in a temporal and a spatial term. The solution based on the first mode approximation, holding for small rotation angle, is extremely useful to highlight general features for the cell design and the selection of the LC material. In particular, the characteristic relaxation and settling times of the LC cell are proportional to the splay elastic

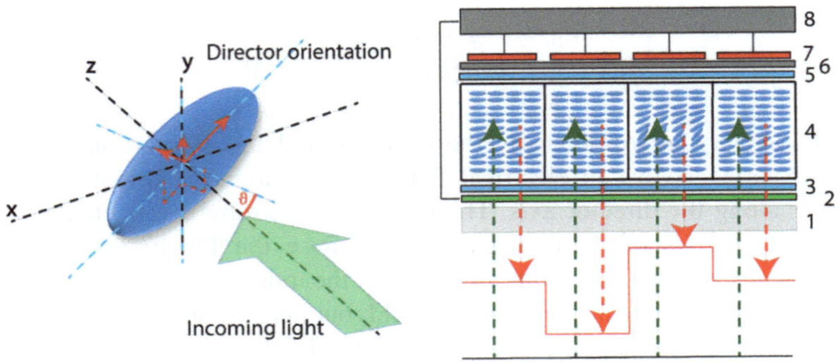

Fig. 2 **a** A schematic cartoon illustrating the spatial arrangement of LC molecules with respect to the light propagation direction and the profile these assume within an active cell of a spatial light modulator. **b** A simplified diagram of the SLM layout showing a set of cell (*4*) with different modulation profiles according to the control voltage applied by the electronic interface (*7* + *8*) and the change in the wave front superimposed to the propagating beam leaving the device

constant K_2, to the square of the cell thickness d, and to the viscosity coefficient of the material γ_1, according the following equations:

$$\tau_{\mathrm{OFF}} = \frac{\gamma_1}{K_2}\left(\frac{d}{\pi}\right)^2 \quad \text{and} \quad \tau_{\mathrm{ON}} = \frac{\gamma_1}{\Delta\varepsilon}\left(\frac{1}{E_z^2 - E_F^2}\right)$$

where E_F represents the threshold or Friedrich electric field. The performances and the modulation schemes achievable with SLM are not only depending on the intrinsic properties of the LC molecules but also depending on the overall design of the device and on configuration of the optical train build around it for the particular application.

3.2 LCoS-SLM: Temporal Features and Overdriving Procedures

The choice of the LC compound and the characteristic sizes of the active cell are fundamental parameters in the definition of the temporal response of device, and these factors should be carefully evaluated in the light of the envisaged application and modulation scheme. For instance, the ferroelectric LC have response times in the order of hundreds of microseconds leading to refresh rate of about 1 kHz, but are suitable for a binary phase modulation only resulting in SLMs with a lower diffraction efficiency compared to devices with an extended phase modulation resolution and thus limiting their utilization in applications where the light utilization efficiency is not the limiting factor. These figures of efficiency are enhanced in the case of devices with nematic LC but at the expense of the response times that are falling down to the range of 5–50 ms due to the relatively higher viscosity characterizing this kind of molecules [28]. To improve the temporal performances for nematic LC devices, different approaches have been proposed and are continuously investigated. Along with the engineering of new LC blends [29] and of specific cell architectures [30, 31], a lot of efforts have been applied on the side of the LC electrical driving circuits and in overdriving techniques to minimize the effective response times. Generally, in the overdrive scheme [32], to speed up the transitions from different cell profiles corresponding to different phase values, the applied voltage to control the electric field across the cell is not simply switched from a level to the next one, but it is switched for an initial short time to the maximal or the minimal value allowed depending on the transition direction and only successively turned to the static level required for the next phase modulation once the cell configuration and the desired phase value has been achieved. Assuming an exponential response [33], the amount of time Δt during which the maximum or minimum voltage is applied depends on the phase difference between the transition states φ_0, φ_1 and the extreme phase modulation or phase stroke φ_m, and from the time constant τ required for switching the voltage between the minimum and the maximum values, according the following formula:

$$\Delta t(\varphi_0, \varphi_1) = -\tau \ln\left(\frac{\varphi_1 - \varphi_m}{\varphi_0 - \varphi_m}\right)$$

It is straightforward to realize that faster transitions are achieved when smaller phase jumps are considered. The performances of theses overdrive methods strongly depend on the maximum phase stroke achievable and the maximum driving voltage accepted, in particular in consideration of the fact that the maximum phase stroke is proportional to the cell thickness and that this one is affecting the temporal response in a quadratic dependence according to the approximated solutions of the models. Moreover, this approach assumes the use of control interfaces with refreshing times in the submillisecond range, significantly smaller with respect to the typical settling times of the LC cell.

3.3 LCoS-SLM: Design of the Device

In order to have a working phase modulation cell, the LC molecules need to be embedded in an optoelectronic functional unit able to manage at the same time the optical features of the material and the electrical circuit required for phase modulation. This is generally achieved with a structure including the following components: (1) an outer antireflection broadband or wavelength-specific protective glass layer; (2) a transparent ground electrode common to the all the cells made of indium tin oxide (ITO) coated with; (3) a SiO2 alignment layer to confer the proper orientation and pretilting angle to the LC molecules; (4) a layer of liquid crystal molecules; (5) an alignment matching layer; (6) a planar dielectric mirror to increase the effective fill factor and to optimize the light utilization efficiency; and (7) a layer of evaporated Al pads.

The matrix of these functional units is matched to driving and addressing electronic circuits realized in complementary metal-oxide-semiconductor technology at very large integration scale (8).

On the market, there are many solutions available that mainly are differing for the LC material adopted, the alignment scheme, the number of the active elements in the SLM matrix, the feature size of the pixel, the electrical driving scheme of the LC cells, and the control interface. Typical matrix sizes are ranging from 256×256 to 1000×1900 elements with pixel diameters from 40 to 10 μm and active area of about 10 mm × 10 mm, resulting in spatial resolutions in the range of 20–33 lp/mm. All these parameters together contribute to define the temporal dynamics, the optical performances, and the modulation scheme of the devices.

3.4 LCoS-SLM: Optical Properties

One of the most important features of a SLM is its phase modulation characteristics, meaning the empirical relation between the voltage applied to the cell and the actual phase change produced. Depending on the working wavelength, the voltage applied to achieve a defined phase modulation, encoded with an 8-bit value, has to be calibrated according to a specific lookup table. These input–output characteristics can be obtained by interferometric methods or measuring the diffraction for binary linear gratings, or Ronchi gratings, with increasing phase modulation depth [34].

Another important figure for SLM is its diffraction efficiency η, defined as the ratio between the integrated intensity redirected in the first diffracted order and the intensity in the zero order due to the non-diffracted components of the light wave front. It strongly depends on the number of phase quantization levels q and the grating period Λ of the device according the following expression:

$$\eta = \frac{\sin c^2\left(\frac{1}{q}\right)}{\sin c^2\left(\frac{1}{[q,\Lambda]}\right)} \times \sin c^2\left(\frac{1}{\Lambda}\right)$$

where [ ] indicates the least common multiplier. Although the dependency in the quantization levels becomes quickly negligible increasing the quantization levels for the phase, the spatial discretization of the devices deriving from the pixelated structure of the matrix results in a non-uniform optical power density depending on the period of the diffraction grating. The method based on Ronchi gratings is less straightforward in the extrapolation of the LUT compared to interferometric methods, relaying in the measurement of the power addressed in the zero and first orders for the different modulations, but gives a more complete characterization of the devices regarding its non-ideal behavior due to different kinds of phenomena such as edge effects, fringe field effects, back plate curvature, and non-uniformity. All these factors contribute in the reduction of the effective diffraction efficiency achievable. The pixel fill factor is also influencing the diffraction efficiency. Indeed, the regions between adjacent pixels of the device are without any control in the phase modulation, so increasing the amount of light contributing in the zero-order component. This contribution, in the case of the modern devices characterized by fill factors greater than 90 %, is hardly to be relevant.

4 Optical Configurations for CGH

4.1 Basic Configuration for CGH

The optical design required for CGH is essentially a Fourier holographic system, where the electric field at the output is given by the Fourier transformation of the electric field distribution at the input of the optical system realized with a single lens, with an objective or with a more elaborated optical path. In order to control the light intensity distribution at the focal volume of an objective, the phase correction map, superimposed by the SLM on the wave front of a propagating beam, has to be projected onto the objective back focal plane (Fig. 3).

In a basic optical setup, the coherent beam generated by a continuous wave or pulsed laser source is passed through an intensity modulation unit (MU), e.g., acousto-optic modulator or Pockell cell, in order to control the power density at the sample. Afterward, a half-wave plate to control the direction of the light polarization is generally placed to match the polarization requirement of the SLM for pure phase modulation dictated by the LC orientation. The original diameter of the beam successively is expanded by means of a telescope or commercial beam expander of 5x–10x factor in order to match the input window of the SLM that is illuminated at near-perpendicular incidence to preserve the optimal performances of the device while at the same time allowing for an easy lateral separation of the reflected beam. At this stage, in order to project the phase modulation map generated by the SLM onto the objective back focal plane of the imaging system, a telescope-based demagnification stage composed by lenses L1 and L2 is inserted along the path to generate a replica of the SLM screen properly fitting in the diameter of the objective back plane according to the following equations [35]:

$$mL_{\mathrm{SLM}} = 2 \times f_0 \times \mathrm{NA} \quad \text{and} \quad m = f_2/f_1$$

where m is the magnification factor of the telescope with L1 and L2 having focal lengths f_1 and f_2 respectively, f_0 is the focal length, NA the numerical aperture of the objective, and L_{SLM} is the size of the SLM. If the SLM image at the back aperture plane is smaller than the aperture diameter, the nominal NA of the

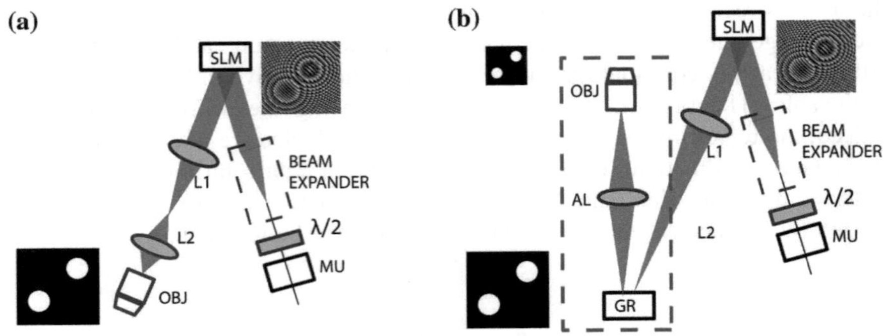

Fig. 3 Basic optical configuration for holographic projection **a** and its expansion to integrate the optical path for temporal focusing. At the SLM is superimposed to a propagating beam a proper phase correction map or diffractive optical element (DOE). At the Fourier-conjugated optical plane (between L1 and L2 in **a** and at the level of the grating GR in **b**), the desired intensity distribution is realized and demagnified at the sample plane

objective is not fully utilized with a consequent reduction in spatial resolution. On the other hand, with SLM image larger than the aperture, light efficiency is compromised.

4.2 Integration of CGH into Imaging Systems

Integration of a CGH optical train into an imaging system is very frequent and, depending on the implementation type, may require some degree of modification in the optical elements in particular when imaging and CGH designs are overlapping in the final segment of the microscope light path. The more common situations are represented by commercial infinity-corrected microscopes. In this case, a good approach is to couple the CGH path at the level of the fluorescence port between the objective and the tube lens. This approach, requiring a proper combining mirror in the microscope optical path and the telescope lenses mounted outside, is very versatile to combine in principle at the same time the phase modulation with other imaging techniques (Fig. 4).

A slight deviation from this scheme is achieved when the microscope tube lens serves as the second lens of the telescope required to image the SLM plane to the back focal plane of the objective. In this case, the integration is requiring minor hardware changes because the CGH path can be easily fed in through a secondary camera port in combination with a proper mirror but, on the other hand, leads to some implications on the CGH train design in terms of focal distances of the lenses and overall length of optical path. In principle, this last configuration compared to the original one allows for a more comfortable integration of the CGH module within a 2-photon or confocal scanning system. An alternative design [36–38] is

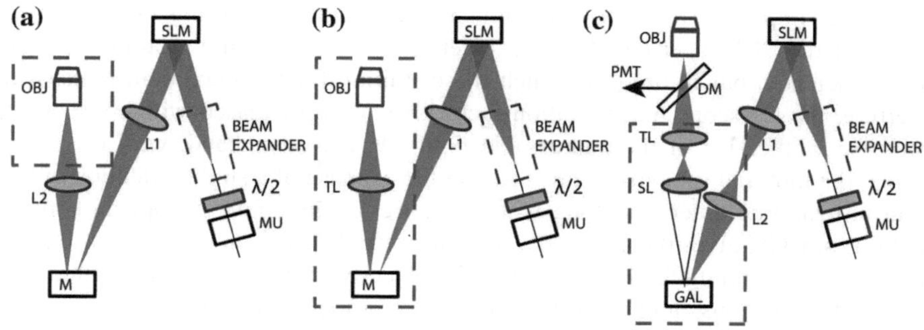

Fig. 4 Different integration schemes of CGH within imaging systems, almost independent optical trains in **a** where CGH shares with the imaging path only the segment after the lens L2; shared tube lens in **b**; configuration with scanning system in **c** with conjugation at the galvo mirror level

represented by the layout merging the CGH and the imaging paths within the scan head by means of a combining optic element placed before the scan lens. This configuration, being the magnification factor dictated by the tube and scan lenses in the last segment of the optical path, generally requires an additional magnification stage to project the SLM image into the plane of the galvo mirrors or to a plane located at the same distance from the scan lens.

4.3 Optical Figures of CGH

One of the optical aspects generally considered to characterize a CGH configuration is the size of the volume that can be addressed by means of the phase modulation, either in the transversal direction with respect to the direction of the light propagation or in the longitudinal direction when the reconstruction plane is shifted from the focal plane of the objective. The diameter of the excitation field in the transversal direction is a figure depending at the same time on the objective focal length f_0 and on the pixel size d_{SLM} of the SLM, on the basis of the following equations:

$$F_{\text{MAX,TRANS}} = \frac{2 \times \lambda \times f_0}{n} \times \left(\frac{f_1}{f_2 d_{\text{SLM}}}\right) = \frac{2 \times \lambda \times f_0}{n} \times \left(\frac{1}{m \times d_{\text{SLM}}}\right)$$

where n is the refraction index of the medium. The second factor on the right side of the equation is representing the size of the pixel obtained at the back focal plane with the magnification performed by the telescope placed after the SLM. It is straightforward to realize that for a defined telescope configuration the addressable field is proportional to the objective focal length and inversely proportional to the size of the SLM pixel at the level of the back focal plane. The equation points also to another important aspect of the optical configuration regarding the degree of filling of the SLM aperture. Indeed, underfilling SLM aperture assures for high light efficiency, but in order to match the condition for the maximum numerical aperture, the telescope magnification factor should be increased with consequent decrease of the size of the addressable field. On the other hand, with the over-filling, the minimal spot size is preserved at the expense of a slight reduction in the light utilization efficiency. Along with a transversal direction, the optical configuration for CGH gives the possibility to control the distribution of the electric field in a volume including the focal plane of the objective, with excitation foci located at different positions along the direction of light propagation [39, 40]. In this case, the longitudinal addressable range is limited by energy loss of rays coming from SLM pixel imaged at the periphery of the pupil. Pixels diffract light in a cone of half-angle in the order of $\lambda \times f_1 / (2 \times f_2 \times d_{\mathrm{SLM}})$, but compared to pixels located at the center of the pupil, those at the periphery can cover a smaller range of the optical axis. So with the axial defocus, progressively less energy is redirected from the peripheral pixels, limiting the effective axial range to about $\lambda \times f_0 \times n / (\mathrm{NA} \times m \times d_{\mathrm{SLM}})$.

The other important figure for a CGH system is the diffraction efficiency. Generally, the phase corrections leading to large displacement of illumination foci in the axial and lateral directions are characterized by high spatial frequencies. The discrete nature of the SLM renders these frequency components in a less accurate manner with respect to the low-frequency ones, leading to position-dependent diffraction efficiency. According to the Fraunhofer propagation theory [41], the diffraction efficiency η can be expressed in the case of a lateral displacement x, y with respect the center of the plane:

$$\eta(x, y) = \left(\sin c\left(\frac{\pi \mathrm{nm} d_{\mathrm{SLM}} x}{\lambda f_0}\right)\right)^2 \left(\sin c\left(\frac{\pi \mathrm{nm} d_{\mathrm{SLM}} y}{\lambda f_0}\right)\right)^2$$

This position-dependent diffraction efficiency leads to a non-uniform optical power at the sample and to an overall reduction of the power density over the addressable region. This scenario is further complicated in the case of combination of lateral and axial displacement.

4.4 *Peculiar Aspects of CGH Beams for Bidimensional Extended Patterns*

When CGH is adopted for the generation of extended regions of illumination in the observation space, the resulting intensity distributions present a few characteristic aspects. The first one is the evident intensity variability across the illuminated area. This phenomenon is mainly due to the cross talk between adjacent points of the plane of the DOE, which, having slightly different optical paths because of the specific phase profile and the non-ideal properties of the overall optical system, sum up together randomly at the image plane, resulting in constructive or destructive mutual interference [42, 43]. This effect can lead to sensible fluctuations in the local intensity of the pattern, with values ranging from 30 to 60 % depending on the absorption mechanism of the excited molecules. Different methods have been adopted to minimize this variability in the applications requiring more uniform illumination profiles. The design of specific algorithms [44] is one approach based essentially on three different schemes: (1) imposing a constraint in the variation of the phase delay between adjacent pixels to smooth the phase profile [45]; (2) compute a set of DOEs including random phase patterns to mitigate the intensity fluctuation by time averaging [46]; and (3) with time averaging, but calculating a single DOE and then applying a procedure named cycling shifting to produce derived DOEs leading to field distributions at the image plane characterized by different phase profiles but sharing the same intensity distribution [47]. All these computational methods, although with different weight, have a common bottleneck in the computational time and load, limiting their application to relative slow experimental protocol or sophisticated computation platform. Other solutions have been proposed [48] based on the introduction along the optical path of a diffusing element to mitigate the beam coherency, but their potentiality cannot be adopted in those applications where light efficiency budget is critical.

A more severe aspect to carefully consider in CGH when generating two-dimensional patterns is the axial extension of the illumination profile along the direction of propagation of the light [49]. The relation of the axial extension to the lateral size of the projected illumination patter is linear and dictated by the rules of the diffraction [28]. This phenomenon leads to a relevant deterioration in the spatial resolution along the light propagation direction. This problem can be mitigated in the case of the 2-photon excitation, by means of temporal focusing [50]. With this method, an increase in the axial confinement of an extended illumination pattern is achieved controlling the dispersion of the pulse in such a way that the different spectral components are recombining effectively together only at the focal plane, while the pulse results dispersed in the regions out of focus. The hardware configuration is generally based on a grating to disperse the different chromatic components of the pulse at different angles and an achromatic lens forming with objective a perfect imaging system. In this way, a two-dimensional pattern projected to the grating is imaged at the sample plane at the same time with the spatial and the temporal focusing [51, 52].

5 Applications and Perspective for CGH

Over the last five years, the application of CGH method to shape the light illumination pattern has been used prevalently in combination with caged neurotransmitters. In 2008, the group of Emiliani [53] first demonstrated the possibility of holographic photolysis of MNI-caged neurotransmitters in brain slices using a 405 nm laser source in combination with a SLM. Evident became the physiological implications in the study of neuronal activity and function deriving from the possibility to extend with the holographic beams the spatial and temporal features of the uncaging patterns in particular with respect to the traditional diffraction-limited point scanning approach. At the end of the same year, by the group of Yuste [36], the combination of spatial light modulators with a pulsed source was demonstrated in brain slices either applied for the simultaneous 2-photon uncaging of groups of dendritic spines with MNI glutamate or applied for a 2-photon parallel imaging approach of neuronal activity in cells loaded with the synthetic calcium indicator Fluo4 in acute brain slice based on the holographic excitation of large ensemble of points in combination with a CCD camera detection.

After the initial characterization of the performances of holographic MNI glutamate photostimulation based on electrophysiological recordings, the development moved toward the integration of the holographic shaping capabilities into systems for high-resolution functional imaging of the network activity. The clear goal was to develop an "all-optical" investigation system for triggering the network activation with high degree of control in the stimulation pattern in combination with the detection of the induced activity pattern using fluorescent probes for changes in calcium concentration. In particular, a CW holographic system [49] was presented for the stimulation of a set of cells in brain slice in combination with Hi-Lo, a widefield imaging scheme with improved depth discrimination to mitigate the tissue scattering. In the same period, an all-optical 2-photon configuration was presented in dal Maschio et al. [37]. In this work, a SLM-based holographic module was integrated within the scan head of a 2-photon imaging system equipped with two independent pulsed sources: one designed for the imaging and the other dedicated for the photostimulation. The engineered optical coupling was designed in order to have the possibility to use alternatively the SLM either for holographic uncaging while at the same time performing the imaging with the traditional scanning approach or to use the traditional diffraction-based uncaging done with the galvo mirror while imaging with a CCD camera and the parallel excitation scheme provided by the holographic beam.

At the end of the same year, the first demonstration of 2-photon excitation of Chr2 by means of wave front modulation approach was published by the Emiliani group [54]. The technique adopted in that case was the generalized phase-contrast GPC in combination with temporal focusing, where SLM is inserted in a common path interferometer and conjugated to the sample plane.

The possibility to use SLMs and the peculiar property of the holographic beams to design complex 3D light pattern was presented during the 2011 in three different works. Simultaneous generation of many uncaging spots at different depths within a brain slice was reported for the study of the integration property of dendritic processes [55, 56]. In dal Maschio et al. [38], it was presented a design to move the imaging focus plane in a 2P system along the light propagation direction without the movement of the objective but with the phase modulation approach. These approaches demonstrated the effective possibility to decouple the imaging plane from the photostimulation plane.

Very recently, CGH beams have been used in combination with temporal focusing for the 2-photon photostimulation of a Chr2 variant, called C1V1, deep in the brain slice tissue [52]. In this study, the peculiar properties of temporal focusing in preserving the shape of the pattern within a scattering tissue were characterized.

In conclusion, only in the last 5 years, the efforts of many research groups led to the engineering effective of optical methods suitable for the investigation of the molecular function and of the dynamics of neuronal circuits. More time is required for the diffusion and the optimization of this kind of techniques, in particular for the application in vivo of these approaches.

References

1. Hell SW (2007) Far-field optical nanoscopy. Science 316(5828):1153–1158. doi:10.1126/science.1137395
2. Martin-Fernandez ML, Tynan CJ, Webb SE (2013) A 'pocket guide' to total internal reflection fluorescence. J Microsc 252(1):16–22. doi:10.1111/jmi.12070
3. Axelrod D, Thompson NL, Burghardt TP (1983) Total internal inflection fluorescent microscopy. J Microsc 29(Pt 1):19–28
4. Betzig E, Lewis A, Harootunian A, Isaacson M, Kratschmer E (1986) Development and biophysical applications. Biophys J 49(1):269–279
5. Aurélie J, Heintzmann R (2013) Superresolution multidimensional imaging with structured illumination microscopy. Annu Rev Mater Res 43:261–282. doi:10.1146/annurev-matsci-071312-121648
6. Santi PA (2011) Light sheet fluorescence microscopy: a review. J Histochem Cytochem 59(2):129–138. doi:10.1369/0022155410394857
7. Grewe BF, Langer D, Kasper H, Kampa BM, Helmchen F (2010) High-speed in vivo calcium imaging reveals neuronal network activity with near-millisecond precision. Nat Methods 7(5):399–405. doi:10.1038/nmeth.1453
8. Ashkin A, Dziedzic JM, Bjorkholm JE, Chu S (1986) Observation of a single-beam gradient force optical trap for dielectric particles. Opt Lett 11:288–290
9. Grier DG (2003) A revolution in optical manipulation. Nature 424:21–27. doi:10.1038/nature01935
10. Kramer RH, Fortin DL, Trauner D (2009) New photochemical tools for controlling neuronal activity. Curr Opin Neurobiol 19(5):544–552. doi:10.1016/j.conb.2009.09.004
11. Ellis-Davies GC (2007) Caged compounds: photorelease technology for control of cellular chemistry and physiology. Nat Methods 4(8):619–628. doi:10.1038/nmeth1072
12. Rial Verde EM, Zayat L, Etchenique R, Yuste R (2008) Photorelease of GABA with visible light using an inorganic caging group. Front Neural Circuits 13(2):2. doi:10.3389/neuro.04.002.2008
13. Heinbockel T, Brager DH, Reich CG, Zhao J, Muralidharan S, Alger BE, Kao JP (2005) Endocannabinoid signaling dynamics probed with optical tools. J Neurosci 25(41):9449–9459. doi:10.1523/JNEUROSCI.2078-05.2005
14. Nagel G, Szellas T, Huhn W, Kateriya S, Adeishvili N, Berthold P, Ollig D, Hegemann P, Bamberg E (2003) Channelrhodopsin-2, a directly light-gated cation-selective membrane channel. Proc Natl Acad Sci USA 100(24):13940–13945. doi:10.1073/pnas.1936192100

15. Zhang F, Wang LP, Brauner M, Liewald JF, Kay K, Watzke N, Wood PG, Bamberg E, Nagel G, Gottschalk A, Deisseroth K (2007) Multimodal fast optical interrogation of neural circuitry. Nature 446(7136):633–639. doi:10.1038/nature05744
16. Han X, Boyden ES (2007) Multiple-color optical activation, silencing, and desynchronization of neural activity, with single-spike temporal resolution. Plos One 212(3):e299. doi:10.1371/journal.pone.0000299
17. Volgraf M, Gorostiza P, Numano R, Kramer RH, Isacoff EY, Trauner D (2006) Allosteric control of an ionotropic glutamate receptor with an optical switch. Nat Chem Biol 2(1):47–52. doi:10.1038/nchembio756
18. Leach J, Wulff K, Sinclair G, Jordan P, Courtial J, Thomson L, Gibson G, Karunwi K, Cooper J, Laczik ZJ, Padgett M (2006) Interactive approach to optical tweezers control. Appl Opt 45(5):897–903. doi:10.1364/AO.45.000897
19. Yang S, Papagiakoumou E, Guillon M, de Sars V, Tang CM, Emiliani V (2011) Three-dimensional holographic photostimulation of the dendritic arbor. J Neural Eng 8(4):046002. doi:10.1088/1741-2560/8/4/046002
20. Gerchberg RW, Saxton WO (1972) A practical algorithm for the determination of the phase from image and diffraction plane pictures. Optik 35:237–246
21. Curtis JE, Koss BA, Grier DG (2002) Dynamic holographic optical tweezers. Opt Commun 207:169–175
22. Piestun R, Shamir J (1998) Generalized propagation-invariant wave fields. J Opt Soc Am A 15:3039–3044. doi:10.1364/JOSAA.15.003039
23. Sinclair G, Leach J, Jordan P, Gibson G, Yao E, Laczik Z, Padgett M, Courtial J (2004) Interactive application in holographic optical tweezers of a multi-plane Gerchberg–Saxton algorithm for three-dimensional light shaping. Opt Express 12(8):1665–1670. doi:10.1364/OPEX.12.001665
24. Whyte G, Courtial J (2005) Experimental demonstration of holographic three-dimensional light shaping using a Gerchberg–Saxton algorithm. New J Phys 7:117. doi:10.1088/1367-2630/7/1/117
25. Shabtay G (2003) Three-dimensional beam forming and Ewald's surfaces. Opt Commun 226(1–6):33–37. doi:10.1016/j.optcom.2003.07.056
26. Wu ST (1986) Phase retardation dependent optical response time of parallel-aligned liquid crystals. J Appl Phys 60:1836–1838. doi:10.1063/1.337228
27. Stewart I (2004) The static and dynamic continuum theory of liquid crystals: a mathematical introduction. Taylor and Francis, UK. ISBN:9780748408962
28. Oron D, Papagiakoumou E, Anselmi F, Emiliani V (2012) Two-photon optogenetics. Progress in brain research. In: Knöpfel T, Boyden E (eds) vol. 196. ISSN:0079-6123
29. Wu YH, Lin YH, Lu YQ, Ren H, Fan YH, Wu J, Wu ST (2004) Submillisecond response variable optical attenuator based on sheared polymer network liquid crystal. Opt Express 12(25):6382–6389. doi:10.1364/OPEX.12.006382
30. Wu ST, Wu CS (1989) High speed liquid crystal modulators using transient nematic effect. J Appl Phys 65:527. doi:10.1063/1.343135
31. Lin XW, Hu W, Hu XK, Liang X, Chen Y, Cui HQ, Zhu G, Li JN, Chigrinov V, Lu YQ (2012) Fast response dual-frequency liquid crystal switch with photo-patterned alignments. Opt Lett 37(17):3627–3629. doi:10.1364/OL.37.003627
32. Hu H, Hu L, Peng Z, Mu Q, Zhang X, Liu C, Xuan L (2012) Advanced single-frame overdriving for liquid-crystal spatial light modulators. Opt Lett 37(16):3324–3326. doi:10.1364/OL.37.003324
33. Thalhammer G, Bowman RW, Love GD, Padgett MJ, Ritsch-Marte M (2013) Speeding up liquid crystal SLMs using overdrive with phase change reduction. Opt Express 21(2):1779–1797. doi:10.1364/OE.21.00177
34. Zhang Z, Lu G, Francis TS (1994) Simple method for measuring phase modulation in liquid crystal televisions. Opt Eng 33:3018–3022
35. Pawley JB (ed) (2006) Handbook of biological confocal microscopy, 3rd edn. Springer, Berlin. ISBN 038725921X

36. Nikolenko V, Watson BO, Araya R, Woodruff A, Peterka DS, Yuste R (2008) SLM microscopy: scanless two-photon imaging and photostimulation with spatial light modulators. Front Neural Circuits 19(2):5. doi:10.3389/neuro.04.005
37. Dal Maschio M, De Stasi AM, Benfenati F, Fellin T (2011) Three-dimensional in vivo scanning microscopy with inertia-free focus control. Opt Lett 36(17):3503–3505. doi:10.1364/OL.36.003503
38. Dal Maschio M, Difato F, Beltramo R, Blau A, Benfenati F, Fellin T (2010) Simultaneous two-photon imaging and photo-stimulation with structured light illumination. Opt Express 18(18):18720–18731. doi:10.1364/OE.18.018720
39. Go MA, Stricker C, Redman S, Bachor HA, Daria VR (2012) Simultaneous multi-site two-photon photostimulation in three dimensions. J Biophotonics 5(10):745–753. doi:10.1002/jbio.201100101
40. Daria VR, Stricker C, Bowman R, Redman S (2009) Arbitrary multi-site two-photon excitation in four dimensions. Appl Phys Lett 95(9):093701–093703. doi: 10.1063/1.3216581. http://ieeexplore.ieee.org/search/searchresult.jsp?searchWithin=p_Authors:.QT.Bachor,%20H..QT.&searchWithin=p_Author_Ids:37283969400&newsearch=true
41. Golan L, Reutsky I, Farah N, Shoham S (2009) Design and characteristics of holographic neural photo-stimulation systems. J Neural Eng 6(6):066004. doi:10.1088/1741-2560/6/6/066004
42. Haupt C, Kolodziejczyk A, Tiziani HJ (1995) Resolution and intensity distribution of output images reconstructed by sampled computer-generated holograms. Appl Opt 34(17):3077–3086. doi:10.1364/AO.34.003077
43. Palima D, Gluckstad J (2008) Comparison of generalized phase contrast and computer generated holography for laser image projection. Opt Express 16:5338–5349. doi:10.1364/OE.16.005338
44. Golan L, Shoham S (2009) Speckle elimination using shift-averaging in high-rate holographic projection. Opt Express 17:1330–1339. doi:10.1364/OE.17.001330
45. Wyrowski F, Bryngdahl O (1988) Iterative Fourier-transform algorithm applied to computer holography. J Opt Soc Am A 5:1058–1065
46. Amako J, Miura H, Sonehara T (1995) Speckle-noise eduction on kinoform reconstruction using a phase-only spatial light modulator. Appl Opt 34(17):3165–3171. doi:10.1364/AO.34.003165
47. Matar S, Golan L, Shoham S (2011) Reduction of two-photon holographic speckle using shift-averaging. Opt Express 19(27):25891–25899. doi:10.1364/OE.19.025891
48. Papagiakoumou E, de Sars V, Emiliani V, Oron D (2009) Temporal focusing with spatially modulated excitation. Opt Express 17:5391–5401. doi:10.1364/OE.17.005391
49. Zahid M, Vélez-Fort M, Papagiakoumou E, Ventalon C, Angulo MC, Emiliani V (2010) Holographic photolysis for multiple cell stimulation in mouse hippocampal slices. PLoS One 5:e9431. doi:10.1371/journal.pone.0009431
50. Oron D, Tal E, Silberberg Y (2005) Scanningless depth-resolved microscopy. Opt Express 13:1468–1476. doi:10.1364/OPEX.13.001468
51. Papagiakoumou E, de Sars V, Oron D, Emiliani V (2008) Patterned two-photon illumination by spatiotemporal shaping of ultrashort pulses. Opt Express 16:22039–22047. doi:10.1364/OE.16.022039
52. Bègue A, Papagiakoumou E, Leshem B, Conti R, Enke L, Oron D, Emiliani V (2013) Two-photon excitation in scattering media by spatiotemporally shaped beams and their application in optogenetic stimulation. Biomed Opt Express 4(12):2869–2879. doi:10.1364/BOE.4.002869
53. Lutz C, Otis TS, DeSars V, Charpak S, David A, DiGregorio DA, Emiliani V (2008) Holographic photolysis of caged neurotransmitters. Nat Methods 5(9):821–827. doi:10.1038/nmeth.1241
54. Papagiakoumou E, Anselmi F, Begue A, de Sars V, Gluckstad J, Isacoff E, Emiliani V (2010) Scanless two-photon excitation of channelrhodopsin-2. Nat Methods 7:848–854. doi:10.1038/nmeth.1505

55. Anselmi F, Ventalon C, Bègue A, Ogden D, Emiliani V (2011) Three-dimensional imaging and photostimulation by remote-focusing and holographic light patterning. Proc Natl Acad Sci USA 108(49):19504–19509. doi:10.1073/pnas.1109111108
56. Yang S, Papagiakoumou E, Guillon M, de Sars V, Tang C, Emiliani V (2011) Three-dimensional holographic photostimulation of the dendritic arbor. J Neural Eng 8:046002. doi:10.1088/1741-2560/8/4/046002

Super-Resolution Fluorescence Optical Microscopy: Targeted and Stochastic Read-Out Approaches

Alberto Diaspro, Francesca Cella Zanacchi, Paolo Bianchini and Giuseppe Vicidomini

Abstract This chapter is dedicated to a general overview of some of the emerging and well-established super-resolution techniques recently developed and known as optical nanoscopy and localization precision method. Due to the way of probing the sample, one can consider them as targeted and stochastic-based techniques, respectively. Here, we stress how super-resolution is obtained without violating any physical law, i.e., diffraction. The strong idea behind such approaches, operating in fluorescence contrast mode, is related to the ability of controlling the states, bright/dark or red/blue, of the fluorescent labels being used in order to circumvent the diffraction barrier. Super-resolution is achieved by precluding simultaneous emission of spectrally identical emission of adjacent (<diffraction limit distance) molecules. Also, the evolution of such techniques toward applications on thick (>50 micron thickness) samples is discussed along with correlative microscopy approaches involving scanning probe methods. Examples are given within the neuroscience framework.

1 Introduction

One of the most relevant optical microscopy approaches toward cell studies, i.e., far-field fluorescence optical microscopy, in the last 30 years benefited of a series of key advances from confocal to multiphoton microscopy [22], from single-molecule detection methods to super-resolution microscopy and optical nanoscopy [23, 24, 41]. Also, advances in labeling techniques, imaging speed, and detection sensitivity greatly expanded the impact of fluorescence optical microscopy in life sciences [66]. Unfortunately, spatial information about packed or adjacent

A. Diaspro (✉) · F. C. Zanacchi · P. Bianchini · G. Vicidomini
NanoBioPhotonics—LAMBS, Nanophysics, Istituto Italiano di Tecnologia, Via Morego 30, 16163 Genoa, Italy
e-mail: alberto.diaspro@iit.it

F. Benfenati et al. (eds.), *Novel Approaches for Single Molecule Activation and Detection*, Advances in Atom and Single Molecule Machines,
DOI: 10.1007/978-3-662-43367-6_3,

(positioned at distances closer than the diffraction limit, i.e., approx. 200 nm) fluorescent molecules—that one should consider as nanoscale sources of light having a size ranging from 1 to 5 nm—is irretrievably lost when the related signatures overlap on a microscopic image [71]. This prevents the ability of following in a quantitative way biological mechanisms molecule by molecule. Such a scenario dramatically changed with the development of a number of methods that are usually grouped as super-resolution optical microscopy approaches. In general, the so-called super-resolution microscopies, realized using focused light, can be separated into two main categories, namely stochastic read-out or localization methods and targeted read-out or engineering point spread function (PSF) methods. However, improvement in spatial resolution is one of the requirements when developing advanced optical methods and should be balanced with others such as low phototoxicity, temporal resolution, penetration depth, and artifacts minimization [73]. As early reported [86], one of the most relevant developments in super-resolution and/or single-molecule detection methods is the one that brings single-molecule observation to the interior of living cells [18]. We would add that a great jump is moving from super-resolution imaging of single cells on glass slides to cell aggregates, organs, and tissues [12, 13, 45]. Specimens are rarely optically transparent, and imaging depth can be hampered by wavefront distortions mainly caused by random scattering and undesired absorption. The basic idea of precluding spectrally identical emitting fluorescent molecules to emit in the very same temporal acquisition window allows circumventing the inherent physical constraint given by the utilization of focusing lenses. Single molecules mapped into an image by an objective lens of numerical aperture NA appear as confused in a unique emission pattern when they are closer than $\lambda\_em/(2NA)$. Likewise, diffraction makes impossible to focus excitation light of wavelength $\lambda\_ex < \lambda\_em$ more sharply than to a spot of $\lambda\_ex/(2NA)$ in size. As a result, features that are spectrally identical and closer than the diffraction limit, say $\lambda/(2NA)$, are difficult to be separated. It turns that part of the modern research in microscopy has been focused to improve spatial resolution and effective far-field optical methods have been designed and implemented to this end (STED/RESOLFT, PALM/STORM and SSIM) [73]. Some of these approaches realize a substantial spatial resolution improvement by precluding simultaneous emission of fluorescent molecules closer than the diffraction limit. Key differences rely on the mechanism by which the fluorescence emission is modulated/controlled and by whether the emission takes place [at specific space coordinates (STED/RESOLFT) or at random coordinates molecule by molecule (PALM, STORM)]. In parallel, new classes of fluorescent dyes and proteins able to selectively target specific molecules in biological samples and suitable for the super-resolution approaches have been developed, becoming an election tool for imaging biological sub-structures at the whole-cell level [30, 44]. Despite the development of brighter and spectrally optimized fluorescent molecules, super-resolution imaging in depth is limited and its application to thicker samples represents one of the most challenging tasks.

2 Toward Super-Resolution Imaging of Thick (>50 Mm) Biological Specimens

Recently, several advances, both in single-molecule localization-based techniques and in stimulated emission depletion (STED) microscopy, brought to a wider application range of super-resolution techniques toward imaging of thick scattering objects [4, 8, 11–13, 27, 50, 55, 59, 83]. Also, comparatively new methods for 3D imaging of large biological samples, such as two-photon excitation (2PE) and selective plane illumination microscopy (SPIM), are now mature for super-resolution completing the framework comprising confocal and computational optical sectioning microscopy [9, 26, 47, 76, 95]. The growing interest in tissue and large organisms imaging has led in the last few years to an increased popularity of fluorescence techniques such as (2PE) microscopy [20] and light-sheet-based microscopy [47]. Nonlinear microscopy is an elegant method, based on the use of nonlinear contrast mechanism [21, 26], to perform non-invasive deep imaging and became in the last decade the election tool for whole organism and tissue imaging. 2PE laser scanning microscopy (LSM) opened a rapidly expanding field of imaging studies in intact tissues and living animals. Two main reasons make 2PE-LSM the matter of choice for deep imaging: First, because for commonly used fluorescent markers, the TPE needs near-infrared light (700–1,000 nm), which not only penetrates deeper into scattering samples but is also less phototoxic; and the second major advantage of 2PE, similarly shared also with all nonlinear contrast mechanisms, is that the signal depends supralinearly on the density of the excitation photons. Thereby, when focusing the laser beam into the sample, the generation of fluorescence signal is spatially confined to the perifocal region, i.e., out-of-focus fluorescent signal, which is intrinsic of one-photon excitation and which is usually rejected by using confocal detection, is not anymore a problem. Further, confinement of the fluorescence is maintained even in strongly scattering sample because the density of scattered excitation photons is too weak to generate TPE. Within this scenario, the use of higher wavelengths and the exploitation of 2PE represent a unbeatable opportunity to improve imaging depth capabilities and performances of far-field super-resolution techniques [8, 12, 13, 27]. STED microscopy has been initially used to image cortical spines in brain slices in depth, and recently, its combination with 2PE represents an optimal opportunity for multicolor imaging in living brain tissue. STED allowed two-color super-resolution imaging in acute brain slices [5] and the study neuronal morphology up to 30-mm deep in living brain tissue [82]. Furthermore, the implementation of adaptive optics has been proven to be a good opportunity to minimize the aberration effects in STED microscopy [35]. 2PE can also be used to improve performances of localization-based techniques through the confinement of the photoactivation process and the higher imaging depth capability provided. On the other side, light-sheet-based fluorescence microscopy has been recently established as one of the most promising tools for live, three-dimensional and fast imaging in depth [47, 49]. In SPIM, a confined illumination volume is used to excite a thin layer orthogonally oriented to the detection path, thus reducing the phototoxicity and allowing fast 3D imaging of biological samples in depth. Light-sheet-based microscopy has been recently proven to be a golden approach for improving imaging capabilities of entire organisms and tissues, thus improving the imaging depth capabilities and reducing scattering effects due to light–sample interactions [64]. Furthermore, light-sheet-based configurations have been successfully combined with 2PE [12, 13, 65, 88], showing a reduction of the scattering effects and improved imaging performances [52]. Here, the use of particular optical arrangements, such as light-sheet-based illumination [12, 13] and array tomography [63], extended the imaging depth capabilities (up to 150 microms) of single-molecule localization-based techniques in cellular cell spheroids and intact brain tissue, respectively. Furthermore, the use of efficient dyes emitting in the near-infrared range [2], the wide range of laser sources currently available, and the improved sensitivity of the new generation of sensors are also contributing to the extension of the applicability of super-resolution techniques toward tissue and whole-animal imaging.

A different optical microscopy approach, out from the scope of this paper but worth to be mentioned and able to reach sub-diffraction spatial resolution using non-uniform illumination, is represented by structured illumination (SIM). This technique is based on a widefield architecture, and periodic illumination patterns with high spatial frequency are applied to the sample. Images corresponding to different orientation and phases are used to reconstruct a super-resolution image with spatial resolution doubled in the focal plane compared to conventional microscopy [36]. Furthermore, the combination of SIM with fluorescence saturation (SSIM) allowed to push the spatial resolution reachable down to 50 nm in the 2D case [37]. SIM has been used for 3D super-resolution imaging at the cellular level in living samples [38, 72, 75]. SIM-based techniques are a challenging task in thick samples and tissues since the patterned illumination and the wavefronts are strongly affected/degraded in inhomogeneous samples. For this reason, SIM imaging is often restricted to cells or thin sections. Recently, a combination of line scanning and SIM microscopy improved the modulation and the quality of the illumination pattern, allowing optically sectioned images of 30-μm-thick fluorescent sample [57]. On the other side, a recent approach based on multifocal SIM microscopy, allowed three-dimensional (3D) super-resolution in live multicellular organisms [99]. A strong improvement in the maximum imaging depth reachable by SIM (>45 μm) has been obtained thanks to the rejection of out-of-focus light by means of sparse multifocal illumination patterns. This super-resolution technique allowed microtubules imaging in live transgenic zebrafish embryos.

In the next paragraphs, we focus on PALM-like and STED-like methods.

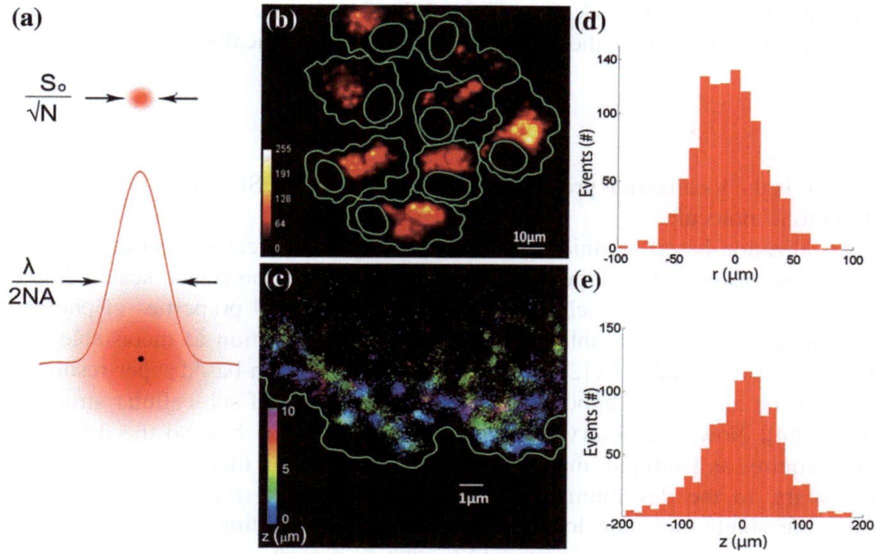

Fig. 1 Schematic representation of the localization process (**a**). Selective plane illumination imaging of the entire mammary cell spheroid (**b**) and super-resolution imaging of connexin43 spatial distribution (**c**) in different planes (*z color-coded map*). The effective localization accuracy estimated directly within thick biological samples in the radial (**d**) and axial (**e**) direction (68 and 141 nm, respectively). (Image modified from [12, 13])

3 Single-Molecule Localization Techniques (Palm-Like)

Detection of single molecules was a pioneering work, carried out more than 20 years ago in condensed phases [58]. The ability to recognize and localize single molecules from pixelated recordings was successful in several cases [6, 10, 16, 17, 31, 51, 85, 96, 97]. When one is confident enough of dealing with a single emitting molecule, then this extra knowledge allows one to interpret the center of the PSF as a measurement of the location of the molecule. Talking about super-resolution, so invoking the ability of discerning to light sources when at distances closer than the diffraction limit, could be risky without a series of "clever" assumptions when forming an image. Within this scenario, in the past few years, several emerging super-resolution techniques, based on localization of single molecules, have been used to provide information of the cellular protein organization with a spatial resolution far beyond the diffraction limit [7, 43, 70]. The basic idea behind these techniques is the repeated imaging and localization with high precision of single fluorophores within the sample. The final image can be rendered by plotting the position of each single emitter that can be precisely determined by measuring the center position of its image or PSF [61, 85]. As long as the number of photons collected for each emitter (N) is sufficient, the fluorophore position is determined with precision higher than the diffraction limit (as schematically shown in Fig. 1a):

$$\sigma \sim \frac{s_0}{\sqrt{N}}$$

where σ is the localization precision, s_0 is the size of the PSF, and N is the number of photons/molecule.

Localization-based techniques allow imaging of sub-cellular structures with nanometer resolution and provide information at the molecular scale. Single-molecule imaging can be achieved exploiting the spectral properties of photoactivatable and photoswitchable proteins [56] or the transition to metastable dark states of conventional dyes [29, 33, 40]. So far, localization-based super-resolution methods are also robust enough for multicolor imaging of sub-cellular structures [3, 77, 84]. Now, the observation of structures with size beyond the diffraction limit requires an isotropic improvement in resolution, and this brought, in the last few years, to the development of 3D super-resolution strategies that allowed achieving single-molecule localization along the three dimensions. The implemented setup based on engineered PSF such as astigmatism [46], double-plane detection [48], interferometric approaches [78], and double helix PSF [53] clearly demonstrated 3D super-resolution imaging at the cell level. Notwithstanding this, imaging of the whole cells is still limited and suffers from extended excitation of out-of-focus signal regions. A strategy to alleviate this effect is represented by 2PE: temporal focusing approaches [90] allowed super-resolution imaging of protein distribution deep inside cells (10 μm). On the other side, the development of new class of fluorophores suitable for two-photon-induced molecular switching [32] allowed super-resolution with optical sectioning. More recently, 2PE and temporal focusing have been also used to confine the activation process in order to perform 3D super-resolution at the whole-cell level [98].

The application of localization-based methods to thick samples (>50 μm), such as tissues and organisms, often represents a challenge since it is limited by the high background, aberrations, and refractive index mismatch. Several approaches contribute to solve this problem, from the sample preparation to the development of new optical approaches and more robust localization algorithms. First, particular attention needs to be addressed to the sample preparation procedure in order to reduce refractive index mismatch and aberrations [46]. Recently, approaches based on single-molecule localization and array tomography, *tomo*STORM [63] allowed imaging of tissue samples with molecular resolution. Similarly, the use of near-infrared lasers enables multicolor super-resolution imaging of complex biological samples, such as cardiac tissue sections and neuronal hippocampal cultures [2]. Efforts have also been addressed to solve the practical problem represented by optical aberrations that distort the single-molecule images affecting the localization precision. To this end, optimal 3D localization methods, taking into account for optical system aberrations in case of arbitrary 3D PSFs in the presence of noise, have been recently presented in literature [68]. Furthermore, the out-of-focus excitation and the high background can be alleviated by axially confining the excitation using an inclined illumination approach [87] or light-sheet-based microscopy [47] instead of a widefield illumination scheme.

Individual molecule localization combined with SPIM (IML-SPIM) has been recently applied to super-resolution imaging of thick mammary cell spheroids (up to 200 μm) [12, 13]. IML-SPIM allowed to reduce out-of-focus background signal and to observe the spatial distribution of connexin43 in selected regions of the cellular spheroid. The entire spheroid can be imaged by conventional SPIM (Fig. 1b) and 3D super-resolution images within a specific volume of interest can be performed at different depths (as represented in the color-coded map Fig. 1c). This approach, under two-photon photoactivation regime [74], has been further improved in terms of robustness to scattering artifacts [11–13, 52].

When dealing with thick scattering media, aberrations and light–sample interactions may decrease the SNR and thus the localization performances. Scattering effects and system instabilities can induce additional errors, and the localization precision can be redefined [1] by considering also the standard deviation σ_{inst} of the overall instabilities:

$$\sigma_{\mathrm{eff}} = \sqrt{\sigma + \sigma_{\mathrm{inst}}}$$

The real localization accuracy is strongly affected by low light conditions and background signal, and this effect is stronger when photons are collected deep within biological samples. When molecules are localized in depth within the sample, particular attention needs to be addressed to the real value of the effective accuracy [12, 13]. An experimental measure of its value can be directly obtained within the real sample by localizing point-like fluorescent objects in depth, both in the radial (63 nm) and axial (140 nm) direction, as shown in Fig. 1d and e, respectively. Similarly, a light-sheet-based approach has been used also for single particle tracking [80] deep within living tissue (up to 200 μm).

On the other side, the development of new robust algorithms for localization [81], able to localize molecule positions within high-density samples [62, 100] featuring high background regimes [79], represents a step forward super-resolution of more complicate and dense samples. In fact, an increasing number of new localization algorithms, based on gaussian fitting, MLE or compressed sensing have been developed to improve localization performances and temporal resolution and allowing single-molecule localization in highly dense samples. The synergy between the development of suitable optical architectures, new far-red or IR dyes and robust localization algorithms will smooth the way to super-resolution imaging of thicker samples or tissues [19].

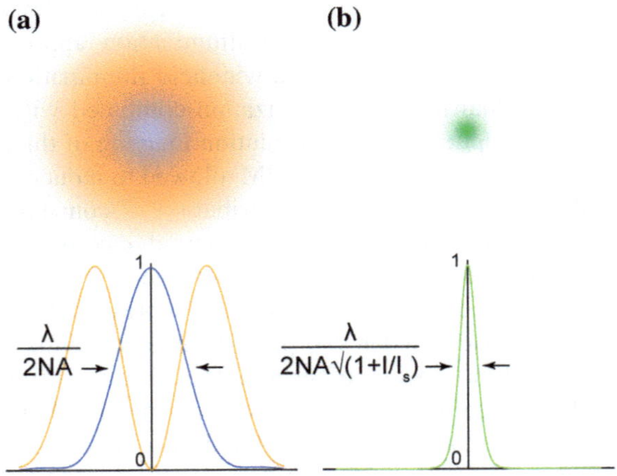

Fig. 2 The STED concept

4 Stimulated Emission Depletion (Sted-Like)

The first technique that made the diffraction barrier "crumbling" was STED microscopy [42]. In a STED microscope, a regularly focused excitation beam shared the focus with a second beam (usually called STED beam) able to de-excite the fluorophores via stimulated emission. Since the STED beam usually forms a doughnut shape, featuring a zero intensity in the center, all the fluorophores eventually located on the excitation spot are kept dark, except those in the proximity of the zero-intensity point, which, on the contrary, spontaneously emit, see Fig. 2. By increasing the intensity of the STED beam, the probability to de-excite the fluorescence via stimulated emission saturates at the outer part of the excitation spot, and the volume from which the fluorophores spontaneously emit decreases to sub-diffraction size. Scanning the two co-aligned beams across the sample and recording the spontaneously emitted fluorescence yield the final image whose spatial resolution can be tuned by the STED beam intensity.

Nevertheless, the spatial resolution of a STED system is theoretically "infinite"; many practical issues limit the ultimate value achievable. First, potential phototoxicity as well as photodamage effects on the sample limits the amount of STED light that is possible to imping. Second, many experimental imperfections in the focal intensity distributions of both STED and excitation beams, such as astigmatism, lateral misalignments, and zero-intensity point defects, reduce the contrast of the inhibition process, that is, so crucial obtaining substantial resolution improvement. The attainable resolution further reduces when imaging is performed depth into the sample, since to the experimental imperfections cited above, it is necessary adding scattering, absorption, and aberration induced by the sample itself.

An extreme way to circumvent the problem of depth imaging is to mimic the strategy adopted in electron microscopy: The sample is sectioned into a series of thin slices, imaged on the microscope, and then dedicated software helps to merging all the slices into a three-dimensional structure. This method can greatly improve the performances of correlative light-electron microscopy [92]. Punge et al. [67] combined this method with STED and demonstrated 80 nm 3D spatial resolution in immunostained cultured rat hippocampal neurons. Under a complete different perspective, 2PE microscopy allows combining 3D non-invasive imaging capabilities with a reduced sensitivity to scattering effects, thus providing a golden standard for tissue imaging. Driven by these advantages, several groups [27, 39, 54, 59] have shown that 2PE can be realized within the STED microscopy umbrella, thus pushing the resolution of 2PE microscopy beyond the diffraction barrier. 2PE microscopy is usually performed with a high-repetition (80 MHz) train of ultrashort ($\sim$200-fs) laser pulses provided by mode-locked Ti:Sapphire lasers. The high concentration of photons in this short time ensures the TPE process takes place. In stark contrast, de-exciting fluorophores by stimulated emission is one-photon process, meaning that the STED beam can be supplied also by continuous-wave beams. In fact, the first combination of STED with two-photon microscopy was realized with continuous-wave STED beams [27, 59]. This combination greatly simplified the combination of 2PE and STED, because the depletion beam does not need to be synchronized with the ultrafast 2PE laser beam pulses. However, despite its simplicity, the classical CW-STED implementation is not able to provide the same imaging performance of the all-pulsed STED implementation [94]. It recently turned out that the development of gated CW-STED (gCW-STED) microscopy is able to provide further advances toward applications on living systems by allowing a significant reduction of the intensity needed by the depletion beam to get a certain super-resolution. gCW-STED essentially relies on a pulse de-excitation beam and time-gated detection [94]. An intuitive formulation to explain the principle governing the gCW-STED is that the stimulated emission (SE) shortens the average time that a fluorophore spends in the excited state, that is, its effective excited-state lifetime s varies inversely with the de-excitation rate of SE, which increases linearly with the STED beam intensity I_{STED} (Fig. 3a). Consequently, in the doughnut-shaped pattern, the excited-state lifetime of a fluorophore changes according to its position. In particular, the effective excited-state lifetime t decreases away from the zero-intensity point, reaching a minimum in the proximity of the doughnut crest where there is the maximum STED beam intensity $I^{\mathrm{m}}_{\mathrm{STED}}$ (Fig. 3b). By collecting the photons after a time delay T_{g} from the excitation events (time-gated detection) that are triggered by the pulses of the excitation beam, it is possible to reject photons emitted by short-lived, excited-state fluorophores, located in the periphery, and highlight photons emitted by long-lived excited-state fluorophores located close to the zero-intensity point. As a result, the effective area from which the fluorescence signal is registered is further confined (Fig. 3b). Theoretically, this area can be tuned to infinitely small size by infinitely delaying the detection, namely the full width at half-maximum of the effective fluorescent area scales as $d_c/\sqrt{\left(1+d_{\mathrm{c}}^2 a^2 I^{\mathrm{m}}_{\mathrm{STED}}/I_{\mathrm{sat}}\left(1+T_{\mathrm{g}}/(\mathrm{t_{fl}}\ln 2)\right)\right)}$, with d_{c} the effective area of the confocal associated confocal microscope, I_{sat} the intensity of the STED beam for which

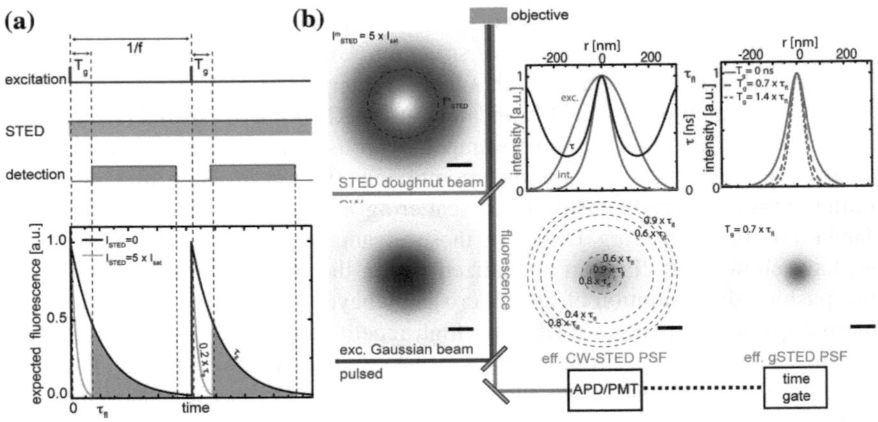

Fig. 3 Principles of gCW-STED microscopy. **a** Time evolution of the fluorescence signal in the absence ($I_{\text{STED}} = 0$) and presence ($I_{\text{STED}} = 5 \times I_{\text{sat}}$) of the STED beam with experimental time sequence (*upper panel*). **b** Schematic drawing of the gCW-STED setup with pulsed excitation (*blue*) and CW-STED (*red*) laser beams. Fluorescence light (*green*) is detected by the objective lens and imaged onto an avalanche photon detector (*APD*) or photomultiplier (*PMT*). A time-gating card discards the early photons (with respect to the excitation events). (**b**, *upper-left panel*) Calculated lateral intensity distribution for the excitation beam (*blue*), the effective region in which the dyes allow fluorescing (*green*) with the relative excite-state lifetime (*black*). (**b**, *lower-left panel*) The two-dimensional contour plot of the relative excite-state lifetime super-imposed with the two-dimensional effective fluorescence area. (**b**, *upper-right panel*) Calculated lateral intensity distribution for the gCW-STED effective point spread function for increasing time delay T_g of the gated detection. (**b**, *lower-right panel*) The two-dimensional effective point spread function. Scale bars 100 nm

the probability of spontaneous emission equals the probability of stimulated emission, t_{fl} the excited-state lifetime of the unperturbed fluorophore, and *a* a constant that depends on the doughnut shape of the STED beam. However, the loss of signal, which is intrinsic to the gated detection, fixes an upper limit on the time delay [91].

Two further approaches have been implemented to improve the 2PE-STED performances in depth. A first method has been recently realized by two groups [5, 54, 82], which have simultaneously elaborated an all-pulsed 2PE-STED implementation based on the synchronization of two femtosecond mode-locked Ti:Saphire lasers, see Fig. 4.

Another effective method performs 2PE-STED imaging using a single wavelength (SW) for 2PE and depletion [39]. The related SW 2PE-STED setup simplifies the image formation scheme, provides the super-resolution advantage of a pulsed STED, and allows maintaining a comparatively low cost, see Fig. 5.

It is worth noting that even if the stimulated emission process is a one-photon process, scattering of stimulating photons will not represent a source of background since, in most of the case, their wavelength is far away from the absorption

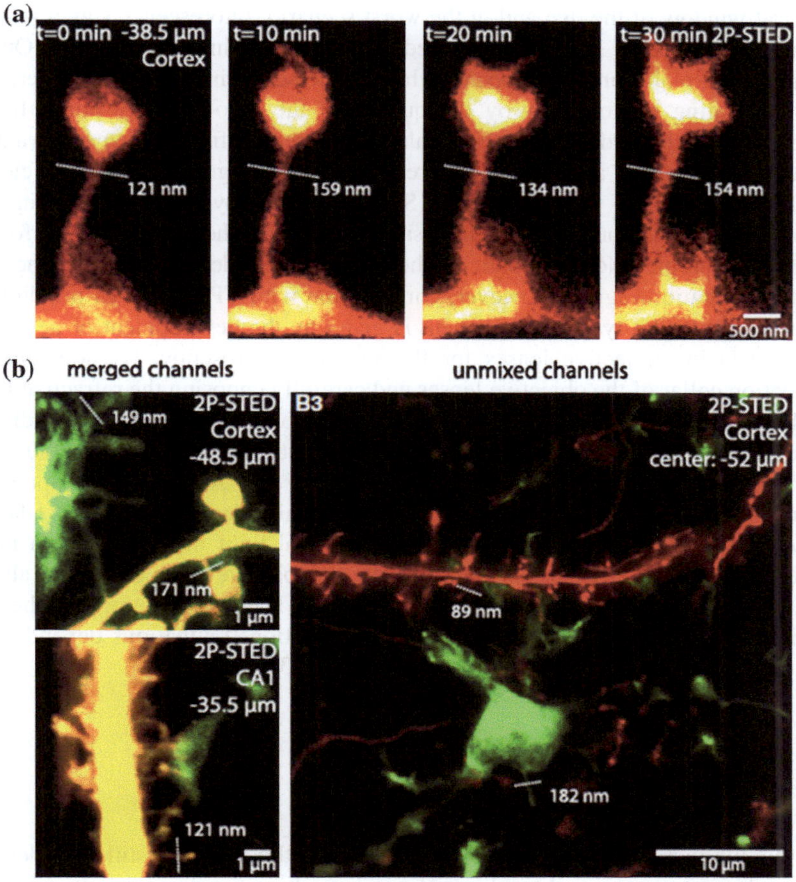

Fig. 4 Time-lapse and dual-color 2P-STED imaging. **a** Time-lapse images of a cortical spine acquired at 38.5 μm below the tissue surface. **b** Two-color 2PE-STED imaging of neurons and microglia (from [5])

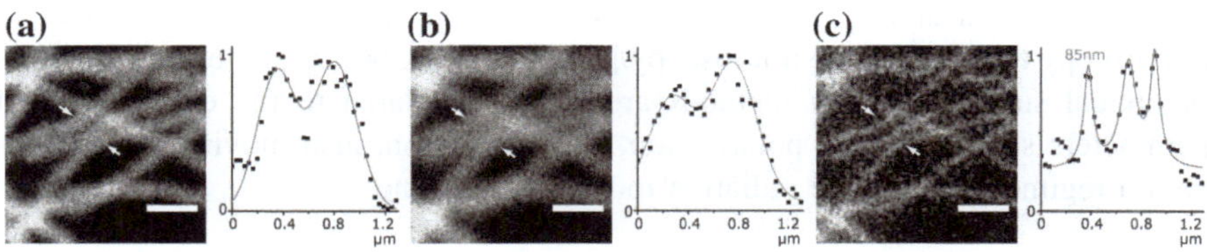

Fig. 5 Immunofluorescence 2PE-STED single-wavelength microscopy shown in comparison with confocal and 2PE microscopy (**a–c**). **a** Confocal (excitation at 633 nm), **b** 2PE (excitation at 770 nm), and **c** 2PE-STED (excitation and depletion at 770 nm) micrographs of microtubules immunostained with the dye ATTO647N. On the right of each image, we plotted the line profile (*black squares*) along the *arrows* indicated in the images, together with a multiple peak fit (*red*). (Scale bars: 1 μm). Modified from (Bianchini, Harke et al. 2012)

spectral window of the dye and in the worst scenario, fluorescence signal induced by the STED beam can be subtracted by lock-in technique [69, 93]. On the contrary, the major problem appears when due to scattering and due to aberration the stimulating photons destroy the quality of the zero-intensity point. In this scenario, stimulated depletion occurs also in the region from which one wants to collect fluorescence and the spatial resolution fast degrade [34]. Two elegant methods have been combined with STED microscopy for compensating the aberrations induced on the STED intensity distribution when imaging is performed deep into sample. Notably, both methods have been demonstrated in the one-photon excitation regime; however, combination with TPE could in principle be reliable. A possibility of improvement is given by the development of new aberration-reducing objective lenses for fluorescence microscopy. By adjusting the correction collar of the objective lenses and carefully choosing the refractive index of the objective embedding medium, Urban and colleagues [89] resolved distinct distribution of actin inside dendrites and spines with a resolution of 60–80 nm in living organotypic brain slices at depths up to 120 μm. Also, adaptive optics can be utilized to compensate for aberration in STED microscopy, which uses adaptive optics. Gould et al. [35] showed that using spatial light modulators both in the excitation and STED beam can substantially compensate sample-induced aberration in a three-dimensional STED implementation. It is worth noting that the SW-2PE-STED approach can improve the above-mentioned situations depending on the availability of suitable fluorescent molecules [39].

5 Perspective Note

Despite the fact that so-called super-resolution methods can be still considered in their infacy, so many technical variations have been implemented that is quite impossible to mention all of them. It is self-evident how targeted and stochastic methods tend to merge their own "technicalities." We think that two emerging classes of new approaches could be very promising, namely smart investigation of the sample toward 3D super-resolution in thick scattering specimens [25, 28] and coupling with scanning probe methods extending the domain of correlative microscopy to atomic force microscopy [14, 15, 28, 39, 60]. Also, further advances in optical super-resolution methods are the ones related to the exploitation of parameters such as phase, polarization, and absorption, also moving the investigation regime from visible radiation to the infrared one.

References

1. Aquino D, Schönle A, Geisler C, Middendorff CV, Wurm CA, Okamura Y, Lang T, Hell SW, Egner A (2011) Two-color nanoscopy of three-dimensional volumes by 4Pi detection of stochastically switched fluorophores. Nat Methods 8(4):353–359
2. Baddeley D, Crossman D, Rossberger S, Cheyne JE, Montgomery JM, Jayasinghe ID, Cremer C, Cannell MB, Soeller C (2011) 4D super-resolution microscopy with conventional fluorophores and single wavelength excitation in optically thick cells and tissues. PLoS One 6(5):e20645
3. Bates M, Huang GTDB, Zhuang X (2007) Multicolor super-resolution imaging with photo-switchable fluorescent probes. Science 317:1749–1753
4. Berning S, Willig KI, Steffens H, Dibaj P, Hell SW (2012) Nanoscopy in a living mouse brain. Science 335(6068):551
5. Bethge P, Chéreau R, Avignone E, Marsicano G, Nägerl UV (2013) Two-photon excitation STED microscopy in two colors in acute brain slices. Biophys J 104(4):778–785
6. Betzig E, Chichester RJ (1993) Single molecules observed by near field scanning optical microscopy. Science 262:1422–1425
7. Betzig E, Patterson GH, Sougrat R, Lindwasser OW, Olenych S, Bonifacino JS, Davidson MW, Lippincott-Schwartz J, Hess HF (2006) Imaging intracellular fluorescent proteins at nanometer resolution. Science 313(5793):1642–1645
8. Bianchini P, Harke B, Galiani S, Vicidomini G, Diaspro A (2012) Single-wavelength two-photon excitation-stimulated emission depletion (SW2PE-STED) superresolution imaging. Proc Natl Acad Sci USA 109(17): 6390–6393
9. Bianco B, Diaspro A (1989) Analysis of the three dimensional cell imaging obtained with optical microscopy techniques based on defocusing. Cell Biophys 15(3):189–200
10. Bobroff N (1986) Position measurement with a resolution and noise limited instrument. Rev Sci Instrum 57(6):1152–1157
11. Cella Zanacchi F, Lavagnino Z, Faretta M, Furia L, Diaspro A (2013) Light-sheet confined super-resolution using two-photon photoactivation. PLoS One 8(7):e67667
12. Cella Zanacchi F, Lavagnino Z, Perrone Donnorso M, Del Bue A, Furia L, Faretta M, Diaspro A (2011) Live-cell 3D super-resolution imaging in thick biological samples. Nat Methods 8(12):1047–1049
13. Cella Zanacchi F, Lavagnino Z, Ronzitti E, Diaspro A (2011) Two-photon fluorescence excitation within a light sheet based microscopy architecture. Proc SPIE 7903(1)
14. Chacko JV, Canale C, Harke B, Diaspro A (2013) Sub-diffraction nano manipulation using STED AFM. PLoS One 8(6):e66608
15. Chacko JV, Zanacchi FC, Diaspro A (2013) Probing cytoskeletal structures by coupling optical superresolution and AFM techniques for a correlative approach. Cytoskeleton 70(11):729–740
16. Chirico G, Cannone F, Beretta S, Baldini G, Diaspro A (2001) Single molecule studies by means of the two-photon fluorescence distribution. Microsc Res Tech 55(5):359–364
17. Chirico G, Cannone F, Beretta S, Diaspro A, Campanini B, Bettati S, Ruotolo R, Mozzarelli A (2002) Dynamics of green fluorescent protein mutant2 in solution, on spin-coated glasses, and encapsulated in wet silica gels. Protein Sci 11(5):1152–1161
18. Coelho M, Maghelli N, Tolic-Norrelykke IM (2013) Single-molecule imaging in vivo: the dancing building blocks of the cell. Integr Biol (Camb) 5(5):748–758
19. Del Bue A, Cella Zanacchi F, Diaspro A (2013). Super-resolution 3D reconstruction of thick biological samples: a computer vision perspective. IEEE international conference on computer vision (ICCV)
20. Denk W, Strickler JK, Webb WW (1990) Two photon laser scanning fluorescence microscopy. Science 248(4951):73–76
21. Denk W, Svoboda K (1997) Photon upmanship: why multiphoton imaging is more than a gimmick. Neuron 18(3):351–357

22. Diaspro A (2001) Confocal and two-photon microscoscopy: foundations, applications, and advances. Wiley-Liss Inc, New York
23. Diaspro A (2010a) Nanoscopy and multidimensional optical fluorescence microscopy. CRC Press, Taylor & Francis
24. Diaspro A (2010b) Optical fluorescence microscopy: from the spectral to the nano dimension. Springer, Berlin
25. Diaspro A (2013) Taking three-dimensional two-photon excitation microscopy further: encoding the light for decoding the brain. Microsc Res Tech 76(10):985–987
26. Diaspro A, Chirico G, Collini M (2006) Two-photon fluorescence excitation and related techniques in biological microscopy. Q Rev Biophys 15:1–70
27. Ding JB, Takasaki KT, Sabatini BL (2009) Supraresolution imaging in brain slices using stimulated-emission depletion two-photon laser scanning microscopy. Neuron 63(4): 429–437
28. Ducros M, Houssen YG, Bradley J, de Sars V, Charpak S (2013) Encoded multisite two-photon microscopy. Proc Natl Acad Sci USA 110(32):13138–13143
29. Egner A, Geisler C, von Middendorff C, Bock H, Wenzel D, Medda R, Andresen M, Stiel AC, Jakobs S, Eggeling C, Schönle A, Hell SW (2007) Fluorescence nanoscopy in whole cells by asynchronous localization of photoswitching emitters. Biophys J 93(9):3285–3290
30. Fernandez-Suarez M, Ting AY (2008) Fluorescent probes for super-resolution imaging in living cells. Nat Rev Mol Cell Biol 9(12):929–943
31. Flors C (2013) Super-resolution fluorescence imaging of directly labelled DNA: from microscopy standards to living cells. J Microsc 251(1):1–4
32. Fölling J, Belov V, Riedel D, Schönle A, Egner A, Eggeling C, Bossi M, Hell SW (2008) Fluorescence nanoscopy with optical sectioning by two-photon induced molecular switching using continuous-wave lasers. ChemPhysChem 9(2):321–326
33. Fölling J, Bossi M, Bock H, Medda R, Wurm CA, Hein B, Jakobs S, Eggeling C, Hell SW (2008) Fluorescence nanoscopy by ground-state depletion and single-molecule return. Nat Methods 5(11):943–945
34. Galiani S, Harke B, Vicidomini G, Lignani G, Benfenati F, Diaspro A, Bianchini P (2012) Strategies to maximize the performance of a STED microscope. Opt Express 20(7):7362–7374
35. Gould TJ, Burke D, Bewersdorf J, Booth MJ (2012) Adaptive optics enables 3D STED microscopy in aberrating specimens. Opt Express 20(19):20998–21009
36. Gustafsson MG (2000) Surpassing the lateral resolution limit by a factor of two using structured illumination microscopy. J Microsc 198(Pt 2):82–87
37. Gustafsson MG (2005) Nonlinear structured-illumination microscopy: wide-field fluorescence imaging with theoretically unlimited resolution. Proc Natl Acad Sci USA 102:13081–13086
38. Gustafsson MG, Shao L, Carlton PM, Wang CJR, Golubovskaya IN, Cande WZ, Agard DA, Sedat JW (2008) Three-dimensional resolution doubling in wide-field fluorescence microscopy by structured illumination. Biophys J 94(12):4957–4970
39. Harke B, Chacko JV, Haschke H, Canale C, Diaspro A (2012) A novel nanoscopic tool by combining AFM with STED microscopy. Opt Nanoscopy 1(1):3
40. Heilemann M, van de Linde S, Mukherjee A, Sauer M (2009) Super-resolution imaging with small organic fluorophores. Angew Chem Int Ed Engl 48(37):6903–6908
41. Hell SW (2007) Far-field optical nanoscopy. Science 316(5828):1153–1158
42. Hell SW, Wichmann J (1994) Breaking the diffraction resolution limit by stimulated emission: stimulated-emission-depletion fluorescence microscopy. Opt Lett 19(11):780–782
43. Hess ST, Girirajan TPK, Mason MD (2006) Ultra-high resolution imaging by fluorescence photoactivation localization microscopy. Biophys J 91(11):4258–4272
44. Hou S, Liang L, Deng S, Chen J, Huang Q, Cheng Y, Fan C (2013) Nanoprobes for super-resolution fluorescence imaging at the nanoscale. Sci China Chem 57(1):100–106
45. Huang B (2011) An in-depth view. Nat Methods 8(4):304–305

46. Huang B, Jones SA, Brandenburg B, Zhuang X (2008) Whole-cell 3D STORM reveals interactions between cellular structures with nanometer-scale resolution. Nat Methods 5(12):1047–1052
47. Huisken J, Swoger J, Bene FD, Wittbrodt J, Stelzer EHK (2004) Optical sectioning deep inside live embryos by selective plane illumination microscopy. Science 305(5686): 1007–1009
48. Juette MF, Gould TJ, Lessard MD, Mlodzianoski MJ, Nagpure BS, Bennett BT, Hess ST, Bewersdorf J (2008) Three-dimensional sub-100 nm resolution fluorescence microscopy of thick samples. Nat Methods 5(6):527–529
49. Keller PJ, Schmidt AD, Wittbrodt J, Stelzer EHK (2008) Reconstruction of zebrafish early embryonic development by scanned light sheet microscopy. Science 322(5904):1065–1069
50. Kempf C, Staudt T, Bingen P, Horstmann H, Engelhardt J, Hell SW, Kuner T (2013) Tissue multicolor STED nanoscopy of presynaptic proteins in the calyx of held. PLoS One 8(4):e62893
51. Kim H, Ha T (2013) Single-molecule nanometry for biological physics. Rep Prog Phys 76:1–16
52. Lavagnino Z, Cella Zanacchi F, Ronzitti E, Diaspro A (2013) Two-photon excitation selective plane illumination microscopy (2PE-SPIM) of highly scattering samples: characterization and application. Opt Express 21(5):5998–6008
53. Lee H-LD, Sahl SJ, Lew MD, Moerner WE (2012) The double-helix microscope super-resolves extended biological structures by localizing single blinking molecules in three dimensions with nanoscale precision. Appl Phys Lett 100(15):153701–1537013
54. Li Q, Wu SSH, Chou KC (2009) Subdiffraction-limit two-photon fluorescence microscopy for GFP-tagged cell imaging. Biophys J 97(12):3224–3228
55. Loew LM, Hell SW (2013) Superresolving dendritic spines. Biophys J 104(4):741–743
56. Lukyanov KA, Chudakov DM, Lukyanov S, Vverkhusha V (2005) Innovation: photoactivatable fluorescent proteins. Nat Rev Mol Cell Biol 6(11):885–889
57. Mandula O, Wicker MKK, Krampert G, Kleppe I, Heintzmann R (2012) Line scan: structured illumination microscopy super-resolution imaging in thick fluorescent samples. Opt Express 20:24167–24174
58. Moerner WE, Kador L (1989) Optical detection and spectroscopy of single molecules in a solid. Phys Rev Lett 62:2535–2538
59. Moneron G, Hell SW (2009) Two-photon excitation STED microscopy. Opt Express 17(17):14567–14573
60. Monserrate A, Casado S, Flors C (2013) Correlative atomic force microscopy and localizationbased superresolution microscopy: revealing labelling and image reconstruction artefacts. ChemPhysChem Comm, pp 1–5 (in press)
61. Mortensen KI, Churchman LS, Spudich JA, Flyvbjerg H (2010) Optimized localization analysis for single-molecule tracking and super-resolution microscopy. Nat Methods 7(5):377–381
62. Mukamel EA, Babcock H, Zhuang X (2012) Statistical deconvolution for superresolution fluorescence microscopy. Biophys J 102(10):2391–2400
63. Nanguneri S, Flottmann B, Horstmann H, Heilemann M, Kuner T (2012) Three-dimensional, tomographic super-resolution fluorescence imaging of serially sectioned thick samples. PLoS One 7(5):e38098
64. Ntziachristos V (2010) Going deeper than microscopy: the optical imaging frontier in biology. Nat Methods 7(8):603–614
65. Palero J, Santos SICO, Artigas D, Loza-Alvarez P (2010) A simple scanless two-photon fluorescence microscope using selective plane illumination. Opt Express 18(8):8491–8498
66. Pawley J (Ed.) (2006) Handbook of biological confocal microscopy, 3rd ed., XXVIII, p 988 Springer
67. Punge A, Rizzoli SO, Jahn R, Wildanger JD, Meyer L, SchÃnle A, Kastrup L, Hell SW (2008) 3D reconstruction of high-resolution STED microscope images. Microsc Res Tech 71(9):644–650

68. Quirin S, Pavani SRP, Piestun R (2012) Optimal 3D single-molecule localization for superresolution microscopy with aberrations and engineered point spread functions. Proc Natl Acad Sci USA 109(3):675–679
69. Ronzitti E, Harke B, Diaspro A (2013) Frequency dependent detection in a STED microscope using modulated excitation light. Opt Express 21(1):210–219
70. Rust MJ, Bates M, Zhuang X (2006) Sub-diffraction-limit imaging by stochastic optical reconstruction microscopy (STORM). Nat Methods 3(10):793–795
71. Sahl SJ, Moerner WE (2013) Super-resolution fluorescence imaging with single molecules. Curr Opin Struct Biol 23(5):778–787
72. Schermelleh L, Carlton PM, Haase S, Shao L, Winoto L, Kner P, Burke B, Cardoso MC, Agard DA, Gustafsson MGL, Leonhardt H, Sedat JW (2008) Subdiffraction multicolor imaging of the nuclear periphery with 3D structured illumination microscopy. Science 320(5881):1332–1336
73. Schermelleh L, Heintzmann R, Leonhardt H (2010) A guide to super-resolution fluorescence microscopy. J Cell Biol 190(2):165–175
74. Schneider M, Barozzi S, Testa I, Faretta M, Diaspro A (2005) Two-photon activation and excitation properties of PA-GFP in the 720-920-nm region. Biophys J 89(2):1346–1352
75. Shao L, Kner P, Rego EH, Gustafsson MGL (2011) Super-resolution 3D microscopy of live whole cells using structured illumination. Nat Methods 8(12):1044–1046
76. Sheppard CRJ (2002) The generalized microscope. In: Diaspro A (ed) Confocal and two-photon microscopy: foundations, applications and advances. Wiley-Liss, New York, pp 1–18
77. Shroff H, Galbraith CG, Galbraith JA, White H, Gillette J, Olenych S, Davidson MW, Betzig E (2007) Dual-color superresolution imaging of genetically expressed probes within individual adhesion complexes. Proc Natl Acad Sci USA 104(51):20308–20313
78. Shtengel G, Galbraith JA, Galbraith CG, Lippincott-Schwartz J, Gillette JM, Manley S, Sougrat R, Waterman CM, Kanchanawong P, Davidson MW, Fetter RD, Hess HF (2009) Interferometric fluorescent super-resolution microscopy resolves 3D cellular ultrastructure. Proc Natl Acad Sci USA 106(9):3125–3130
79. Smith CS, Joseph N, Rieger B, Lidke KA (2010) Fast, single-molecule localization that achieves theoretically minimum uncertainty. Nat Methods 7(5):373–375
80. Spille J-H, Kaminski T, Königshoven H-P, Kubitscheck U (2012) Dynamic three-dimensional tracking of single fluorescent nanoparticles deep inside living tissue. Opt Express 20(18):19697–19707
81. Starr R, Stahlheber S, Small A (2012) Fast maximum likelihood algorithm for localization of fluorescent molecules. Opt Lett 37(3):413–415
82. Takasaki KT, Ding JB, Sabatini BL (2013) Live-cell superresolution imaging by pulsed STED two-photon excitation microscopy. Biophys J 104(4):770–777
83. Testa I, Urban NT, Jakobs S, Eggeling C, Willig KI, Hell SW (2012) Nanoscopy of living brain slices with low light levels. Neuron 75(6):992–1000
84. Testa I, Wurm CA, Medda R, Rothermel E, von Middendorf C, Fölling J, Jakobs S, Schönle A, Hell SW, Eggeling C (2010) Multicolor fluorescence nanoscopy in fixed and living cells by exciting conventional fluorophores with a single wavelength. Biophys J 99(8):2686–2694
85. Thompson RE, Larson DR, Webb WW (2002) Precise nanometer localization analysis for individual fluorescent probes. Biophys J 82(5):2775–2783
86. Tinoco I Jr, Gonzalez RL Jr (2011) Biological mechanisms, one molecule at a time. Genes Dev 25(12):1205–1231
87. Tokunaga M, Imamoto N, Sakata-Sogawa K (2008) Highly inclined thin illumination enables clear single-molecule imaging in cells. Nat Methods 5(2):159–161
88. Truong TV, Supatto W, Koos DS, Choi JM, Fraser SE (2011) Deep and fast live imaging with two-photon scanned light-sheet microscopy. Nat Methods 8(9):757–760
89. Urban NT, Willig KI, Hell SW, Nagerl UV (2011) STED nanoscopy of actin dynamics in synapses deep inside living brain slices. Biophys J 101(5):1277–1284

90. Vaziri A, Tang J, Shroff H, Shank CV (2008) Multilayer three-dimensional super resolution imaging of thick biological samples. Proc Natl Acad Sci USA 105(51):20221–20226
91. Vicidomini G, Coto Hernandez I, d'Amora M, Cella Zanacchi F, Bianchini P, Diaspro A (2013) Gated CW-STED microscopy: a versatile tool for biological nanometer scale investigation. Methods
92. Vicidomini G, Gagliani MC, Cortese K, Krieger J, Buescher P, Bianchini P, Boccacci P, Tacchetti C, Diaspro A (2010) A novel approach for correlative light electron microscopy analysis. Microsc Res Tech 73(3):215–224
93. Vicidomini G, Moneron G, Eggeling C, Rittweger E, Hell SW (2012) STED with wavelengths closer to the emission maximum. Opt Express 20(5):5225–5236
94. Vicidomini G, Moneron G, Han KY, Westphal V, Ta H, Reuss M, Engelhardt J, Eggeling C, Hell SW (2011) Sharper low-power STED nanoscopy by time gating. Nat Methods 8(7):571–575
95. Wilson T, Sheppard CJR (1984) Theory and practice of scanning optical microscopy. Academic Press, Waltham
96. Xie XS (1996) Single-molecule spectroscopy and dynamics at room temperature. Acc Chem Res 29:598–606
97. Yildiz A, Selvin PR (2005) Fluorescence imaging with one nanometer accuracy: application to molecular motors. Acc Chem Res 38(7):574–582
98. York AG, Ghitani A, Vaziri A, Davidson MW, Shroff H (2011) Confined activation and subdiffractive localization enables whole-cell PALM with genetically expressed probes. Nat Methods 8(4):327–333
99. York AG, Parekh SH, Nogare DD, Fischer RS, Temprine K, Mione M, Chitnis AB, Combs CA, Shroff H (2012) Resolution doubling in live, multicellular organisms via multifocal structured illumination microscopy. Nat Methods 9(7):749–754
100. Zhu L, Zhang W, Elnatan D, Huang B (2012) Faster STORM using compressed sensing. Nat Methods 9(7):721–723

Superhydrophobic Devices Molecular Detection

Tania Limongi, Lorenzo Ferrara, Gobind Das, Manola Moretti, Monica Marini, Ermanno Miele, Angelo Accardo, Raffaella Raimondo, Francesco Gentile and Enzo Di Fabrizio

Abstract Recent advances in the single-molecule detection and manipulation provided unexpected solutions for the understanding of the physio-pathological behavior of individual biological macromolecules. Modern techniques of patterning at the micro- and nanometer scale combined with chemical treatments are being used to create surfaces that stretch the hydrophobic behavior to the limit. The ability to create surfaces with high static water contact angles (usually greater than 150°) is essential for a variety of applications, ranging from the development of biosensors to the implementation of sensitive and reliable single-molecule collection and sample preparation methods. Thus, superhydrophobic devices could be considered as nano-biotechnological single-molecule detection tools that can be applied to a wide range of high-resolution studies. To outline the paper, single-molecule detection topics and theoretical principles of superhydrophobicity are first introduced. A comprehensive overview is then given, describing how different types

T. Limongi (✉) · L. Ferrara · M. Moretti · E. Miele · A. Accardo · F. Gentile
Nanostructures, Istituto Italiano di Tecnologia, via Morego 30, 16163 Genoa, Italy
e-mail: tania.limongi@iit.it

L. Ferrara
e-mail: lorenzo.ferrara@iit.it

M. Moretti
e-mail: manola.moretti@Kaust.edu.sa

E. Miele
e-mail: ermanno.miele@iit.it

A. Accardo
e-mail: angelo.accardo@iit.it

F. Gentile
e-mail: francesco.gentile77@gmail.com

T. Limongi
Neuroscience and Brain Technology, Istituto Italiano di Tecnologia,
via Morego 30, 16163 Genoa, Italy

F. Benfenati et al. (eds.), *Novel Approaches for Single Molecule Activation and Detection*, Advances in Atom and Single Molecule Machines,
DOI: 10.1007/978-3-662-43367-6_4,

of devices with superhydrophobic surfaces are realized. Finally, the usefulness of the presented devices for a wide range of applications and the concluding comments are proposed.

1 Introduction

1.1 Single-Molecule Detection

The interest in single-molecule characterization and detection spans from chemistry, biomedicine, and nanoscience. Single-molecule detection properties normally averaged out and therefore hidden in bulk experiments, reveal essential information about the interaction of the molecule with its nano-environment. Among all the approaches for the study of single molecules, scanning probe microscopy (SPM) represents a powerful solution thanks to its ability to accurately probe and control chemical synthesis at the single-molecule level [1]. Versatile experimental condition (air, liquid, vacuum)-allowed atomic force microscope (AFM) was used to image and change conformational states of nucleic acid and protein [2, 3]. Experiments with near-field scanning optical microscopy (NSOM) led to a scan, at a resolution higher than the diffraction limit, of the near field generated by a sample, such as a single molecule on a surface, upon illumination [4]. An alternative but equivalent single-molecule detection approach is given by confocal microscopy; the ability to track and identify single molecules plays a very important part in biology and functional genomic [5]. By exploiting the characteristics of fluorescence, various

T. Limongi · G. Das · M. Marini · Enzo Di Fabrizio (✉)
Physical Science and Engineering Division, King Abdullah University of Science and Technology, Bldg. 3, 3rd floor, Office # 3235, Thuwal 23955-6900, Kingdom of Saudi Arabia
e-mail: enzo.difabrizio@kaust.edu.sa

G. Das
e-mail: gobind.das@kaust.edu.sa

M. Marini
e-mail: monica.marini@kaust.edu.sa

M. Moretti
Computer, Electrical and Mathematical Sciences and Engineering (CEMSE) Division, King Abdullah University of Science and Technology, Thuwal 23955-6900, Kingdom of Saudi Arabia

R. Raimondo · F. Gentile
Lab. BIONEM, Dipartimento di Medicina Sperimentale e Clinica, Università degli Studi Magna Graecia di Catanzaro, viale Europa, loc. Germaneto, 88100 Catanzaro, Italy
e-mail: r.raimondo@unicz.it

techniques have been developed that enable the visualization and analysis of slow dynamics of molecular machines and even establish spatial distributions of fluorescently tagged macromolecules. Nanometer-localized multiple single-molecule (NALMS) fluorescence microscopy, fluorescence imaging with one-nanometer accuracy (FIONA), fluorescence resonance energy transfer (FRET), fluorescence recovery after photobleaching (FRAP), fluorescence loss in photobleaching (FLIP), and stochastic optical reconstruction microscopies (STORM) could be applied to monitor fluorescent molecules in time and space with high sensitivity and contrast [6–8].

The most critical aspect of performing single-molecule experiment is the choice of the correct sample preparation in terms of methods and substrates coupled to each specific detection technique (fluorescent microscopy, AFM, Raman spectroscopy, electron microscopy, and so on).

Single-molecule techniques typically use tools that consist of essential parts at nano- and micrometer scale, which is in general determined by the size of single molecule or cell to be investigated. Microfabrication techniques are largely contributing to design and fabricate nano- and micropatterned devices for single-molecule manipulation, probing, and sensing. For instance, nanopores [9], nanochannels [10–12], nanomechanical sensors [13], nanoantennas [14, 15], and more generally laboratory-on-chips [16, 17] are the ultimate developments of micro-nanotechnology aimed at single-molecule detection. In this paper, we describe how our micro- and nanofabricated superhydrophobic surfaces can be used to concentrate and vehiculate to specific detection points the analyte to achieve single-molecule detection with a high signal-to-noise ratio, allowing X-ray diffraction, Raman spectroscopy, and electron microscopy characterizations.

1.2 Superhydrophobic Surfaces

In surfaces and interfaces sciences, superhydrophobicity represents a well-known and studied phenomenon that may be referred to the ability of a surface of repelling water. In particular, a drop of water in contact with a superhydrophobic surface tends to remain confined, assuming a quasi-spherical shape, pursuing low surface interaction. This behavior and the solid–liquid interaction can be quantitatively measured by the contact angle (CA) value.

The relationship between equilibrium contact angle and the interfacial tensions at interface is described by the Young equation:

$$\cos\theta = \frac{\gamma_{sv} - \gamma_{sl}}{\gamma_{lv}} \tag{1}$$

γ is generally a tension (in N/m) at the interface, described by the subscript s, v, and l, for solid, vapor, and liquid phase, respectively. A surface showing a water CA larger than 150° and low hysteresis (the difference between the advancing and receding contact angle) is considered to be superhydrophobic. The superhydrophobic effect is related to surface morphology and chemistry. To illustrate the effect of roughness on contact angle of liquid drops, a few models have been proposed; Wenzel [18] and Cassie-Baxter [19] are the two universally accepted wetting theories. The first one predicts that the surface roughness increases the surface area of the solid and that consequently modifies the surface contact angle, according to:

$$\cos\theta^* = r\cos\theta \tag{2}$$

where $\theta*$ is the apparent CA, while r is the surface roughness, and θ is the contact angle on a smooth surface made by the same material, i.e., with the same surface chemistry. The second model, i.e., Cassie-Baxter, describes a configuration of the interface between solid and liquid in which the formation of air pockets occurs; thus, it is valid in the study of composite wetting states that involve the presence of solid, liquid, and vapor phases:

$$\cos\theta^* = -1 + \phi_s(1 + \cos\theta) \tag{3}$$

Here, ϕ_s represents the fraction of solid in contact with the liquid. In Cassie-Baxter wetting regime, a contact angle $\theta^* > 90°$ is allowed even if $\theta < 90°$. This phenomenon can be explained in terms of surface energy by thermodynamic consideration such as the presence of metastable wetting states [20].

The application and the effectiveness of the two models depend on the physical nature of the solid–liquid interface. In particular, the Wenzel model is effective in describing superhydrophobicity on surfaces with low surface energy and pinning effects of the evaporating droplet. On the other hand, the Cassie-Baxter model is more adequate to the case of interfaces composed by a hierarchical roughness or in general when the droplet does not pin the asperities but remains in a suspended state. In this case, air remains trapped between the asperities, creating an equilibrium condition between evaporative phenomena and tensions at the interfaces. The transition between Wenzel and Cassie-Baxter regimes may occur in case of break in surface symmetry and perturbation of the metastable wetting [20]. Despite the robustness of the proposed superhydrophobicity models, many research activities are conducted in order to better explain and describe the wetting phenomena, especially the transition between Wenzel and Cassie states [20].

Superhydrophobic surfaces hold a paramount importance in a variety of industrial and scientific applications. For this reason, researchers are engaging in profuse and ever-growing efforts in order to develop efficient fabrication methods for robust surfaces, targeted to different application fields: functional coatings, fluidic drag reduction, improved water-supporting force, and controlled fluidics in heat and mass transfer, just to name a few [21].

Among others, one of the most useful features of superhydrophobic surfaces is the possibility to tune the evaporative phenomena triggering and driving nano-object assembly processes and concentrating solutes in a precise point [22–26]. In particular, the latter functionality has been successfully exploited as strategy to design efficient sensors with extra-low sensitivity (attomolar) [23] and tune surface energy to control over force field at the interfaces between solid, liquid, and vapor [24].

2 Fabrication

2.1 Superhydrophobic Pillared Substrates

Among the possible fabrication approaches of superhydrophobic surfaces, textured pillars ones undoubtedly hold a wide popularity. Pillared substrates realization is based on robust microfabrication techniques, mainly based on optical lithography and inductively coupled plasma reactive ion etch (ICPRIE). The parameters of the surface (i.e., pillars' gap, height, diameter, and area fraction ϕ_s) can be easily tuned, and this fact offers the possibility to modify the superhydrophobicity in terms of stability, CA value, and hysteresis, remaining in a Cassie-Baxter wetting regime. Here, we will describe the fabrication steps of a device composed by hexagonal lattice of pillars equipped with holes embedded on the pillars lattice that can be used in conventional transmission electron microscopy (TEM) experiments [24]. As regards the holes realization double polished, ⟨100⟩, 50 μm thick, p-type thin silicon wafers were purchased from Si-Mat (Silicon Materials, Kaufering, Germany); they were cleaned with acetone and isopropanol to remove possible contaminant and then etched with a 4 % wet HF solution. The wafers were then rinsed with DI water and dried with N_2. A 100-nm layer of chrome was deposited upon the back side of the substrates using a sputter coater (Q150TES Quorum Technologies, Dixon, CA, USA). Standard optical lithography techniques (Suss Microtec MA6/BA6, Sunnyvale, CA, USA) were employed to realize a regular pattern of disks within a layer of positive resist (SPR220, from Rohm and Haas) that was spin-coated onto the chrome layer. Afterward, the chrome was removed from the disks by dipping the substrate for 50 s in a standard chrome-etching solution (ETCH 18 from Organo Spezial Chemie (OSC), Bitterfeld, Germany). Upon removal of the residual resist with acetone and oxygen plasma, a deep reactive ion etching (DRIE) process (SI 500 Sentech Instruments Gmbh, Berlin, Germany) was used to etch holes passing through the substrate, and in this process, the patterned chrome layer served as a mask. The 6-μm-diameter holes were sufficiently large to allow a practical use in TEM imaging. A number of alignment markers were conveniently positioned around the holes to be used as a reference frame in the fabrication of the pillars paired to the holes. The samples were then immersed in a chrome-etching solution to dissolve the remaining chrome. The substrate was then turned upside down, and the fabrication process was pursued on the front side of the samples.

As regards the pillars fabrication, a second lithography exposure step was used to realize a regular hexagonal pattern of disks within a layer of negative resist (AZ5214 from MicroChemicals GmbH, Ulm, Germany). Due to an accurate alignment procedure with the alignment markers, obtained in the precedent fabrication step, the disks were positioned exactly on the middle point of each subset of 6 holes; thereby, a highly symmetric design was obtained. A DRIE process was therefore used, whereby the final pillars were obtained with a height of about 10 μm, a diameter of 10 μm, and a pitch of 30 μm (Fig. 1). These values are consistent with a criterion of optimal design as described in [23], and hydrophobicity with contact angles approaching 170° is guaranteed.

2.2 Polymeric Superhydrophobic Substrates

The need to have a material more flexible and cheaper than Si motivated the development of novel processes for the fabrication of poly(methyl methacrylate) (PMMA) superhydrophobic surfaces [27, 28]. These PMMA substrates can be also thermally shaped acquiring a parabolic curvature, which allow the deposition of the droplets on an exact point without oscillations. Here we present two different kind of polymeric superhydrophobic substrates. The first one is characterized by micropillar structures realized with a two-step fabrication process composed by an optical lithography step and a following plasma treatment. The second one is characterized by PMMA nanofibrils realized by means of a single-step plasma roughening [27, 28].

The process developed to fabricate a micrometric topology is shown schematically in Fig. 2a. 500 μm PMMA sheets were coated with a 100-nm-gold layer by a sputtering process using a microplasma system (I). The gold film served as a physical mask to create the desired geometries by final plasma etches. An optical lithography process was used to create the pattern array that defined the position of the pillars. In a first step, a layer of AZ5214 positive tone resist was spin-coated at 4000 rpm for 60 s to reach a final thickness of about 1.5 μm (II). In order to keep a protective layer of gold during the deep reactive ion etch (DRIE) step, the tone of AZ5214 had to be inverted. After spin coating, a prebaking step (60 s at 95 °C on a hot plate) was necessary to make the solvent evaporate and to accomplish a better adhesion between the resist and the Au layer. The following step was the exposure to UV light ($\lambda \sim 365$ nm) for 19 s by means of a proximity contact mask aligner. The post-exposure bake was the most crucial phase since it was responsible for the "image reversal." The optimal bake-out was found to be 3–4 min at 95 °C. A 60-s blank UV exposure (i.e., without the optical mask) was employed to make the previously unexposed areas soluble. Then, an AZ400 K:H_2O = 1:4 development was used to create the resist array pattern (III). To remove the exposed gold areas, an isotropic wet etch based on aqueous KI/I_2 solution was needed (IV).

The final step was composed of a two-step plasma process spaced out by a gold etch removal. A hybrid etch process based on a gas mixture of O_2/Ar (25 sccm/30 sccm) allowed obtaining both an ion-enhanced chemical and a physical sputter etching (V). By optimizing the parameters of pressure, electrical power of the platen and of the coil, etch period, and gas ratios, a good aspect ratio and slightly

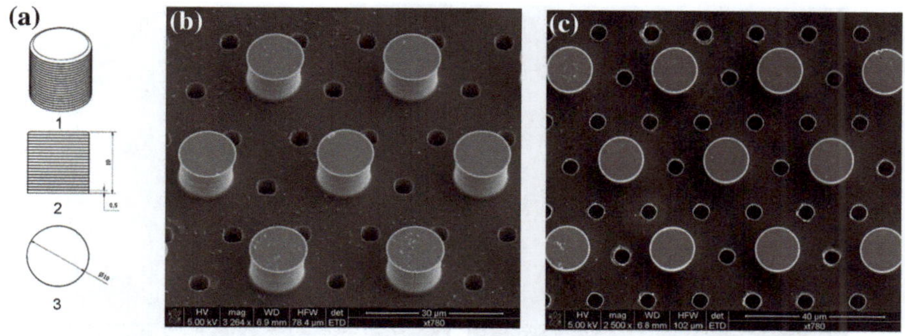

Fig. 1 Pictorial drawing (**a**) of a silicon pillar in axonometric projection (*1*), *side view* (*2*), and *top view* (*3*). Scanning electron micrographs in *tilted view* (**b**) and *top view* (**c**) of a pillars' textured superhydrophobic surface equipped with hexagonal lattice of holes

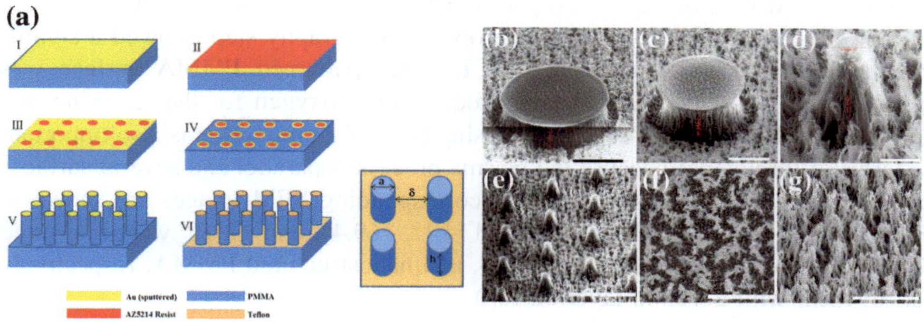

Fig. 2 **a** Microfabrication process steps to develop micropatterned superhydrophobic surfaces: (*I*) gold sputtering; (*II*) AZ5214 spin coating; (*III*) resist baking, exposure, tone inversion, and development; (*IV*) gold etch; (*V*) DRIE plasma process; (*VI*) gold stripping and Teflon coating. The parameters describing the dimensions and separation of the microcylinders are shown schematically to the right. Typical parameters are: a = 10 μm, $\delta = 20$ μm, and h = 15 μm. **b** O_2/Ar plasma etch phases for micro- and nanostructured PMMA surfaces 10 min (with Au mask, 5-μm scale bar); **c** 20 min (with Au mask, 5-μm scale bar); **d** 40 min (Au mask stripped, 5-μm scale bar). **e** overall PMMA micropillars-based SHS (40-μm scale bar); **f**, **g** 12-min O_2 plasma etch of flat PMMA substrate (10-μm scale bar). (kindly adapted from [27])

bent sidewalls could be obtained. The introduction of Ar and the reduction of the oxygen component indeed avoided the problem of extreme tapering at the cost of a slower etch rate (~0.3/0.4 μm/min). The optimal etch period in terms of aspect ratio was found to be 40 min. After that, a final gold wet etch was used to remove the Au mask caps from top of the pillars and a 5s C_4F_8 (48 sccm)-based plasma process coated the whole micropatterned surface with a thin (~5/10 nm) amorphous Teflon layer (VI). The formation of micropillars during the plasma etch process is shown in Fig. 2b–e, showing a coexistence of micro- and nanopatterned features.

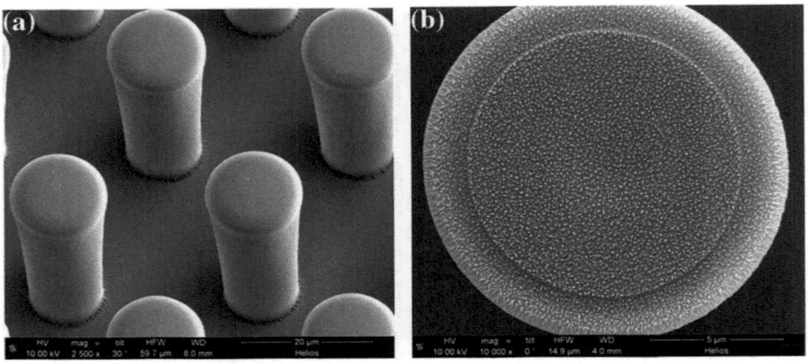

Fig. 3 **a** SU8 pillars; **b** *top view* of an SU8 pillar further processed by plasma nanoroughening on *top*

The second kind of superhydrophobic PMMA substrate included the presence of nanostructured fibers. Avoiding the optical lithography step allowed a significant speed-up of the whole process. The nanopatterned PMMA surface was obtained through a two-step plasma process (pure oxygen for the texturing and C_4F_8 for the Teflon layer) with a processing time of 12 min. This is also important in terms of throughput because the plasma process is parallel and several surfaces can be simultaneously processed. The corresponding SEM images are shown in Fig. 2f and g. Initial contact angles of 114.1°, 169.4°, and 171.1° were measured for the flat PMMA Teflon layer, micro-, and nanostructured PMMA, respectively (all of them with a Teflon layer on top).

Another type of microfabricated polymeric substrate for the study of molecule aggregation and characterization, in particular DNA, has been recently developed using the SU8 photosensitive resin. The great advantage from the use of this photoresist consists in the fact that the formation of the micrometric pillars is obtained directly through the optical lithography process; a further plasma processing is required only for the final Teflon deposition and eventually for the obtainment of nanofibrils on top of the pillars (Fig. 3).

3 Applications

3.1 *Molecular Detection of Organic and Inorganic Sample's Structural Information Through In-situ X-ray Diffraction*

The polymeric nanostructured superhydrophobic surfaces, just described in the previous section, have been extensively used for accurate in situ detection of conformational changes of biological and inorganic compounds [29, 30]. The general

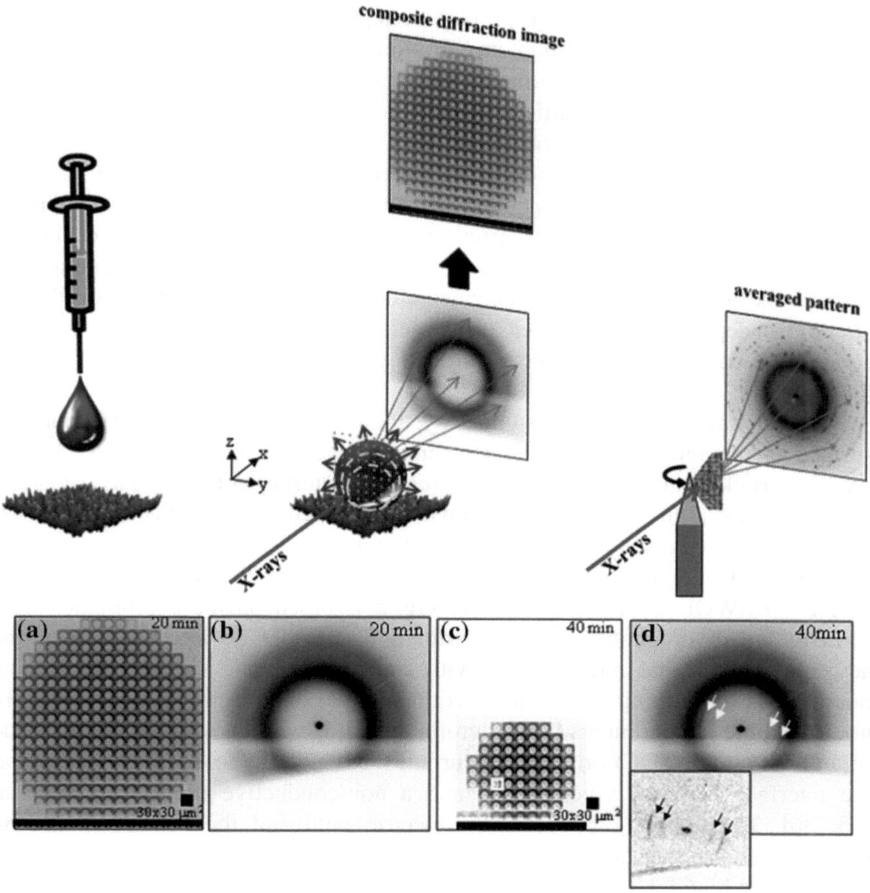

Fig. 4 *Up* Schematics of the deposition of a solution droplet by a syringe on a superhydrophobic nanostructured surface for in situ X-ray diffraction analysis. **a** CDIs of a 10 mg/ml lysozyme solution droplet containing calcium bicarbonate at 20 min and **c** 40 min evaporation time. **b** Single pattern from raster scan at 20 min and **d** at 40 min. The location of the pattern is evidenced in (**c**) by a *white* rectangle. The zoom of pattern (**d**) reveals more clearly the fibrous scattering after subtraction of aqueous scattering. (kindly adapted from [28] and [29])

experimental approach is shown schematically in Fig. 4. After deposition, the evaporating droplet is probed by consecutive raster-diffraction scans. Each raster scan allows assembling a composite diffraction image (CDI), composed of "pixels" of individual small and wide angle X-ray scattering (SAXS/WAXS) patterns that corresponds to a projection of the volume scattering onto a plane. The precipitate is optionally detached from the substrate, glued to a glass tip, and further analyzed by a raster-diffraction or powder-diffraction approach. In the case of lysozyme fibrillation analysis, the azimuthally integrated patterns at different stages of evaporation, reflect lysozyme solution scattering that overlaps with aqueous solution scattering (Fig. 4a and b). With increasing concentration, a rapid deposition of fibrillar lysozyme with amyloidic cross-β structure was observed at the interface of the droplet with the substrate under conditions of weak acidity (Fig. 4c and d) [29].

The speed of lysozyme conversion into a cross-β structure is surprisingly high as compared to amyloid formation under acidic conditions [31]. Shear-flow enhancement formation of amyloid-β in aqueous solution has been tentatively related to aggregation although partial unfolding of the native protein cannot be discarded [32, 33]. In conclusion, using these superhydrophobic nanostructured surface, the transformation of lysozyme solution into a cross-β structure fibrillar amyloidic deposit occurs under weak acidity and high Ca^{2+} ionic strength conditions, on a timescale shorter (around 1 h) than that reported in literature [34].

Another interesting subject studied by the aid of an *active* superhydrophobic surface is the monitoring of the conversion of amorphous calcium carbonate (ACC) particles into more stable phases (calcite, vaterite) at the level of the single crystallites [35]. In this context, a novel electrowetting on dielectric (EWOD) system was used as a microreactor for the mixing of two drops, containing respectively $CaCl_2$ and Na_2CO_3 to monitor the biomineralization process (Fig. 5).

The SHEWOD device presented in Fig. 5 is based on the combination of a PMMA superhydrophobic substrate with a Si EWOD device. Combining the advantages of digital microfluidics with superhydrophobic surfaces opens new possibilities for probing ultrasmall reaction volumes and reaction products that are not accessible to continuous flow microfluidics. It is well known that the presence of a direct current (DC) or alternate current (AC) electric field gathers charges at the interface between a conductive and a non-conductive (dielectric) material [35, 36]. The coupling between this electric field and the interfacial electric charges exerts an interfacial force, which can distort a deformable interface such as a conductive liquid. Using this electrowetting principle, it was possible to study the onset of calcite crystallites precipitation, probing the liquid–air interface during the mixing (Fig. 5d) [29, 35]. Calcite crystallites were observed already at 2.87 min after the onset of mixing (Fig. 5e), while calcite crystallites are only observed at about 20 min under conditions of homogeneous droplet evaporation (Fig. 5f).

3.2 Superhydrophobic Substrate for Raman Detection

In this section, the use of silicon-pillared superhydrophobic substrates for Raman spectroscopy characterization of highly diluted ($10^{-15}/10^{-18}$ mol l^{-1}) molecules is presented. Spectroscopic experiments, based on Raman scattering, will be presented to demonstrate the extreme sensing capability and wide applicability of methods using this particular types of microfabricated superhydrophobic devices. We reported [23] the detection of very diluted molecules from water solutions below a concentration of 1 fM. Evaporation of the drop (on bare superhydrophobic

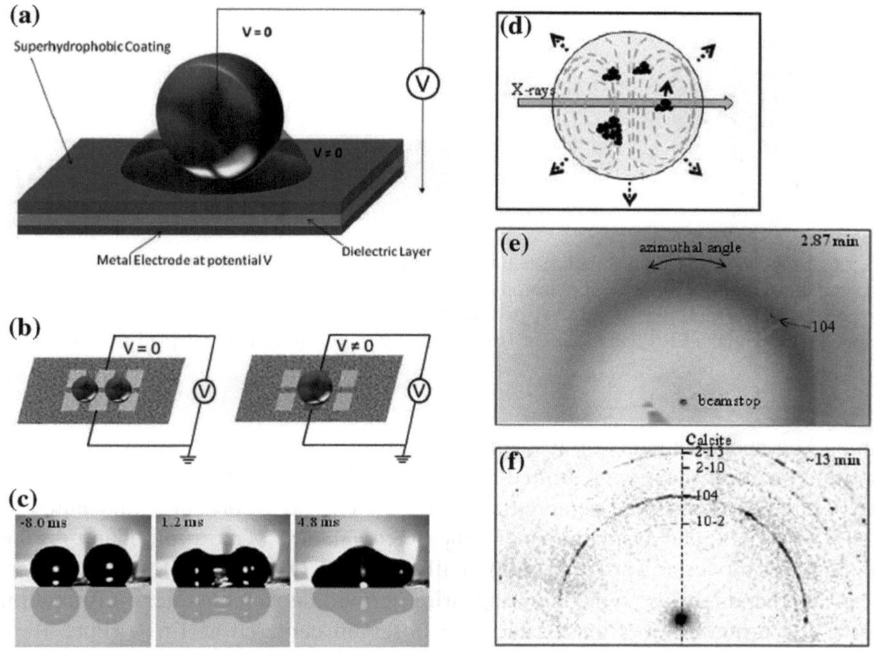

Fig. 5 **a** Principle of the superhydrophobic EWOD (SHEWOD) device showing a droplet in the high contact angle (suspended) state (V = 0). Once the voltage (V ≠ 0) is applied, the contact angle decreases and the droplet spreads. **b** Multiple electrode SHEWOD device showing schematically the coalescence of two droplets. **c** Selected video frames during coalescence of two 4-μl water droplets of d = 1 mm diameter on the SHEWOD device at 45 V AC (1 kHz). **d** Schematic design of convective flow transporting calcite crystallites through the X-ray microbeam. **e** Diffraction pattern at t = 2.87 min of mixing showing a diffuse ring due to water and calcite crystallites via the 104 reflection. **f** Diffraction patterns at advanced stages of precipitation showing textured calcite powder rings at t ≈ 13 min. Miller indices for the calcite and vaterite reflections are indicated. (kindly adapted from [35])

surfaces) was followed over time until precipitation of the solute. The solute molecules were confined into a small area of a few tens of micrometers square. To fully exploit the advantage of our approach, we designed superhydrophobic surfaces with a silver plasmonic cone directly embedded in the micropillar array (Fig. 6a); the cone was fabricated at the center of the array as a replacement for a micropillar. For symmetry reasons, the cone can be viewed as a spatial defect that breaks the spatial symmetry defined by the pillars and the hydrophobic radial forces cause the collapse of the drop on the cone tip. Molecules initially dispersed in solution will be progressively guided toward the plasmonic tip, where they will be deposited at the end of the evaporation process. A 160-nl droplet, contained lysozyme at a concentration of 1 fM, was deposited on the surface of the device using a microinjector system; the entire process of evaporation was followed over

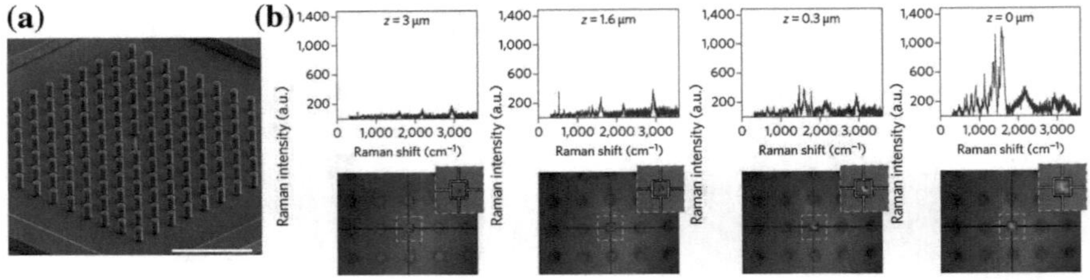

Fig. 6 Embedded plasmonic nanostructure in superhydrophobic device for single-lysozyme-molecule detection. **a** FIB grown plasmonic antenna realized in the center of a superhydrophobic array of pillars (scale bar is 50 μm). **b** Raman spectra at different Z illumination along the nanocone. Optimal SERS enhancement is reached at Z = 0 μm. (kindly adapted from [23])

time. The spectra taken for lysozyme (Fig. 6b) when focused 3 μm out of the optimal position show very limited information about the substance, as only three weak bands are observed centered at 2.940, 2.184, and 1.600 cm^{-1}, attributed to C–Hx, –S–CHx, and aromatic bands (Phe, phenylalanine; Trp, tryptophan). As the focal point moves toward the optimal illumination ($z \leq 1.6$ μm), various characteristic bands of lysozyme start appearing, providing rich chemical information about the protein. Vibrational bands are observed centered at 1.610 (Phe, Trp), 1.555–1.568 (amide II), 1.450 (C–Hx), 1.350 (Trp), 1.230–1.295 (Amide III), 1.069 and 1.126 (C–N stretching), 990 (a-helix), 895 (Trp), 650 (C–S stretching), and 620 cm^{-1} (Phe breathing), in addition to the broadband centered around 3.000 cm^{-1} (C–Hx stretching), which confirm the presence of proteins at the tip. Furthermore, the most significant band centered at 2.140 cm^{-1}, attributed to the –S–CHx, can be clearly observed. This band is relatively less visible in the protein spectrum from the liquid sample. The broadband around 3.350 cm^{-1}, which is related to the N–Hx stretching (amine groups), ($z = 0$), is also observed. Here, it should be noted that even if the scattering crosssection for N–H vibration is very low, this band is observed in our measurement thanks to surface plasmon generation on the metal surface and thereafter the enhancement of the electric field on the nanocone tip.

In brief, the combination of superhydrophobic surfaces and nanoplasmonic devices allows us to overcome the limit dictated by diffusion when sensors approach a nanoscale length. This new technique provides chance to detect few or single molecules after standard purification protocols normally used in biology and medicine, thus overcoming the need for expensive and time consuming methods to concentrate very diluted samples. When applied to biomedicine, the present tools, combined with already available purification methods, suggest the possibility of improving the early detection of several diseases, including cancer, where the number of clinically significant molecules at the onset of the pathology is very small and often generated by a single cell.

3.3 Superhydrophobic Substrates for Background-Free TEM DNA Characterization and for Nucleic Acid Pattern Generation

DNA and RNA are two of the most famous individual molecules known in the scientific world; for this reason, single-molecule experiments with nucleic acids are undoubtedly useful. Techniques for their preparation and handling include the use of laboratory-on-a-chip devices [37, 38], AFM nanotips, optical, and magnetic tweezers [39]. Various are the methods permitting the extension of DNA molecules to obtain nanowires of several micrometers length [40–42]; however, none of them has been exploited for further characterization down to few molecules in substrate-free configuration [43, 44].

By using the holed pillared superhydrophobic surface presented in the Sect 2.1, we implemented a novel sample preparation technique that allows, obtaining very thin and stable lambda phage DNA fibers suspended between micropillars and over-passing holes, to image them by transmission electron microscopy (TEM) without any substrate contribution (Fig. 7a and c) [24]. The superhydrophobic surface was treated in a way that the adhesion force between it and the water was very low (friction coefficient about 0.02) and, during the evaporation, the pinning of the drop was avoided. Under this condition, after the evaporation, DNA molecules could be deposited, suspended, and well tensed between the pillars (Fig. 7b), and more importantly, several DNA bundles resulted suspended in correspondence of the holes (inset Fig. 7b). The suspended DNA bundles have well-reproducible diameter, between 8 and 200 nm at our salinity conditions. Compared to all the previous DNA preparation and characterization methods, the use of these superhydrophobic devices shows unique and innovative results for the nucleic acid detection, because TEM images are completely background free and the images are then much clearer and simpler to interpret. Thus, it opens new research perspectives; one of its major advantages is that it can provide a technique for studying how different molecules (drugs, transcriptional factor, etc.) affect single DNA fiber structures.

Generation of polymer patterns of λ DNA was achieved by using the same superhydrophobic devices. As clearly seen in Fig. 8a and b, very well-ordered patterns were generated with lengths of several hundreds of micrometers. As long as superhydrophobicity is maintained also at different side-to-side pillar distance, given relative parameters accordingly, it was possible to extend dsDNA also between pillars at, e.g., 4 μm side-to-side distance (Fig. 8a), with very reproducible results. Of course, because the λ dsDNA length in A conformation equals to 12 μm (calculated from [45]), the longer the distance between the pillars, the more improbable the well-ordered extension of DNA patterns would be; conversely, the shorter the distance between two pillars and the more probable the bridging of a single dsDNA molecule would be. In Fig. 8b, it is shown that even for a 20 μm distance between two pillars, the λ DNA can still form very long and ordered wires. As an example, we also report the experimental result of a suspended λ dsDNA in between two pillars at 3 μm side-to-side distance (Fig. 8c), in which a

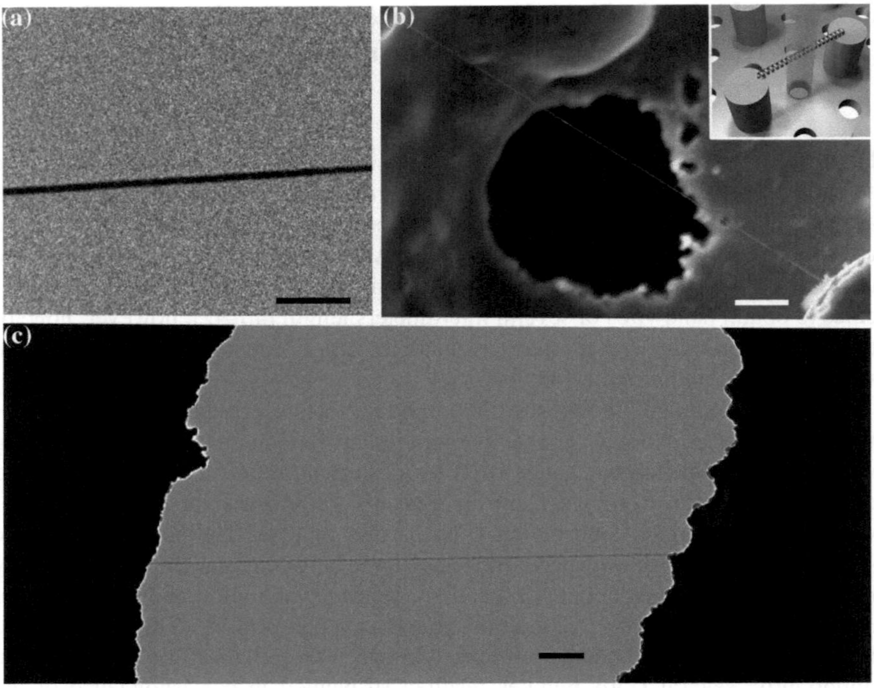

Fig. 7 **a** High-resolution (scale bar 100 nm) TEM image of λ DNA molecules stretched and suspended between two pillars. **b** SEM image of λ DNA suspended in front of one of the holes embedded on the pillars lattice (scale bar 1 μm). The inset in B elucidates the concept exploited for TEM direct imaging: the passing through etched hole allows the suspended DNA fibers to be orthogonally crossed by the electron beam. In **c** TEM low-magnification image of free-standing DNA bundle (scale bar 200 nm)

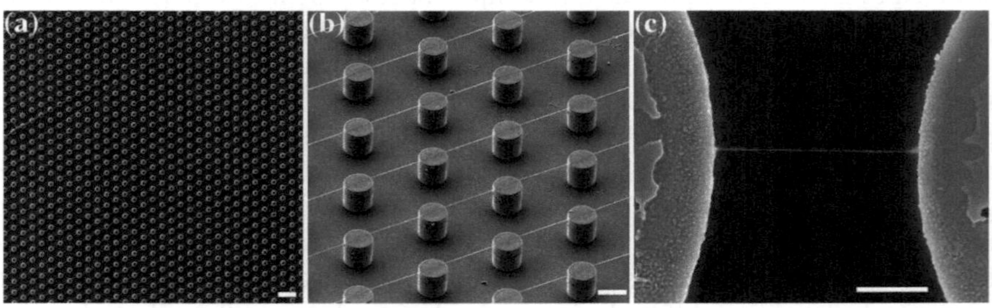

Fig. 8 SEM images of ordered substrate-free generation of DNA nanowire arrays upon superhydrophobic substrate. The pillars are fabricated at 4 μm side-to-side distance (**a**) and 20 μm side-to-side distance (**b**). In **c**, a single dsDNA extended between two pillars at 3 μm side-to-side distance (Pt/Pd coating is present) is shown. Scale bars are 10 μm (**a** and **b**) and 1 μm (**c**)

single DNA molecule is evident, being the diameter of the filament 12 nm with a Pt/Pd coating of 5 nm at each side. Thus, the eventual issue could be overcome by changing the distance between the pillars, the type/size of DNA extended (longer or shorter) as well as buffer type, and/or salt concentration.

To conclude, regarding the use of these pillared superhydrophobic substrate, we reported about two different approaches that aim at: (i) characterization of few/ single molecules in a substrate-free environment, without any tagging or modification; (ii) exploit of biomolecules for generation of well-ordered nanowire arrays. These superhydrophobic substrates have been recently implemented and are currently used in our laboratory to characterize single-molecule assay for enzymatic activity on stretched DNA [46] and to obtain suspended gold nanowires circuitry. Further functionality has been added by decorating gold nanowires with rod-shaped hydrothermally grown ZnO nanocrystals.

References

1. Hou J, Zhao A (2006) Detecting and manipulating single molecules with STM. NANO 01:15–33
2. Poma A et al (2005) Interactions between saporin, a ribosome-inactivating protein, and DNA: a study by atomic force microscopy. J Microsc 217:69–74
3. Muller D et al (2006) Single-molecule studies of membrane proteins. Curr Opin Struct Biol 16:495–498
4. Mingqian H, Jiye C (2008) Scanning near-field optical microscope and its applications in the field of single molecule detection. Prog Chem 20:984–988
5. Moerner W, Fromm D (2003) Methods of single-molecule fluorescence spectroscopy and microscopy. Rev Sci Instrum 74:3597–3619
6. Huang B, Babcock H, Zhuang X (2010) Breaking the diffraction barrier: super-resolution imaging of cells. Cell 143:1047–1058
7. Roy R, Hohng S, Ha T (2008) A practical guide to single-molecule FRET. Nat Methods 5:507–516
8. Ishikawa-Ankerhold H, Ankerhold R, Drummen G (2012) Advanced Fluorescence Microscopy Techniques—FRAP, FLIP, FLAP, FRET and FLIM. Molecules 17:4047–4132
9. Moretti M et al (2008) An ON/OFF biosensor based on blockade of ionic current passing through a solid-state nanopore. Biosens Bioelectron 24:141–147
10. Fernandez-Cuesta I et al (2011) Fabrication of fluidic devices with 30 nm nanochannels by direct imprinting. J Vac Sci Technol B 29:06F801–1/7
11. Thamdrup L, Klukowska A, Kristensen A (2008) Stretching DNA in polymer nanochannels fabricated by thermal imprint in PMMA. Nanotechnology 19:125301
12. Lesoine J et al (2012) Nanochannel-based single molecule recycling. Nano Lett 13:3273–3278
13. Dong M, Sahin O (2011) A nanomechanical interface to rapid single-molecule interactions. Nat Commun 2:247
14. De Angelis F et al (2008) A Hybrid Plasmonic–Photonic Nanodevice for Label-Free Detection of a Few Molecules. Nano Lett 8:2321–2327
15. Schäfer C et al (2013) A single particle plasmon resonance study of 3D conical nanoantennas. Nanoscale 5:7861–7866
16. Zhang C et al (2013) A nanofluidic device for single molecule studies with in situ control of environmental solution conditions. Lab Chip 13:2821–2826
17. Zhao Y et al (2013) Lab-on-a-chip technologies for single-molecule studies. Lab Chip 13:2183–2198

18. Wenzel R (1936) Resistance of solid surfaces to wetting by water. Ind Eng Chem 28:988–994
19. Cassie A, Baxte S (1944) Wettability of porous surfaces. Trans Faraday Soc 40:546–551
20. Tuteja A et al (2008) Robust omniphobic surfaces. PNAS 105:18200–18205
21. Zhang X et al (2008) Superhydrophobic surfaces: from structural control to functional application. J Mater Chem 18:621–633
22. Cui Y et al (2004) Integration of colloidal nanocrystals into lithographically patterned devices. Nano Lett 4:1093–1098
23. De Angelis F et al (2011) Breaking the diffusion limit with super-hydrophobic delivery of molecules to plasmonic nanofocusing SERS structures. Nat Photonics 5:682–687
24. Gentile F et al (2012) Direct imaging of DNA fibers: the visage of double helix. Nano Lett 12:6453–6458
25. Han W, Byun M, Lin Z (2011) Assembling and positioning latex nanoparticles via controlled evaporative self-assembly. J Mater Chem 21:16968–16972
26. Su B et al (2012) Small molecular nanowire arrays assisted by superhydrophobic pillar-structured surfaces with high adhesion. Adv Mater 24:2780–2785
27. Accardo A et al (2010) In situ X-ray scattering studies of protein solution droplets drying on micro- and nanopatterned superhydrophobic PMMA surfaces. Langmuir 26:15057–15064
28. Accardo A et al (2011) Ultrahydrophobic PMMA micro- and nano-textured surfaces fabricated by optical lithography and plasma etching for X-ray diffraction studies. MNE 88:1660–1663
29. Accardo A et al (2011) Lysozyme fibrillation induced by convective flow under quasi contact-free conditions. Soft Matter 7:6792–6796
30. Lakshmanan A et al (2013) Aliphatic peptides show similar self-assembly to amyloid core sequences, challenging the importance of aromatic interactions in amyloidosis. PNAS 110: 519–524
31. Yagi N, Ohta N, Matsuo T (2009) Structure of amyloid fibrils of hen egg white lysozyme studied by microbeam X-ray diffraction. Int J Biol Macromol 45:86–90
32. Hill E et al (2006) Shear flow induces amyloid fibril formation. Biomacromolecules 7:10–13
33. Dunstan D et al (2009) Shear flow promotes amyloid-β fibrilization. PEDS 22:741–746
34. Lee J, Um E, Park J, Park C (2008) Microfluidic self-assembly of insulin monomers into amyloid fibrils on a solid surface. Langmuir 24:7068–7071
35. Accardo A et al (2013) Fast, active droplet interaction: coalescence and reactive mixing controlled by electrowetting on a superhydrophobic surface. Lab Chip 13:332–335
36. Berthier J (2008) Microdrops and digital microfluidics. William Andrew Publishing, Norwich
37. Marie R et al (2013) Integrated view of genome structure and sequence of a single DNA molecule in a nanofluidic device. PNAS 110:4893–4898
38. Rasmussen K et al (2011) A device for extraction, manipulation and stretching of DNA from single human chromosomes. Lab Chip 11:1431
39. Neuman K, Nagy A (2008) Single-molecule force spectroscopy: optical tweezers, magnetic tweezers and atomic force microscopy. Nat Methods 5:491–505
40. Oshige M et al (2010) A new DNA combing method for biochemical analysis. Anal Biochem 400:145–147
41. Guan J et al (2010) Large laterally ordered nanochannel arrays from DNA combing and imprinting. Adv Mater 22:3997–4001
42. Li B et al (2013) Macroscopic highly aligned DNA nanowires created by controlled evaporative self-assembly. ACS Nano 7:4326–4333
43. Cerf A, Alava T, Barton R, Craighead H (2011) Transfer-printing of single DNA molecule arrays on graphene for high-resolution electron imaging and analysis. Nano Lett 11:4232–4238
44. Cabin-Flaman A et al (2011) Combed single DNA molecules imaged by secondary ion mass spectrometry. Anal Chem 83:6940–6947
45. Watson J, Crick F (1953) The structure of DNA. Cold Spring Harb Symp Quant Biol 18:123–131
46. Marini M, Das G, La Rocca R, Gentile F, Limongi T, Santoriello S, Scarpellini A,Di Fabrizio E (2014) Raman spectroscopy for detection of stretched DNAs on superhydrophobic surfaces. Microelectron Eng 119:151–154

Tip-Assisted Optical Nanoscopy for Single-Molecule Activation and Detection

Denys Naumenko, Damiano Cassese, Marco Lazzarino and Alpan Bek

Abstract In this paper, we discuss on tip-assisted optical techniques that represent a set of modern and promising tools for investigation of surfaces which combine the ultimate performances of scanning probe microscopies in achieving unprecedented lateral and vertical resolution, with the highly informative contribution of optical spectroscopy in terms of material characterization and chemical analysis.

1 An Overview

Optical investigation of material surfaces based on both elastic and inelastic scattering of light can be performed at a very high lateral spatial resolution, down to a few nanometers, by assistance of sharp conducting tips [1–5]. An important physical principle is common to all tip-assisted optical imaging and spectroscopies. Free electrons in conductors support collective oscillations, named localized surface plasmon resonance (LSPR), which in the case of noble metal particles—mainly gold and silver—have their resonance frequency in the visible spectrum. Due to a mismatch between energy and momentum, these optical excitation do not propagate in air/vacuum and therefore remain confined at the metal surface, and decay exponentially over a distance comparable with the particle size, and thus can be as small as few nanometers constituting a very intense and tightly confined light

D. Naumenko · M. Lazzarino (✉)
CBM scrl Area Science Park—Basovizza, 34149 Trieste, Italy
e-mail: lazzarino@iom.cnr.it

D. Naumenko · D. Cassese · M. Lazzarino
IOM-CNR-Laboratorio TASC, Area Science Park—Basovizza, 34149 Trieste, Italy

A. Bek
Middle East Technical University—Dumlupinar Bulvari 1, 06800 Cankaya Ankara, Turkey
e-mail: bek@metu.edu.tr

F. Benfenati et al. (eds.), *Novel Approaches for Single Molecule Activation and Detection*, Advances in Atom and Single Molecule Machines,
DOI: 10.1007/978-3-662-43367-6_5,

source. Such a small yet intense light source can significantly improve the interaction strength and at the same time limit the strong interaction volume spatially, which makes it ideal for investigating light interaction with the nanometer-scale objects such as molecules or quantum dots, etc., at the limit of individual entity rather than ensemble, even in the case of dense surface coverage. If such an LSP resonant particle is placed at the apex of a scanning probe microscope tip and a suitable illumination system is designed, this nm-scale light source can be scanned over a sample surface and information about optical properties, chemical composition, as well as surface topography can be collected simultaneously at unmatched spatial resolution. Similarly, intense and confined near-field strengths can also be produced by vertically elongated, typically conical or cylindrically shaped, metal tips acting simultaneously as scanning probes and plasmonic antennas (PAs). In addition, propagating surface plasmon polaritons (SPPs) of noble metals when launched on extended tapered conical geometries also give rise to a very intense EM field in the vicinity of its nm-sized apex when adiabaticity of the taper is ensured. The SPP tip, too, can be utilized as the LSPR particle tip or the PA tip for achieving nm-scale resolution.

In energy-conserving optical investigations which rely on Rayleigh scattering of light from the material surface, the contribution of the tip to the light–matter interaction takes form of intensification of the EM field due to strong field enhancement in the vicinity of the tip apex, that is, locally increasing the number of photons impinging on the surface, thereby increasing the number of Rayleigh-scattered photons. In non-energy-conserving light–matter interactions such as Raman scattering and fluorescence, the existence of the sharp metal tip does affect the interaction not only by increasing the number of photons as in the Rayleigh case but also by introducing other effects such as charge transfer or modified radiative rates. Although a complete theoretical treatment that can account for all the introduced physical and chemical effects of a metal tip in the vicinity of any arbitrary surface still lie afar, there are well-established methods for the theoretical computation of local field strength and its distribution in the vicinity of tip–surface system. The revelation of the near-field landscape belonging to a given experimental scenario is essential for unambiguous interpretation of the experimental findings since the highly resolved data originate from the near-field interaction region in tip-assisted optical investigations.

Many implementations of tip-assisted optical systems have been proposed in the last decade, with spatial resolution down to 7 nm [6]. The principle technique is tip-enhanced Raman scattering (TERS), in which the field enhancement produced by the particle resonance is used to generate a strong Raman signal, in a close analogy with the more widely known surface-enhanced Raman scattering. Several optical configurations have also been proposed, and a dominant majority of serious atomic force microscope manufacturers now propose instruments suitable for tip-assisted optics experiments. Tip-assisted optical techniques have been applied on many different kinds of samples, in air and in liquid, ranging from solid-state nanowires and carbon nanotubes, to self-assembled monolayers of organic molecules and proteins [7–11]. TERS has been proposed as an alternative technique for DNA sequencing [12]. The tip-assisted optical techniques have been successfully applied on nanoscale materials with semi-quantitative to quantitative characterization capabilities. Currently, the main limitation of the tip-assisted techniques is the low reproducibility of the tip enhancement. In the large majority of the experiments, two main strategies are adopted in probe fabrication: (i) gold or silver coating of silicon or silicon nitride AFM probes and (ii) cutting or etching a full metal probe from a pure gold or silver wire. In both cases, the tip properties are not controlled at the nanometer scale. Alternative strategies are emerging, which involve an accurate design of the plasmonic response of the tip by fabricating, e.g., bow tie antennas or plasmon polariton waveguides [6, 13–16]. With the achievement of an accurate control and reproducibility of tip-induced field enhancement, we believe that TAO will experience a rapid and dramatic diffusion in surface science laboratories as a powerful surface analytical technique.

1.1 A Brief History of the Tip-Assisted Technique

The tip-assisted optical microscopic and spectroscopic technique is also dubbed the apertureless technique owing to its historical evolution starting from the idea of aperture-based near-field optical microscopy as first proposed by Synge in 1928 [17]. This proposal was long forgotten until in 1956 O'Keefe proposed a similar experiment [18] which was not realized until in 1972 Ash and Nicholls performed it in the microwave regime [19]. It was the invention of scanning probe microscopy (SPM) by Binnig et al. [20] in 1982 which made a breakthrough not only in the high-resolution electrical and mechanical surface characterization but also in the high-resolution optical characterization of surfaces. An essential feedback control mechanism that could keep a sub-wavelength aperture controllably in the close vicinity of a surface was implemented in 1984 by Pohl et al. [21] and independently and simultaneously by Lewis et al. [22] so that a sub-wavelength image of a surface was acquired in the visible spectrum for the first time. This concept has reached the maturity of commercialization upon introduction of metal-coated tapered optical fiber tips for aperture probes and was named the scanning near-field (or near-field scanning) optical microscope (SNOM/NSOM).

The concept of near-field enhancement was associated with surface plasmon excitation first by Weber and Ford [23] in 1981. It was Wessel who has integrated the idea of field enhancement due to a small metal particle with the idea of scanning probe microscopy in 1985 [24]. He proposed an optical probe based on optical field confinement by surface plasmons of a nanometer-sized metal particle which serves as an antenna for an incoming field and enhances the interaction of a nearby lying surface with its near field. The metal nanoparticle-based optical microscopy was proposed by raster scanning this particle above the surface. In his paper, he also mentioned that such a probe could be used to generate a Raman, a two-photon, or a second-harmonic spectrum in its nanoscale vicinity. In this proposal, the necessity of an optical aperture was dropped and an apertureless scheme was introduced for the very first time. Possible beneficial use of apertureless technique for obtaining highly resolved and sensitive Raman spectroscopic–microscopic images was a straightforward extension because surface-enhanced Raman spectroscopy (SERS) of metal-adsorbed molecules had already become an actively researched field 10 years after its accidental discovery by Fleischmann et al. [25] in 1974. Today, tip-enhanced Raman spectroscopy (TERS), as named by Stockle et al. [26] in 2000, has become a field of active research. The spatial resolution limit is being ever pushed down with the hope of producing Raman images of surfaces with the resolution of a single molecule. The wording for the phrase "tip-assisted optical investigation" is in fact chosen to reflect the necessarily interwoven nature of apertureless optical investigation technique with SPM. The words such as "tip" and "probe" are used very often along with the more recently introduced phrase "optical antenna" by Farahani et al. [27] in 2005.

Apertureless probes (metal tips) have several advantages over optical fiber-based near-field aperture probes. The most important advantage is in the spatial extent of the near-field interaction. Since the tips can have apex radii down to a few nanometers, the optical resolution expected to be achieved in tip-assisted technique can be an order of magnitude better. Secondly, the optical field is introduced in free space and therefore does not suffer from optical fiber cutoff losses. By using metal tips, the evanescent near field of the surface under investigation is not only scattered (converted to traveling wave) but also enhanced (amplified). And finally, an additional advantage lies in the fact that the free-space transmission of optical fields allows for a broader range of wavelengths from ultraviolet (UV) to infrared (IR), while the optical fibers do have limitations due to waveguiding and material properties. Tip-assisted scheme is especially beneficial where the scattering efficiency of metals are higher in the IR region [3].

2 Theoretical Background

2.1 The Evanescent Field

When an optical field impinges on a material boundary such as a solid or liquid surface or their interface, it evolves into a scattered field upon interaction with matter with an evanescing part bound to the surface and a traveling part transmitted or reflected from the surface. The field in the surface's immediate vicinity has an evanescent nature with an exponentially decaying dependence on the distance from the surface. The decay constant of the evanescent field is $\gamma = k_2\sqrt{\frac{n_1^2}{n_2^2}\sin^2\theta_1 - 1}$, where $\theta_1 > \theta_c$. The evanescent field can propagate only in direction parallel to the interface. Figure 1a depicts the evanescent field decaying in medium n_2.

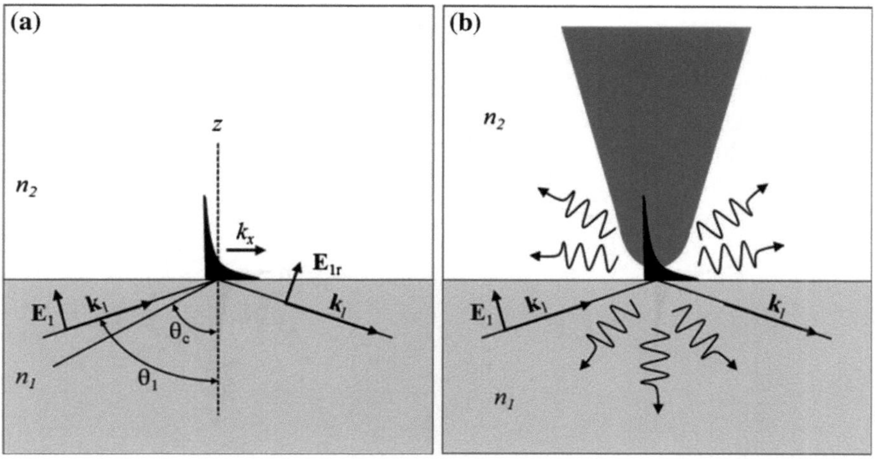

Fig. 1 **a** Evanescent field created in medium n_2 for $\theta_1 > \theta_c$ **b** Scattering of the evanescent field by a metal tip in the vicinity of the surface

When the material on one side of the interface is a metal and the other one is a dielectric, the evanescent field extends couple of nanometers into the dielectric in the visible spectrum. The confined feature of the evanescent field existing in optically excited inhomogeneous media is the most essential ingredient of tip-assisted optical investigation of surfaces. It is this evanescent optical field from the material surface which carries information of the optical properties of the surface at a high spatial resolution since the physical extent of the field is in the sub-wavelength regime. A metal tip brought in the vicinity of the evanescent field can be used to scatter this normally surface-bound field and couple it to detectors at its far field, performing the role of an impedance-matching antenna between the near field of the surface and the far field detector as shown in Fig. 1b.

2.2 *Localized Surface Plasmon Tip*

In an alternative kind of application of the tightly space-confined nature of evanescent field for high-resolution optical investigation of surfaces, the evanescent field is created only at the metal tip of an SPM probe and not on the material surface under investigation as in Fig. 2a. This strongly tip-localized evanescent field is raster scanned over the material surface illuminating it with its nanometer-scale light source. The interacting field with the surface couples to the far field due to surface corrugation and contains information of surface properties such as reflectivity or Raman activity on the nanoscale as in Fig. 2b. This kind of tip-assisted surface investigation relies on efficient creation of a strong evanescent near field in the vicinity of metal tip apexes.

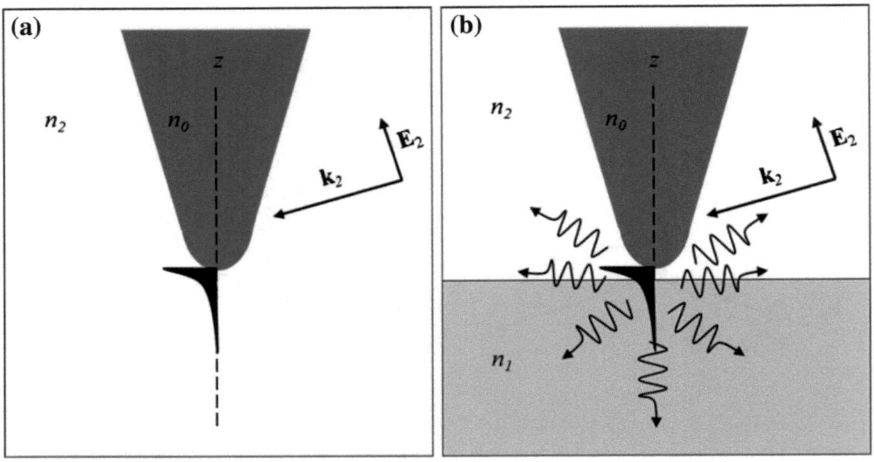

Fig. 2 **a** Upon optical excitation of the metal tip, a strong tip surface-bound evanescent near field evolves. **b** The near field of the tip interacts with the surface and scatters to far field

It is found that when illuminated by external beam of light as in Fig. 2b, the interaction of light with the free electrons on the tip surface causes strong excitation of surface plasmon polaritons which in turn causes a zone of high energy density just as an antenna does in microwave region. Here, the effect of the sharp tip can resemble a simplistic rod-shaped antenna similar to a lightning rod. However, one crucial difference is that the tip is an antenna for SPPs rather than photons alone, or in other words, these photons are non-propagative.

SPP is an electromagnetic excitation (surface wave) existing on the surface of Drude materials. It is an intrinsically two-dimensional excitation whose electromagnetic field decays exponentially with a distance from the surface. SPPs can propagate, scatter, interfere, and localize, etc., depending on either local (nanometer scale) or macro (micrometer-scale) characteristics of the surface profile. Moreover, the enhanced electromagnetic fields can be localized not only at the apex of bulk metal tips but also on the metalized (by evaporation of thin films or linking of single nanoparticles) dielectric structures (Fig. 3a and b). Finally, the enhanced fields at the tip cause enhanced excitation rates that correspond to an increase in the excitation density provided by the light source. The numerical simulations of enhanced field distribution for a metal sphere and cone in close proximity to the surface are shown in Fig. 3c and d.

2.3 Plasmonic Antenna Tip

The latest procedures for creating such a small light source involves injection of surface plasmon polaritons (SPPs) from the base or shaft of the metal tip and letting them propagate toward the tip apex. If the change in the tip shaft diameter fulfills the

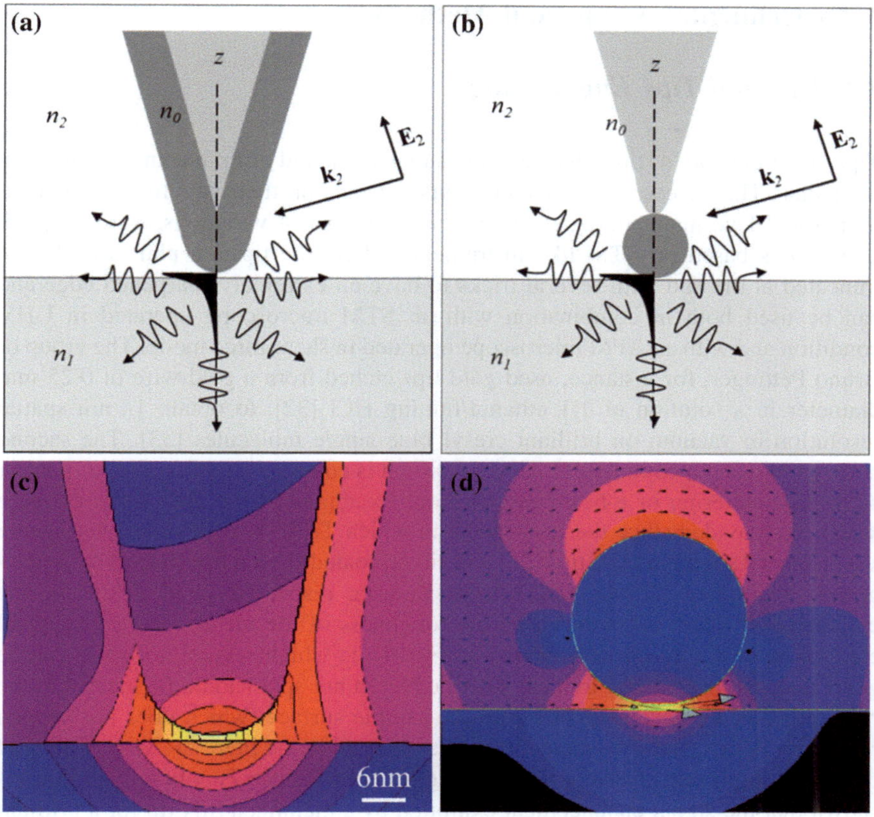

Fig. 3 **a** An alternative to *conical solid metal tip* is the metal-coated conical dielectric tip. **b** A second alternative tip is a *spherical metal* nanoparticle-attached dielectric tip. **c** The calculated near-field distribution in the gap between a *conical metal tip* and a surface (reprinted with permission from Estebanet et al. [28]. Copyright (2007) by the American Physical Society) and **d** between a metal nanoparticle and a surface (adapted with permission from Zhang et al. [29]. Copyright (2007) American Chemical Society)

adiabaticity conditions, the SPPs compress or focus toward the apex along the tip shaft as the field amplitude ever grows. This phenomenon leads to a giant concentration of energy on the nanoscale. The SPPs are adiabatically transformed into localized surface plasmons that are purely electric oscillations that can and do nanolocalize leading to the three-dimensional nanofocusing [30, 31].

3 Experimental Setups and Methods

3.1 Tips and Tips Engineering

Tips are at the core of tip-enhanced optics in general and of tip-enhanced Raman in particular. The success of the experiments depends on their structure and optical properties. Two main strategies, with a large number of variations, were adopted. The first is based on STM-like instruments: Here, the tip is a plain metal wire truncated at the end with several tricks to have an extremely sharpened edge and can be used both in combination with an STM microscope operated in UHV condition and with an AFM microscope operated in shear force mode. The group of Bruno Pettinger, for instance, used gold tips etched from a gold wire of 0.25 mm diameter in a solution of 1:1 ethanol/fuming HCl [32], to obtain 14 nm spatial resolution in vacuum on brilliant cresyl blue single molecules [33]. The second consists in using a commercial AFM tip which is coated with a thin film of gold or silver by sputtering or e-beam evaporation. Examples of the use of tips of these kinds are provided since the first papers in which TERS was demonstrated where silver-coated silicon nitride tips [26] and silver-coated silicon tips [34] were used by groups of Renato Zenobi and Satoshi Kawata, respectively. In this case, the enhancement effect is produced by the roughness of the metal coating (a typical electron microscopy image is shown in Fig. 4) on which several hot spots can be generated. Unfortunately, to have TERS effect, a hot spot should be located at the right position as close as possible to the sample surface, which does not happen regularly. In fact, the enhancement factors reported in the literature vary of several orders of magnitude. In a very illuminating paper by Renato Zenobi group [35], it is shown that the signal enhancement exhibited by a metalized SiN tip for a brilliant cresyl blue sample can be increased by a factor 5 by slight ($<10°$) adjustment of the tip axis–surface plane angle. The cause has been postulated to be a result of the presence of hot Raman-enhancing sites on the randomly shaped metal particle, which is made available for TERS enhancement upon reorientation of the tip on the surface. They also argued that tips made of materials with lower refractive index, such as SiN ($n = 2.05$), present better substrates for TERS probes. Preliminary TERS experiments using Ag-coated SiO_2 ($n = 1.5$) or AlF_3 ($n = 1.4$) tips demonstrate that even larger enhancements can be obtained [36] (Fig. 4c–f).

Currently, as far as we know, a metallization of commercial AFM probes for the TERS needs is still being developed. Some promising results are obtained by Satoshi Kawata [37] group in 2012 by thermal oxidization and subsequent metallization of commercial silicon tips. It was shown that highly reproducible tips exhibit an enhancement factor of >100 with a 100 % yield providing an important step of 'tip-enhanced Raman spectroscopy for everyone.' However, the fabricated probes have an apex diameter of about 50 nm that limits a number of the samples to be characterized by TERS.

A promising strategy to engineer TERS probes with controlled plasmonic response joined to high-aspect ratio tips for topography measurements takes

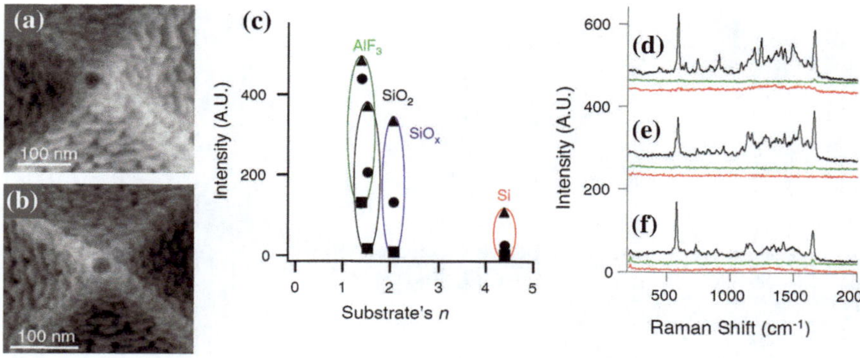

Fig. 4 SEM images of **a** SiO_x- and **b** AlF_3-precoated pyramidal tips vapor-coated with Ag nanoparticles. **c** The integrated SERS intensities of rhodamine 6G (*triangles*), brilliant *cresyl blue* (*circles*), and *malachite green* (*squares*) as a function of refractive index of the underlying surfaces. **d–f** TERS spectra of brilliant *cresyl blue* acquired with a tip in contact (*black*) and retracted (*green trace*) for Ag-coated SiO_2 (**d**), SiO_x/SiN (**e**), and AlF_3/SiN (**f**) tips. The *red curves* are the tips' spectra collected after the experiments (with kind permission from Springer Science + Business Media: Yeo et al. [36])

advantage of the vapor–solid–liquid (VLS) growth of semiconductor nanowires. The growth of nanowires has been deeply investigated during the last 10 years, and many materials geometry and applications have been demonstrated [38]. Although several catalytic processes have been proposed, including self-catalytic ones [39], the most accepted one is the VLS process. In a nutshell, a thin layer of gold ($\sim$1 nm) is deposited on a surface, which could be amorphous or clean with defined crystallographic planes, and then, it is annealed to reach a liquid phase, at about 400 °C. At that point, in order to minimize surface tension, small droplets are formed, with diameter in the 10–100-nm range, depending on the initial thickness and temperature annealing. The substrate is maintained at high temperature and exposed to a flux of precursor gases which diffuse into the Au droplet, increase the concentration to saturation, and precipitate in crystal form at the droplet/substrate interface. The whole process is performed in ultra-high vacuum conditions to prevent carbon contamination. A side effect of the process described above is that the metal particle maintains size and shape and stays at the top of the nanowire, as shown in Fig. 5a–c. Christiansen and coworkers demonstrated this approach by growing Si nanowires on a substrate, picking one isolated NW with a commercial AFM tip and acquiring TERS spectra of malachite green with an overall sevenfold enhancement [40, 41] (Fig. 5d and e). New fabrication strategies to obtain a single NW on an AFM cantilever have been recently demonstrated [42, 43], and the application of this approach to TERS imaging application is just a matter of time.

A novel strategy which has been introduced recently that in principle get rid of all the problems associated with the random location of plasmonic hot spots in metal-etched and metal-coated tips is based on plasmonic localization and compression. The theory of plasmonic compression dates back to 2000 with a paper by

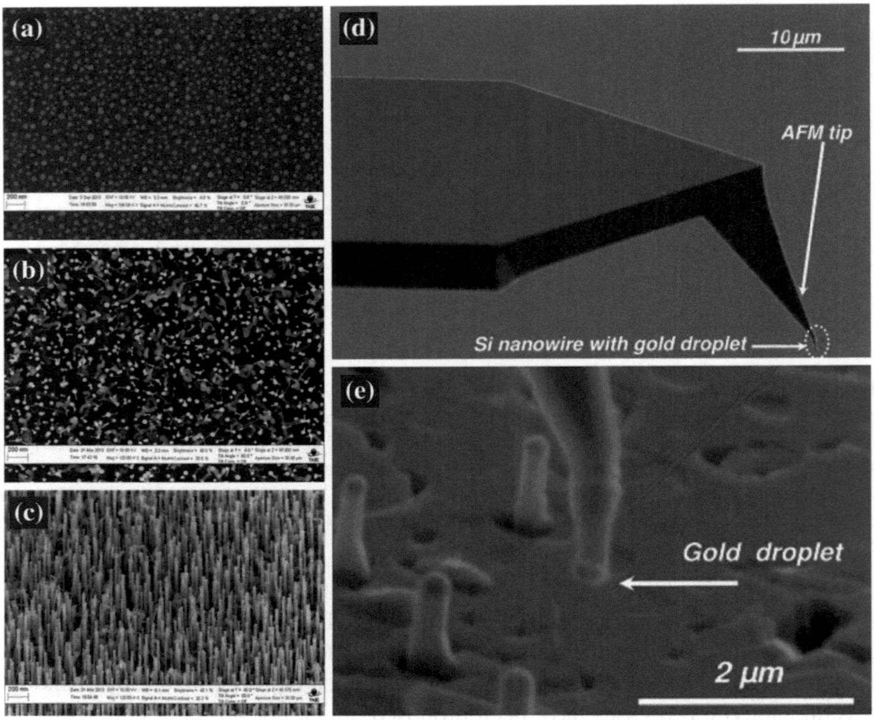

Fig. 5 SEM images of **a** gold NPs on flat GaAs surface formed by annealing of a few monolayers of gold before NW growth, **b** and **c** as-grown 400-nm-long gold-catalyzed ZnSe NWs (*top view* and *tilted view*, respectively), **d** AFM tip with the attached Si nanowire and **e** the magnified NW tip region with the attached nanowire (**a–c** are original images kindly provided by V. Zannier and Dr. S. Rubini, IOM-CNR Laboratorio TASC, Italy, **d** and **e** are adapted with permission from Becker et al. [40]. Copyright (2007) American Chemical Society.)

Babadjanyan and coworkers [44] which unfortunately remained rather unnoticed. A similar approach was reformulated independently 4 years later by Mark Stockman [45] and led to several important applications. The basic concept introduced by Stockman is the adiabatic compression of plasmon polaritons traveling along a metallic cone. A metallic cone propagates surface plasmons from the base to the apex of the cone itself. Due to the reduction of the cone in-plane diameter, the effective dielectric constant increases and the group velocity of the plasmon is reduced, while the frequency remains unchanged. If the variation of the diameter along the cone is slow enough, the decrease of the plasmonic speed is adiabatic and all the energy injected at the base of the cone propagates to the tip. On the contrary, if the diameter is reduced too fast or, equivalently if it is reduced in steps, part of the energy is backreflected at each step, and very little energy reaches the cone tip. This means that special care should be taken during the fabrication process, and plasmonic cones should be made smooth and with the smallest possible cone angle. At the very end of an ideal cone, the radius is zero, as is the group velocity and the

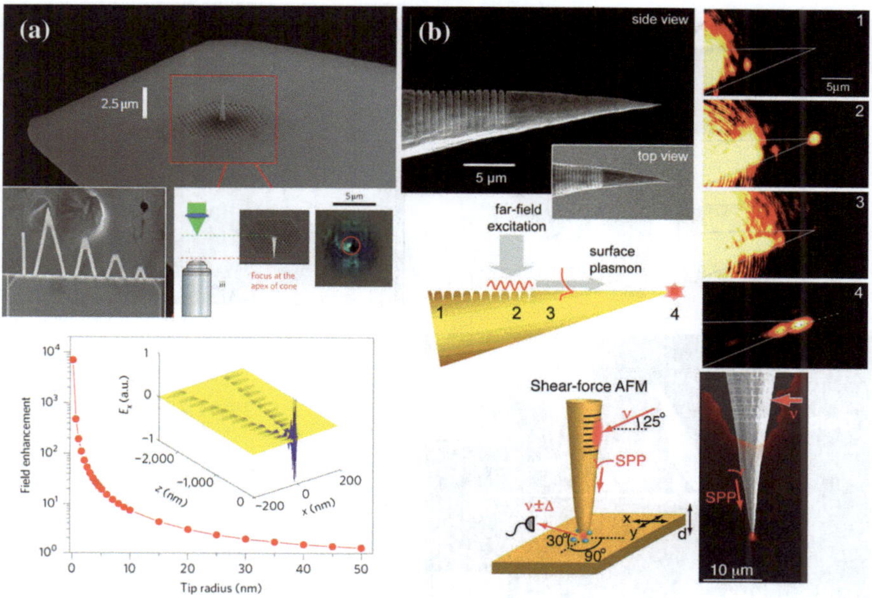

Fig. 6 **a** Tapered waveguide on an AFM cantilever, experimental test of SPP generation and propagation along the waveguide, and theoretical field enhancement in the plasmonic tapered waveguide. Reprinted by permission from Macmillan Publishers Ltd: [Nature Nanotechnology] De Angelis et al. [6]. Copyright (2010). **b** SEM images of a *conical metallic tip* with a grating coupler on the shaft, principle of the non-local excitation of the tip apex, series of microscope images recorded for illumination of the tip at the four positions, and schematic of the experiment using electrochemically etched Au tips mounted onto the quartz tuning fork of a shear force AFM. Adapted with permission from Ropers et al. [14]. Copyright (2007) American Chemical Society and with permission from Berweger et al. [15]. Copyright (2010) American Chemical Society

intensity of the electromagnetic (EM) field associated with the surface plasmons diverges. The problem of coupling a propagating EM wave to the surface plasmons in the cone was theoretically and experimentally solved by the group of Enzo di Fabrizio and by the group of Markus Raschke. De Angelis et al. [6, 13] designed a photonic crystal with a mid-gap defect which is at the same frequency of the maximum of surface plasmon nanofocusing efficiency. A metallic cone is fabricated in correspondence of the defect and acts as a plasmonic waveguide to concentrate the plasmonic fields at the tip extremity (PCPW). In practice, the tips are fabricated using focused ion beam (FIB) milling for the realization of the photonic crystal and chemical vapor deposition (CVD) induced by a focused electron beam for the fabrication of the plasmonic waveguides. In order to withstand the large forces occurring during AFM scans, a more robust shape of the cones is chosen with respect to those used in [13]. A scanning electron microcopy (SEM) image of an AFM cantilever with integrated PCPW tip is shown in Fig. 6a. Ropers et al. [14] and later Berweger et al. [15, 16] fabricated a plasmonic grating on an electrochemically etched tip via focused ion beam (FIB) milling, with the grating period a_0

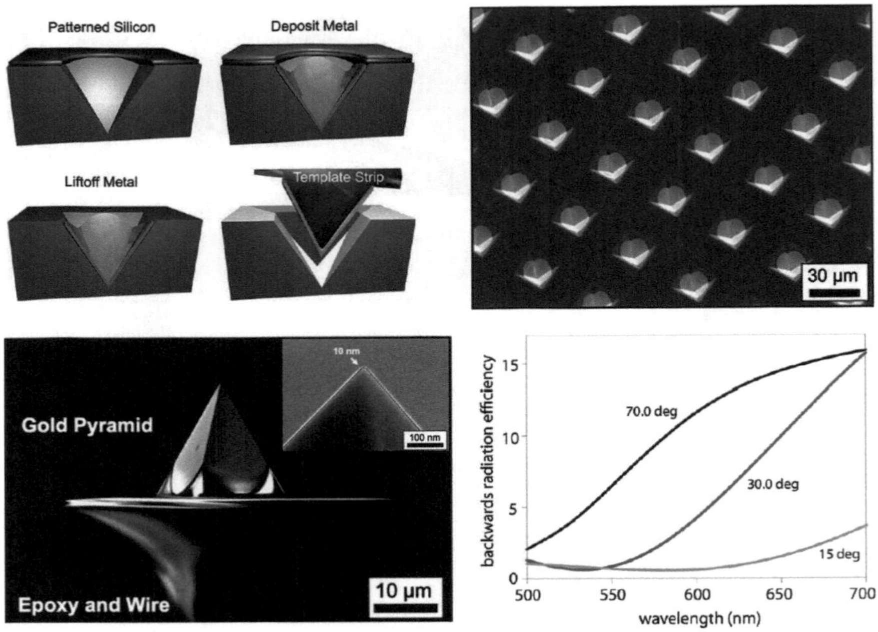

Fig. 7 The schematic of template-stripping fabrication technique; SEM images of template-stripped metallic tips and backward radiation efficiency of a dipole in front of a pyramidal tip as a function of wavelength λ and opening angle of pyramid. Adapted with permission from Johnson et al. [46]. Copyright (2012) American Chemical Society

determined by the in-plane momentum conservation condition $k_{SPP,z} = k_{in,z} \pm nG$, with n integer and $G = 2\pi/a_0$. The tip cone angle of ~ 15 corresponds to a maximum nanofocusing efficiency at =800 nm. The tips so modified are then mounted onto an AFM quartz tuning fork [15] and used in shear force to control the tip to sample distance and to obtain spectroscopic images. A SEM image of the tip with that FIB-fabricated grating and a map of optical excitation are displayed in Fig. 6b.

Another approach to fabricate the metallic probes suitable for near-field imaging techniques such as single-molecule fluorescence imaging and TERS is a template-stripping fabrication technique [46]. In that case (Fig. 7), silicon nitride is used as a mask for anisotropic etching of silicon with KOH that leads to the apex angle of 70.52°. After deposition of 200-nm-thick gold and the following liftoff procedure, the pyramidal tips are ready for template stripping. By placing a small droplet of quick-setting epoxy on a tungsten wire, a single pyramidal tip can be plucked from the mold. Then, the wire is mounted to the prong of a quartz tuning fork and implemented into an NSOM platform. A curvature of fabricated tips typically is about 10 nm. This approach together with a use of a tightly focused radially polarized optical excitation allows to achieve reproducible single-molecule imaging with sub-20-nm spatial resolution and an enhancement in the detected fluorescence

Fig. 8 3D schematic of a campanile structure at the end of a gold-coated conical tapered NSOM fiber and SEM image of the fabricated probe. The image is kindly provided by Dr. Mauro Melli, Molecular Foundry, Lawrence Berkeley National Laboratory, USA

signal of up to 200. However, only 10-fold enhancement in Raman signal was observed on CNT bundle with the same probes. The authors believe that an enhancement in stimulated backward radiation efficiency corresponds to an increased radiative decay rate. This prevents a quantum emitter from complete quenching and is the reason for their ability to perform high-quality near-field fluorescence imaging on samples with single molecules. Thus, a pyramidal tip is well suited to scatter the near-field optical signal into the far field, leading to larger emission enhancement and hence to a larger quantum yield. The high yield (>95 %), reproducibility, durability, and massively parallel fabrication of the probes (1.5 million identical ones over a wafer) make them promising 'for reliable optical sensing and detection and for cementing near-field optical imaging and spectroscopy as a routine characterization technique.' Another advantage of proposed approach is that each gold (or silver) pyramidal probe is used on demand, one at a time, and the unused tips can be stored for extended times without degradation or contamination.

It should be noted as well about promising perspectives of a new generation of photonic–plasmonic transducers, the 'campanile' probes [47, 48] shown in Fig. 8. The campanile probe structure significantly enhances the radiative rate of any dipole emitter located near the probe apex, demonstrating that over 90 % of the light radiated by the emitter is "captured" by this probe. These advantages can be crucial for many optical techniques, including TERS, but still require further investigations. Briefly, the campanile tips couple the photonic to the plasmonic field and then adiabatically compress the plasmon mode over a broad bandwidth. The confinement of the optical near field is determined by the gap size between the two antenna arms, which can be well below 10 nm given the appropriate resolution of the dielectric deposition method. Based on excitation through the back of

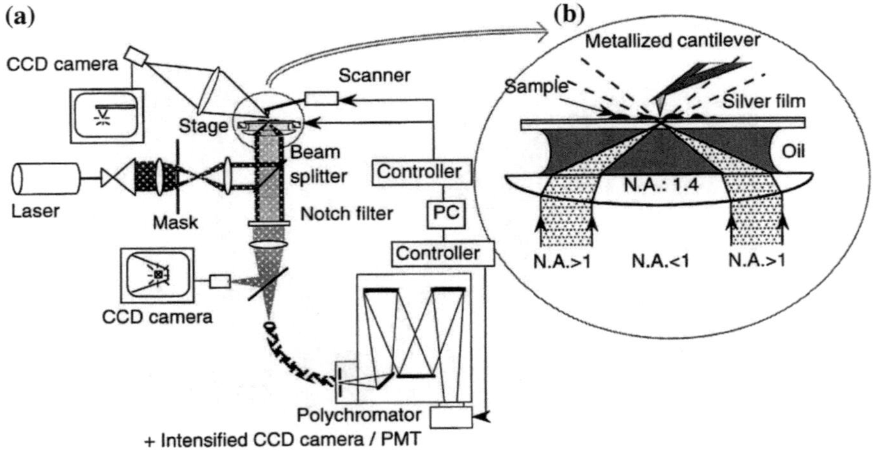

Fig. 9 **a** The configuration of the experimental setup of the near-field Raman microscope and **b** zoomed sample-tip area. Reprinted from Hayazawa et al. [34]. Copyright (2000), with permission from Elsevier

the tip similar to traditional aperture-based NSOM tips, the campanile tips enable nearly background-free nanoscale imaging and spectroscopy, even on dielectric, non-transparent substrates.

3.2 Microscopes Design

The first experiments showing tip enhancement [26, 34] were both performed using an inverted optical microscope and illuminating the tip from below. A simplified scheme of the setup used in [34] is displayed in Fig. 9a. An inverted optical microscope was used. A 488-nm argon ion laser line was aligned with the microscopic optical axis with the help of a beam splitter and was focalized on the sample using a high-numerical aperture oil immersion objective. The scattered light was first filtered with a notch filter to remove the very intense Raleigh scattering and then was focalized on the entrance slit of a monochromator, dispersed in wavelength and recorded with a cooled CCD. The position of an AFM metalized tip was controlled in x, y, and z using a piezodriven AFM controller. A detail of the objective used, shown in Fig. 9b, evidences the advantage of using a high-numerical aperture objective in which only the external ring of the objective is used, as in TIRF microscopy: first, the optical path of the exciting and scattered light results spatially separated, second the light transmitted has a large polarization along the z-axis, which provides the maximum coupling with the metalized tip. Similar setups were adopted in many other laboratories with slight modifications [49, 50]. In the paper of Zenobi group [26], for instance, a tuning fork, instead of an

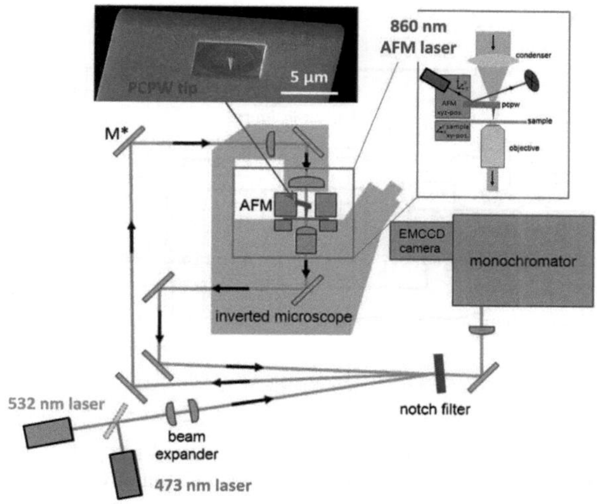

Fig. 10 Schematics of the PCPW-TERS setup with details of the cantilever, illumination, collection alignment, and SEM image of AFM cantilever with a photonic crystal plasmonic waveguide tip used in TERS experiments

AFM in contact mode, is used to control the tip–sample distance in shear force mode, which has the advantage of a weaker interaction and thus a minimal tip wear upon scanning. Other setups are equipped with a double-scanning system which allows at first a precision positioning of the tip within the microscope optical axis and then to scan the sample, thus maintaining the excitation geometry unchanged. The configuration described above offers the wider collection angle possible, thus maximizing the sensitivity, at the cost of operating only with transparent samples or molecular samples dispersed on glass or other transparent substrates.

Modifications based on an inverted microscope system have been introduced to benefit from the fabrication of engineered tips with tailored pasmonic tips. Thus, Bek et al. [51] in order to use the PCPW system described above [6] coupled an optical setup with a commercial AFM mounted on top of an inverted optical microscope. Figure 10 illustrates the optical configuration used. Two DPSS lasers at 532 and 473 nm with variable power up to 50 mW are first expanded and then reflected by a suitable notch filter. The beam is then lifted by a periscope and directed at the entrance of the illumination condenser (NA 0.35) of the microscope. The scattered light is collected from the bottom by a microscope objective (oil immersion NA 1.4) and directed onto the same notch filter through a similar optical path via another periscope. The Raman-shifted light is then focused on the entrance of a 750-mm-long spectrometer and dispersed on a cooled CCD. In practice, a coarse alignment of the excitation path on the photonic crystal using the last mirror of the periscope (*M** in Fig. 11) is firstly completed. Due to the focusing capability of the photonic crystal, a nanometric positioning is not required. Then, the tip with the high-numerical aperture objective of the collection path using the *XY* piezopositioning system of the atomic force microscope is aligned. At this step, an accuracy at the nm level is required, since the focal spot of the objective at the wavelength used is about 200 nm wide. The AFM laser, at

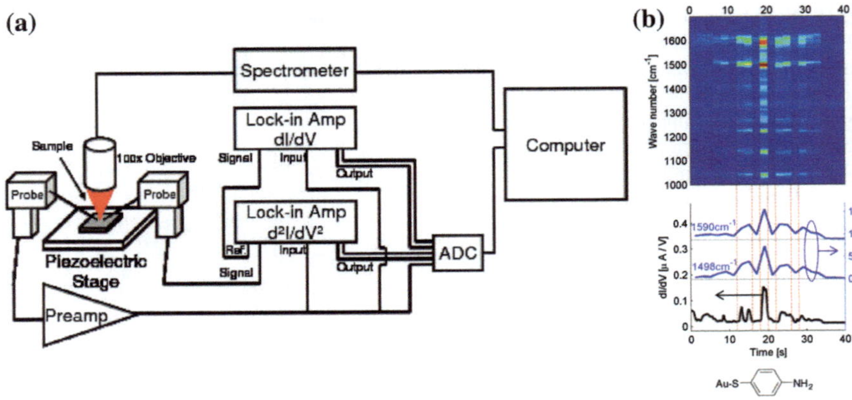

Fig. 11 **a** Schematic of the 'electromigration' setup. **b** Waterfall plot of Raman spectrum (2-s integrations) and positively correlated differential conductance measurements for a p-mercaptoaniline sample. All Raman modes that are visible exhibit this behavior as illustrated by the 1,498 and 1,590 cm^{-1} modes. The 1,590 cm^{-1} mode has been shifted upward on the lower graph with the gray line indicating zero CCD counts. Vertical red lines indicate points of rapidly changing Raman intensity and conduction. Adapted with permission from Ward et al. [55]. Copyright (2008) American Chemical Society

860 nm wavelength, is focused on the back of the cantilever, approximately displaced toward the cantilever chip body by a couple of tens of microns from the PCPW, to avoid cross talk in the two optical paths. Finally, to record an image, the sample is raster scanned using a second *XY* piezoscanner, while maintaining the tip position fixed with respect to the optical axis of the system.

In the case of the metallic tips in which surface plasmon polaritons (SPPs) are excited with the help of a lateral grating (Fig. 6b), the measurements are performed with side-on illumination of the tip–shaft grating with little deviations in the implementation part [14–16, 52]. Typically, gratings with 780 nm period, width of 150 nm and depth of 70–100 nm are focused-ion-beam milled onto the shaft of an electrochemically etched gold tip at a distance of 15 μm from the tip apex. The resulted tip cone angle of $\sim 15°$ corresponds to a maximum nanofocusing efficiency at $\lambda = 800$ nm, yet with broad wavelength range. Then, the grating on the tip shaft is illuminated with light from either a spectrally tunable continuous-wave or a mode-locked Ti:sapphire laser centered at 780 nm. The laser is linearly polarized in the direction perpendicular to the grooves in order to maximize the SPP excitation efficiency. The light is focused onto the tip shaft to a spot size of about 6 μm at close to normal incidence by a microscope objective with a *NA* of 0.35 and a working distance of 20 mm. The excited SPP wave packets travel adiabatically along the taper surface toward the tip apex. The Raman (or Rayleigh)-scattered light from the tip is collected by a second microscope objective ($NA = 0.5$) and imaged onto a photodetector and a CDD camera. For regulation of the distance between tip and sample, a shear-force-based feedback scheme is used in which the gold tip is mounted along the side of one of the arms of a quartz tuning fork. A hardware-linearized *XYZ* piezostage moves the sample with nanometric precision.

When operating with non-transparent samples, the optics should be placed on the same side of the tip. Nice examples of TERS implementation on non-transparent samples are provided by the group of Bruno Pettinger [53] in which the tip is illuminated by a long working distance and high-numerical aperture objective at 60° from the sample normal. The same objective is used to collect a Raman-scattered light. The tip-to-sample distance in those experiments is controlled by keeping constant the tunneling current in a scanning tunneling microscopy (STM) setup. More details of Pettinger setup can be found in [53].

Another solution for non-transparent samples is proposed by Alfred Meixner [54] using a parabolic mirror with high *NA*. In that case, a radially polarized laser beam is focused on reflection on the sample. The gold tip is brought into the center of the focus through a hole in the parabolic mirror and kept at a distance in the order of 2 nm from the sample surface by shear force control. Due to the reflection at the gold surface, a standing wave is formed with two axial intensity maxima of the electric field strictly polarized along the z-axis and a nodal plane about half a wavelength above the sample surface. This well-defined field distribution allows for an effective excitation of a longitudinal plasmon in the front section of the tip, as required for efficient near-field enhancement at the tip apex.

A third, alternative, system design was proposed by Ward and coworkers [55] (Fig. 11). In their experiments, the Raman enhancement is provided by a couple of electrodes whose gap, nanometric in size, is created by electromigration starting from a continuous structure fabricated by in-plane lithographic techniques. Different kinds of molecules are then adsorbed on the gap, and single-molecule current spectroscopy is performed simultaneously with single-molecule Raman spectroscopy, by using a confocal scanning microscope. Up to 120 gaps are fabricated on the same substrate. Both Raman spectra and conductance signal arise from the same single-molecule gap as demonstrated by correlation experiments. This approach is not suitable for Raman mapping but demonstrate the extreme flexibility of single-molecule Raman spectroscopy when coupled with tip enhancement strategies.

4 Applications of TERS

Nowadays, applications of TERS can be divided into two main groups: (i) nanomaterials characterization and (ii) bioapplications.

First group mainly involves high-Raman-active materials, such as dyes, low-dimensional carbon-based or semiconductor materials. These materials have a quite high Raman cross section that is a cause to quantify the structures based on them with relatively low laser powers and acquisition times. The intensity of Raman scattering is a function of the tip–sample distance, Raman polarizability tensor components, magnitude of near EM field, incident laser beam configuration, and tip orientation relative to the sample plane [56]. In particular, the intensity

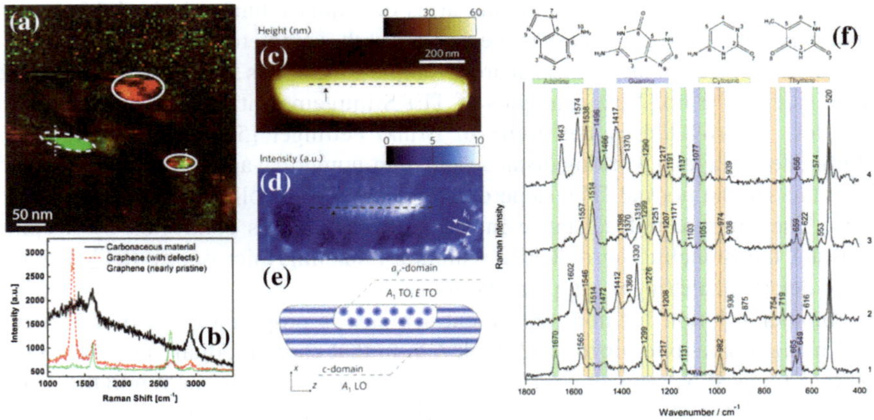

Fig. 12 *Graphene study* **a** A 100 × 100 pixel, 400 × 400 nm^2 Raman image of graphene on template-stripped gold, produced by the scotch-tape technique, 0.3 s/pixel. Areas with strong defects are visible (*white circles*, 75 × 45 nm^2 and 55 × 25 nm^2) as well as a small area with intense graphene signals (*dashed white line*). The weaker graphene signals in most of the background are not visible due to the enormous contrast. **b** Characteristic Raman spectra from the single-layer graphene (*green*), one defect area (*red*), and the graphene-free background (*black*). Resolution has been determined by a Gauss fit on the white dotted lines as 10.6 and 11.8-nm FWHM, respectively. Reprinted with permission from Stadler et al. [62]. Copyright (2011) American Chemical Society. *Ferroelectric domain imaging* **c** Topography of $BaTiO_3$ nanorod. **d** The spectrally integrated TERS signal for ferroelectric domain imaging. **e** Domain assignment based on the Raman selection rules for the TERS geometry used. Reprinted by permission from Macmillan Publishers Ltd: [Nature Nanotechnology] Berweger et al. [64]. Copyright (2009). *Distinction of nucleobases* **f** TERS spectra measured on single-stranded calf thymus DNA immobilized non-specifically on a mica substrate. Unambiguously assignable bands are highlighted. Reprinted from Treffer et al. [67] in accordance with the provisions of the Beilstein-Institut Open Access License Agreement

dependence on the EM field can be described as proportional to the 4th power of the field enhancement factor. Therefore, Raman response is significantly increased in the enhanced-by-tip (or surface) EM fields.

The dyes in a 'TERS history' are used from its beginning. Some illustrative results on technique, tips, microscopes used, etc., can be found in [15, 26, 32, 34, 35, 40] etc.

Since the modern electronics is still based on Silicon technology, but the length scales are constantly being reduced, the characterization of devices at nanoscale is strongly required from technological point of view. Silicon is a well-studied material also by Raman spectroscopy, and such characteristics as intrinsic stress, heating of the structures, the presence of amorphous silicon phase that lead to the recognized shifts in Raman bands can be easily quantified by TERS [6, 51, 57, 58] etc.

Another direction is carbon-based electronics and chemistry that are growing up very fast. The Raman modes of carbon nanotubes and graphene are quite strong and well defined [59]. The peak position of the breath G-mode is very sensitive to the electron doping by electrical means or by chemical functionalization due to a charge transfer. The stress-induced changes in those materials can be also quantified.

Moreover, clearly recognized features in graphene Raman spectra caused by the presence of the defects in its honeycomb structure make a TERS very promising technique to characterize the changes of carbon-based nanostructure properties [49, 60–63]. An example of graphene TERS mapping is shown in Fig. 12a and b.

Finally, among many representative TERS data in material science, the following studies should be noticed: the extension of TERS to optical crystallography on the nanoscale by identifying intrinsic ferroelectric domains of individual $BaTiO_3$ nanocrystals through selective probing of different transverse optical phonon modes in the system [64] (Fig. 12c–e) and high-resolution spectroscopic mapping of the chemical contrast from nanometer domains in P3HT:PCBM organic blend films for solar-cell applications [65].

The second group, bioorganic compounds, due to their complex chemical composition has lower Raman cross sections and lower Raman responses, correspondingly. This requires higher acquisition times. On the other hand, a direct characterization of living cells is obstructed by their motion and possible changes of their shape resulting in the spectral and intensity fluctuations. That is why TERS study of bioobjects mostly operates with relatively simple structures which can be fixed on the substrates, such as DNA and RNA, viruses, membranes, etc. Another impediment is a possible contamination of the tip scanning a sample or tip-induced heating leading to the physical and chemical changes including photochemical reactions as well. However, some examples of TERS investigation of living organisms are present in the literature.

A very representative example of TERS importance in biology is a label-free DNA or RNA sequencing approach (Fig. 12f) based on spectral differences recorded by a lateral tip displacement within the size of a base-to-base distance [12, 66, 67]. Another good example is a demonstration of spectroscopic fingerprint of single viruses without the need of bulk material [68, 69] that makes TERS promising for virus identification on the single-particle level. An important step toward the description of biological processes on the molecular scale is shown in [70], where TERS investigation of streptavidin-labeled lipid films allowed to determine a size distribution of protein aggregations in lipid membranes without the need for extensive labeling that can affect the mobility and functionality of membrane proteins. Moreover, the lipid distribution in a mixed supported lipid monolayer was directly measured with high lateral resolution resulting in the full-spectral images on phase-separated lipid domains [71].

To date, a few works on the cells transferred to the appropriate substrates are performed. The human cells in cancer research [72], cell organelles [73], hemozoin crystals in the digestive vacuole of a sectioned malaria parasite-infected cell [74], bacteria [75], and yeast cells [76] were studied performing the TERS measurements in air. In that aspect, a modification of the tips by self-assembled monolayer of a very short alkanethiols opens a way of TERS use in an aqueous environment [77] that is definitely a huge advantage to study the bioobjects in their natural environment.

A good example of TERS applications and its spread between material science and biology in terms of 'TERS enhancement factor' is summarized and shown in

[78]. The diagram presented there demonstrates mostly the usage of non-biosamples as TERS reporters. The only one biorepresentative reflected there is Staphylococcus epidermidis. However, the most significant advantage of TERS technique is that it is a label-free technique; thus, in addition to clear advantage of TERS study of nanomaterials, chemically undisturbed biological systems can be investigated.

An important characteristic of TERS, as a combination of Raman spectroscopy and scanning probe microscopy, is a spatial resolution on chemical or structural information of the objects. Definitely, a resolution is caused by the tip apex diameter that confirmed as well by theoretical simulations [79]. To date, the best resolution obtained is 4 nm, but an additional effect of local pressure in NSOM-based microscopy is used to achieve super-high spatial resolution [80]. The modern technological approaches allow to fabricate the tips with plasmon-active element of about 10 nm in diameter [6, 46] that is not enough to make the next breakthrough to atomic/molecular resolution in a surface science and biology challenging more research efforts to realize it in the future.

References

1. Novotny L, Sanchez EJ, Xie XS (1998) Near-field optical imaging using metal tips illuminated by higher-order Hermite-Gaussian beams. Ultramicroscopy 71:21
2. Sanchez EJ, Novotny L, Xie XS (1999) Near-field fluorescence microscopy based on two-photon excitation with metal tips. Phys Rev Lett 82:4014
3. Kawata S, Inouye Y, Kataoka T, Okamoto T (2002) CH 4: Apertureless Near-Field Probes. In: Kawata S, Ohtsu M, Irie M (eds) Nano-optics. Springer series in optical sciences, vol 84. Springer, Berlin, p 75
4. Courjon D (2003) Near field microscopy and near field optics. Imperial College Press, London
5. Novotny L, Hecht B (2006) Principles of nano-optics. Cambridge University Press, Cambridge
6. De Angelis F, Das G, Candeloro P, Patrini M, Galli M, Bek A, Lazzarino M, Maksymov I, Liberale C, Andreani LC, Di Fabrizio E (2010) Nanoscale chemical mapping using three-dimensional adiabatic compression of surface plasmon polaritons. Nat Nanotechnol 5:67
7. Bailo E, Deckert V (2008) Tip-enhanced Raman scattering. Chem Soc Rev 37:921
8. Hartschuh A (2008) Tip-enhanced near-field optical microscopy. Angew Chem Int Ed 47:8178
9. Domke KF, Pettinger B (2010) Studying surface chemistry beyond the diffraction limit: 10 years of TERS. Chem Phys Chem 11:1365
10. Pahlow S, Marz A, Seise B, Hartmann K, Freitag I, Kammer E, Bohme R, Deckert V, Weber K, Cialla D, Popp J (2012) Bioanalytical application of surface-and tip-enhanced Raman spectroscopy. Eng Life Sci 12(2):131
11. Schmid T, Opilik L, Blum C, Zenobi R (2013) Nanoscale chemical imaging using tip-enhanced raman spectroscopy: A critical review. Angew Chem Int Ed 52:5940
12. Bailo E, Deckert V (2008) Tip-enhanced Raman spectroscopy of single RNA strands: Towards a novel direct-sequencing method. Angew Chem Int Edit 47(9):1658
13. De Angelis F, Patrini M, Das G, Maksymov I, Galli M, Businaro L, Andreani LC, Di Fabrizio E (2008) A hybrid plasmonic-photonic nanodevice for label-free detection of a few molecules. Nano Lett 8(8):2321
14. Ropers C, Neacsu CC, Elsaesser T, Albrecht M, Raschke MB, Lienau C (2007) Grating-coupling of surface plasmons onto metallic tips: A nanoconfined light source. Nano Lett 7(9):2784

15. Berweger S, Atkin JM, Olmon RL, Raschke MB (2010) Adiabatic tip-plasmon focusing for nano-raman spectroscopy. J Phys Chem Lett 1:3427
16. Berweger S, Atkin JM, Olmon RL, Raschke MB (2012) Light on the tip of a needle: Plasmonic nanofocusing for spectroscopy on the nanoscale. J Phys Chem Lett 3:945
17. Synge EH (1928) A suggested method for extending microscopic resolution into the ultra-microscopic region. Phil Mag 6:356
18. O'Keefe JA (1956) Resolving power of visible light. J Opt Soc Am 46:359
19. Ash EA, Nicholls G (1972) Super-resolution Aperture Scanning Microscope. Nature 237:510
20. Binnig G, Rohrer H, Gerber C, Weibel E (1982) Tunneling through a controllable vacuum gap. Appl Phys Lett 40:178
21. Pohl DW, Denk W, Lanz M (1984) Optical stethoscopy: Image recording with resolution λ/20. Appl Phys Lett 44:651
22. Lewis A, Isaacson M, Harootunian A, Murray A (1984) Development of a 500 Å spatial resolution light microscope. I. Light is efficiently transmitted through λ/16 diameter apertures. Ultramicroscopy 13:227
23. Weber WH, Ford GW (1981) Optical electric-field enhancement at a metal surface arising from surface-plasmon excitation. Opt Lett 6:122
24. Wessel J (1985) Surface-enhanced optical microscopy. J Opt Soc Am B 2:1538
25. Fleischmann M, Hendra PJ, McQuillan AJ (1974) Raman spectra of pyridine adsorbed at a silver electrode. Chem Phys Lett 26:163
26. Stockle RM, Suh YD, Deckert V, Zenobi R (2000) Nanoscale chemical analysis by tip-enhanced Raman spectroscopy. Chem Phys Lett 318:131
27. Farahani JN, Pohl DW, Eisler HJ, Hecht B (2005) Single quantum dot coupled to a scanning optical antenna: A tunable superemitter. Phys Rev Lett 95:017402
28. Esteban R, Vogelgesang R, Kern K (2007) Tip-substrate interaction in optical near-field microscopy. Phys Rev B 75:195410
29. Zhang W, Cui X, Yeo BS, Schmid T, Hafner C, Zenobi R (2007) Nanoscale roughness on metal surfaces can increase tip-enhanced raman scattering by an order of magnitude. Nano Lett 7:1401
30. Bouhelier A, Renger J, Beversluis MR, Novotny L (2003) Plasmon-coupled tip-enhanced near-field optical microscopy. J Microsc 210:220
31. Bouhelier A, Novotny L (2007) Near-field optical excitation and detection of surface plasmons. In: Brongersma ML, Kik PG (eds) Surface plasmon nanophotonics. Springer series in optical sciences, vol 131. Springer, Berlin, p 139
32. Ren B, Picardi G, Pettinger B (2004) Preparation of gold tips suitable for tip-enhanced Raman spectroscopy and light emission by electrochemical etching. Rev Sci Instrum 75:837
33. Steidtner J, Pettinger B (2008) Tip-enhanced Raman spectroscopy and microscopy on single dye molecules with 15 nm resolution. Phys Rev Lett 100:236101
34. Hayazawa N, Inouye Y, Sekkat Z, Kawata S (2000) Metallized tip amplification of near-field Raman scattering. Opt Commun 183:333
35. Yeo B-S, Zhang W, Vannier C, Zenobi R (2006) Enhancement of Raman signals with silver-coated tips. Appl Spectrosc 60(10):1142
36. Yeo B-S, Schmid T, Zhang W, Zenobi R (2007) Towards rapid nanoscale chemical analysis using tip-enhanced Raman spectroscopy with Ag-coated dielectric tips. Anal Bioanal Chem 387:2655
37. Hayazawa N, Yano T, Kawata S (2012) Highly reproducible tip-enhanced Raman scattering using an oxidized and metallized silicon cantilever tip as a tool for everyone. J Raman Spectrosc 43:1177
38. Wang N, Cai Y, Zhang RQ (2008) Growth of nanowires. Mat Sci Eng R 60:1
39. Jabeen F, Grillo V, Rubini S, Martelli F (2008) Self-catalyzed growth of GaAs nanowires on cleaved Si by molecular beam epitaxy. Nanotechnology 19(27):275711
40. Becker M, Sivakov V, Andra G, Geiger R, Schreiber J, Hoffmann S, Michler J, Milenin AP, Werner P, Christiansen SH (2007) The SERS and TERS effects obtained by gold droplets on top of Si nanowires. Nano Lett 7(1):75

41. Becker M, Sivakov V, Gosele U, Stelzner T, Andra G, Reich HJ, Hoffmann S, Michler J, Christiansen SH (2008) Nanowires enabling signal-enhanced nanoscale Raman spectroscopy. Small 4(4):398
42. Jenke MG, Lerose D, Niederberger C, Michler J, Christiansen S, Utke I (2011) Toward local growth of individual nanowires on three-dimensional microstructures by using a minimally invasive catalyst templating method. Nano Lett 11:4213
43. Engstrom DS, Savu V, Zhu X, Bu IYY, Milne WI, Brugger J, Boggild P (2011) High throughput nanofabrication of silicon nanowire and carbon nanotube tips on AFM probes by stencil-deposited catalysts. Nano Lett 11:1568
44. Babadjanyan AJ, Margaryan NL, Nerkararyan KhV (2000) Superfocusing of surface polaritons in the conical structure. J Appl Phys 87(8):3785
45. Stockman MI (2004) Nanofocusing of optical energy in tapered plasmonic waveguides. Phys Rev Lett 93:137404
46. Johnson TW, Lapin ZJ, Beams R, Lindquist NC, Rodrigo SG, Novotny L, Oh S-H (2012) Highly reproducible near-field optical imaging with sub-20-nm resolution based on template-stripped gold pyramids. ACS Nano 6(10):9168
47. Bao W, Melli M, Caselli N, Riboli F, Wiersma DS, Staffaroni M, Choo H, Ogletree DF, Aloni S, Bokor J, Cabrini S, Intonti F, Salmeron MB, Yablonovitch E, Schuck PJ, Weber-Bargioni A (2012) Mapping local charge recombination heterogeneity by multidimensional nanospectroscopic imaging. Science 338:1317
48. Bao W, Staffaroni M, Bokor J, Salmeron MB, Yablonovitch E, Cabrini S, Weber-Bargioni A, Schuck PJ (2013) Plasmonic near-field probes: A comparison of the campanile geometry with other sharp tips. Opt Express 21(7):8166
49. Hartschuh A, Anderson N, Novotny L (2003) Near-field Raman spectroscopy using a sharp metal tip. J Microsc 210(3):234
50. Rasmussen A, Deckert V (2006) Surface- and tip-enhanced Raman scattering of DNA components. J Raman Spectrosc 37:311
51. Bek A, De Angelis F, Das G, Di Fabrizio E, Lazzarino M (2011) Tip enhanced Raman scattering with adiabatic plasmon focusing tips. Micron 42:313
52. Sadiq D, Shirdel J, Lee JS, Selishcheva E, Park N, Lienau C (2011) Adiabatic nanofocusing scattering-type optical nanoscopy of individual gold nanoparticles. Nano Lett 11:1609
53. Pettinger B, Domke KF, Zhang D, Picardi G, Schuster R (2009) Tip-enhanced Raman scattering: Influence of the tip-surface geometry on optical resonance and enhancement. Surf Sci 603:1335
54. Sackrow M, Stanciu C, Lieb MA, Meixner AJ (2008) Imaging nanometre-sized hot spots on smooth Au films with high-resolution tip-enhanced luminescence and Raman near-field optical microscopy. Chem Phys Chem 9:316
55. Ward DR, Halas NJ, Ciszek JW, Tour JM, Wu Y, Nordlander P, Natelson D (2008) Simultaneous measurements of electronic conduction and raman response in molecular junctions. Nano Lett 8(3):919
56. Maximiano RV, Beams R, Novotny L, Jorio A, Cancado LG (2012) Mechanism of near-field Raman enhancement in two-dimensional systems. Phys Rev B 85:235434
57. Hayazawa N, Motohashi M, Saito Y, Ishitobi H, Ono A, Ichimura T, Verma P, Kawata S (2007) Visualization of localized strain of a crystalline thin layer at the nanoscale by tip-enhanced Raman spectroscopy and microscopy. J Raman Spectrosc 38:684
58. Hermann P, Hecker M, Chumakov D, Weisheit M, Rinderknecht J, Shelaev A, Dorozhkin P, Eng LM (2011) Imaging and strain analysis of nano-scale SiGe structures by tip-enhanced Raman spectroscopy. Ultramicroscopy 111:1630
59. Hoffmann GG, de With G, Loos J (2008) Micro-Raman and tip-enhanced Raman spectroscopy of carbon allotropes. Macromol Symp 265:1
60. Cancado LG, Hartschuh A, Novotny L (2009) Tip-enhanced Raman spectroscopy of carbon nanotubes. J Raman Spectrosc 40:1420
61. Snitka V, Rodrigues RD, Lendraitis V (2011) Novel gold cantilever for nano-Raman spectroscopy of graphene. Microelectron Eng 88:2759

62. Stadler J, Schmid T, Zenobi R (2011) Nanoscale chemical imaging of single-layer graphene. ACS Nano 5(10):8442
63. Anderson N, Hartschuh A, Novotny L (2007) Chirality changes in carbon nanotubes studied with near-field Raman spectroscopy. Nano Lett 7(3):577
64. Berweger S, Neacsu CC, Mao Y, Zhou H, Wong SS, Raschke MB (2009) Optical nanocrystallography with tip-enhanced phonon Raman spectroscopy. Nat Nanotechnol 4:496
65. Wang X, Zhang D, Braun K, Egelhaaf H-J, Brabec CJ, Meixner AJ (2010) High-Resolution spectroScopic mapping of the chemical contrast from nanometer domains in P3HT:PCBM organic blend films for Solar-Cell applications. Adv Funct Mater 20:492
66. Zhang D, Domke KF, Pettinger B (2010) Tip-enhanced Raman spectroscopic studies of the hydrogen bonding between adenine and thymine adsorbed on Au (111). Chem Phys Chem 11:1662
67. Treffer R, Lin X, Bailo E, Deckert-Gaudig T, Deckert V (2011) Distinction of nucleobases – a tip-enhanced Raman approach. Beilstein J Nanotechnol 2:628
68. Cialla D, Deckert-Gaudig T, Budich C, Laue M, Moller R, Naumann D, Deckert V, Popp J (2009) Raman to the limit: Tip-enhanced Raman spectroscopic investigations of a single tobacco mosaic virus. J Raman Spectrosc 40:240
69. Hermann P, Hermelink A, Lausch V, Holland G, Moller L, Bannert N, Naumann D (2011) Evaluation of tip-enhanced Raman spectroscopy for characterizing different virus strains. Analyst 136:1148
70. Bohme R, Cialla D, Richter M, Rosch P, Popp J, Deckert V (2010) Biochemical imaging below the diffraction limit - probing cellular membrane related structures by tip-enhanced Raman spectroscopy (TERS). J Biophoton 3(7):455
71. Opilik L, Bauer T, Schmid T, Stadler J, Zenobi R (2011) Nanoscale chemical imaging of segregated lipid domains using tip-enhanced Raman spectroscopy. Phys Chem Chem Phys 13:9978
72. Richter M, Hedegaard M, Deckert-Gaudig T, Lampen P, Deckert V (2011) Laterally resolved and direct spectroscopic evidence of nanometer-sized lipid and protein domains on a single cell. Small 7(2):209
73. Bohme R, Mkandawire M, Krause-Buchholz U, Rosch P, Rodel G, Poppac J, Deckert V (2011) Characterizing cytochrome c states - TERS studies of whole mitochondria. Chem Commun 47:11453
74. Wood BR, Bailo E, Khiavi MA, Tilley L, Deed S, Deckert-Gaudig T, McNaughton D, Deckert V (2011) Tip-enhanced raman scattering (TERS) from hemozoin crystals within a sectioned erythrocyte. Nano Lett 11:1868
75. Neugebauer U, Rosch P, Schmitt M, Popp J, Julien C, Rasmussen A, Budich C, Deckert V (2006) On the way to nanometer-sized information of the bacterial surface by tip-enhanced Raman spectroscopy. Chem Phys Chem 7:1428
76. Naumenko D, Snitka V, Serviene E, Bruzaite I, Snopok B (2013) In vivo characterization of protein uptake by yeast cell envelope: Single cell AFM imaging and μ-tip-enhanced Raman scattering study. Analyst 138:5371
77. Schmid T, Yeo B-S, Leong G, Stadler J, Zenobi R (2009) Performing tip-enhanced Raman spectroscopy in liquids. J Raman Spectrosc 40:1392
78. Pettinger B, Schambach P, Villagomez CJ, Scott N (2012) Tip-enhanced raman spectroscopy: Near-fields acting on a few molecules. Ann Rev Phys Chem 63:379
79. Downes A, Salter D, Elfick A (2008) Simulations of tip-enhanced optical microscopy reveal atomic resolution. J Microsc 229:184
80. Yano T, Verma P, Saito Y, Ichimura T, Kawata S (2009) Pressure-assisted tip-enhanced Raman imaging at a resolution of a few nanometres. Nat Photonics 3:473

DNA as Nanostructuring Element for Design of Functional Devices

Dennis M. Bauer, Dania M. Kendziora, Ishtiaq Ahmed, Yu-Chueh Hung and Ljiljana Fruk

Abstract This chapter gives an overview of the recent developments in the use of natural and synthetic DNA for design of functional devices such as biosensors or optoelectronic platforms. In addition, different strategies for preparation of DNA–protein and DNA–nanoparticle conjugates and their use for DNA origami decoration are discussed. Finally, a summary of DNA-based hybrid materials such as DNA hydrogels are presented with particular emphasis on the current and possible future applications.

1 Introduction

Discovery of the double-helix structure has not only brought the revolution to the field of molecular biology and genetics, but opened up a route toward the use of its molecular architecture to design novel devices, enable structuring of different species or design new programmable materials. DNA's structural properties such as specific base pairing, stability of the double helix as well as the powerful combination of stiffness, and flexibility have attracted the interest of the scientists working in different scientific fields ranging from physics and plasmonics to biochemistry and engineering. Their efforts led to the development of DNA nanotechnology field where DNA, rather than being genetic code carrier, is used as a versatile building block. The structuring potential of DNA was hinted in a theoretical paper written by Ned Seeman in 1982, which then led to the

D. M. Bauer · D. M. Kendziora · I. Ahmed · L. Fruk (✉)
Center for Functional Nanostructures, Karlsruhe Institute of Technology, Wolfgang-Gaede-Str. 1a, 76131 Karlsruhe, Germany
e-mail: Ljiljana.Fruk@kit.edu

Y.-C. Hung
Institute of Photonics Technologies, National Tsing Hua University, Hsinchu, Taiwan

F. Benfenati et al. (eds.), *Novel Approaches for Single Molecule Activation and Detection*, Advances in Atom and Single Molecule Machines,
DOI: 10.1007/978-3-662-43367-6_6, © Springer-Verlag Berlin Heidelberg 2014

establishment of the simple rules employed to build different structures out of DNA strands [1, 2]. The field was further advanced by the development of DNA-directed immobilization (DDI) and conjugation of single-stranded DNA to various molecular species [3] and a DNA-folding technique called DNA origami [4]. Of course, this advances would not be possible if it were not for the development of the chemical synthesis of short DNA strands [5–7], resulting in design of automated synthesizer, which enabled production of short DNA strands of desired sequences at affordable costs. In this paper, we wanted to give an overview of the use of DNA in nanotechnology with special emphasis on the design of functional nanodevices such as biosensors, catalytic platforms, or novel biohybrid materials. It should be noted that two types of DNAs are used today for design of such devices—short chemically synthesized DNA and DNA from natural sources such as viral or salmon sperm DNA. This chapter will give a summary of the devices made of both DNA types.

2 DNA-Directed Immobilization: Use of Chemically Synthesized Short DNA

Short DNA strands (up to 80 bases) can today be synthesized in a relatively straightforward manner using phosphoramidite chemistry and commercially available DNA synthesizers. Such synthetic oligonucleotides are one of the most powerful tools in biotechnology and nanotechnology. The successful phosphoramidite solid-phase synthesis of DNA used today is a product of the extensive research over the past 50 years and is the combination of effective solid-phase chemistry and phosphoramidite methodology first time used by Beaucage and Caruthers [6]. They have described the reaction of two modified nucleosides, which resulted in a 3′-5′ linkage, present in the natural DNA, in high yield. The chemical strategy, which formed the basis of the chemical synthesis of DNA, was also a result of the discovery that P(III) was considerably more reactive than corresponding P(V) acylating agents [8] as well as the work by Sinha et al. [9] with 2-cyanoethyl phosphoramidites used in solid-phase oligonucleotide synthesis. Using phosphoramidite chemistry, numerous modifications can be introduced to the oligonucleotides on both (5′ and 3′) ends or internally, which additionally opened up a range of applications (design of DNA-based biosensing devices such as molecular beacons). Such ability also led to the development of the DNA microarrays and resulted in a range of chemical methodologies, which can be employed to attach DNA to various surfaces or different molecular species. Once immobilized, single-stranded DNA can be hybridized with complementary strand bearing different molecules, and it can act as an immobilization tool within a process called DDI (DNA Directed Immobilization, Fig. 1). Developed in mid-1990s [3], it has been used in various applications such as the preparation of biosensing platforms, but also for precise positioning of biomolecules to enable study of their function.

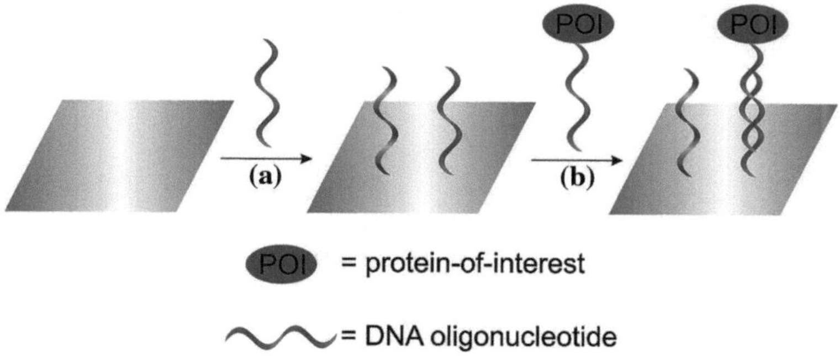

Fig. 1 Schematic illustration of DNA-directed immobilization (DDI). Shown is the formation of a protein microarray in 3 steps. After immobilization of DNA oligonucleotides on a suitable surface (**a**), a DNA-modified protein-of-interest (*POI*) containing complementary sequence can be hybridized (**b**). Complementary sequences are shown in the same color

2.1 DNA as an Immobilization Tool

To enable the use of DNA as immobilization tool, on the one side, capture DNA needs to be anchored to the desired surface (i.e., magnetic bead, nanoparticle (NP), or planar surface). On the other side, a suitable techniques need to be developed for the conjugation of oligonucleotides to the species, which need to be immobilized (biomolecules and NP) [10, 11]. As the biosensing devices are functional devices of huge importance for the investigation of basic biological principles of molecular interaction within the cell but also for diagnostics, the capture and immobilization of important biomolecules such as various proteins are of tremendous significance. The first step, however, is finding appropriate chemical methodology (non-covalent or covalent) to prepare DNA–biomolecule conjugates and vast efforts have gone into developing strategies that enable tight binding and preservation of both DNA and biomolecular functions.

Most common non-covalent coupling methodologies are based on streptavidin (STV)–biotin interaction [12] or use of protein tags such as His tag [13], through which DNA can be attached specifically onto a pre-defined position (Fig. 2a). Another approach applicable to a class of cofactor-containing proteins involves the replacement of the natural cofactor with DNA-modified cofactor, during the process known as cofactor reconstitution [14, 15] (Fig. 2b). Site-specific and stoichiometric heme protein–DNA conjugates were prepared by reconstitution of heme into apoenzymes [15] and successfully used to design electrochemical biosensors [16] and artificial, protein-based photocatalysts [17]. Along the similar line, Simon et al. [18] used flavin-modified DNA to reconstitute apoflavin reductase and designed the system where enzymatic activity can be controlled by DNA hybridization. However, non-covalent approaches often result in the protein–DNA complexes sensitive to the environmental conditions and prone to dissociation, which renders them unsuitable for studies under cellular matrix conditions.

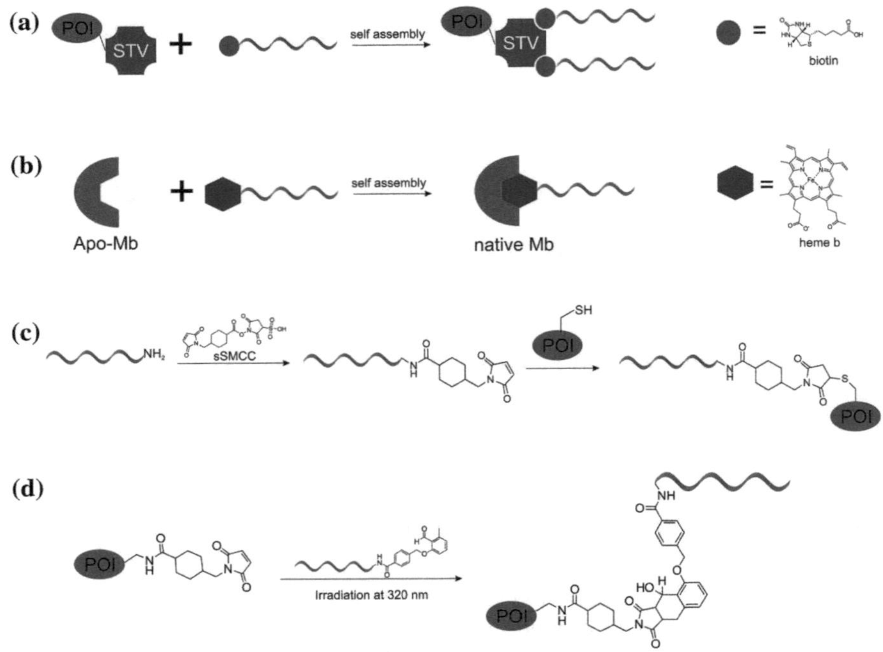

Fig. 2 Overview of different coupling methods for the synthesis of protein–DNA conjugates. **a** Use of STV–biotin interactions to combine protein-of-interest–STV conjugates with biotinylated DNA. **b** Reconstitution of apomyoglobin with DNA-modified cofactor heme. **c** Covalent coupling method with the help of the bifunctional linker sSMCC by modifying a Cys residue inside the protein-of-interest. **d** Light-triggered reaction of maleimide-modified protein-of-interest and photoreactive caged DNA [22]

More robust covalent linkage provides chemical stability and the most widely used methodology for preparation of DNA–protein conjugates involves the use of bifunctional linkers to enable cross-linking of thiol- or amine-modified single-strand DNAs with appropriate amino acid residues on the protein surface (Fig. 2c) [19]. One of the most widely used coupling reagents is heterobifunctional cross-linker sulfosuccinimidyl-4-(*N*-maleimidomethyl)cyclohexane-1-carboxylate (sSMCC) (Fig. 2c) containing a maleimide moiety (thiol-reactive group) and NHS active ester (amine reactive), which then react with naturally occurring or genetically introduced lysines (Lys) and cysteines (Cys) and thiol- or amino-modified DNA [20, 21].

Maleimide-reactive group introduced to the protein using sSMCC can not only be used for modification of Lys with thiolated DNA, but also used for other mild conjugation strategies. In a recent example, Lys groups of horse heart myoglobin (Mb) were converted to maleimide groups to which DNA could be attached via a mild photochemical approach representing a first example of DNA–protein conjugation under light-triggered bioorthogonal conditions [23]. This was achieved by using an oligonucleotide containing a photoreactive caged diene, which can be activated by irradiation and coupled to the maleimide-modified Mb to afford Mb–DNA conjugate (Fig. 2d).

Recently, bioorthogonal approaches have been developed, which allow DNA coupling independently of specific amino acid functional moieties present in a native protein. Using bioorthogonal expressed protein ligation, in which a recombinant protein containing thioester can react with Cys-modified nucleic acids, different DNA–protein conjugates were prepared [24]. Other approaches are based on Huisgen 1,3 dipolar cycloaddition of azide and alkyne [25], Staudinger ligation [26], keton–aminooxy [27], or enzymatically aided reactions [28, 29]. However, these methods usually include additional modifications of the protein either by addition of bioorthogonal functional groups through Lys and Cys residues or by introduction of artificial amino acids [30]. However, for certain classes of proteins, the introduction of common affinity protein tags or artificial amino acids is often a complicated process from a molecular biology side, which results in low yields of desired protein and often in the loss of activity. On the other side, modification through Cys might not be feasible due to the highly conserved Cys regions making thiol unavailable for further modifications and there might be simply too many Lys on the surface, the modification of which effects the distribution of the surface charges and might influence the inherent protein–protein interactions. Therefore, there is often a need to employ other amino acids for attachment of DNA. Along this line, Bauer et al. [22] designed specific tyrosine-binding bifunctional linkers, which were used to modify STV and Mb proteins using mild click chemistry procedures.

In general, DDI methodology can be used to immobilize different species onto a range of surfaces and it is in particular interesting for design of biosensing devices. In one such example, Bano et al. [31] designed a protein nanoarray for detection of proteins using atomic force microscopy (AFM) nanografting by employing DDI of protein–DNA conjugates and using topographic AFM height measurements to characterize the interactions within the nanoarray. Recently, a protease-sensing device was developed employing DDI immobilization of protease trypsin onto nanoporous silica substrate [32]. Silica platform was tuned in such a way to contain nanopores of a particular size, and the DNA–protease conjugate was immobilized onto the surface, while pores remain empty. When protein mixture is exposed to such device, degradation of proteins occurs in the presence of trypsin and small peptide proteolytic products can be captured within the pores. In the case of this sensor, the use of DNA for immobilization enables the mild removal of proteases after digestion and the successive analysis of the fragments captured within pores by mass spectrometry (Fig. 3)

The combination of DDI with mass spectrometry can also be used to determine protein–protein interaction of the Ypt family of small GTPases, which are modified by native chemical ligation to yield protein–DNA conjugates [33]. The successful immobilization of these conjugates could be visualized by exchanging the cofactor GDP with a fluorescently labeled GTP analogue. To test for nucleotide-dependent activity of immobilized Ypts, fluorescence experiments were conducted

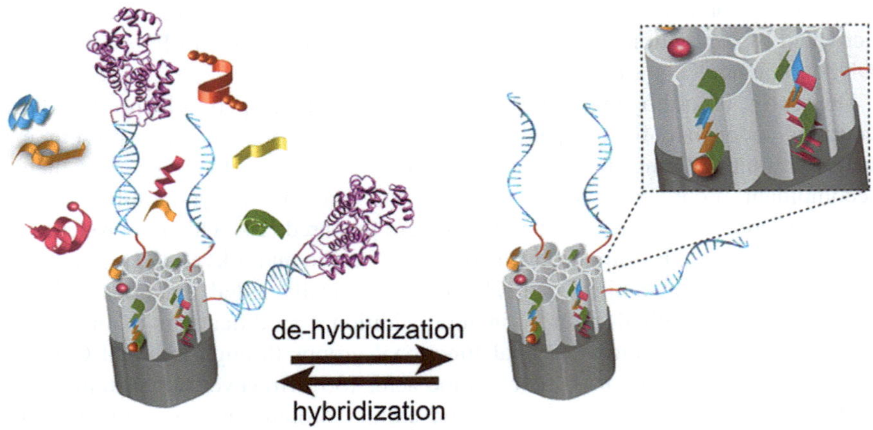

Fig. 3 Use of trypsin–DNA conjugates for the development of a biosensor for detection of the products of protein digestion by mass spectrometry. Reprinted with permission from [32]. Copyright 2013 American Chemical Society

with the labeled guanine nucleotide exchange factor DSS4, which binds to a specific Ypt causing the release of its cofactor GDP. Binding of DSS4 to immobilized Ypt1–DNA conjugates could also be detected after coating the microarray with matrix by conduction MALDI-MS measurements on the slide.

Besides the fabrication of chips for mass spectrometry, immobilization of protein–DNA conjugates was also used to build up electrochemical biosensors for detection of DNA hybridization. Capture DNA probes were bound to nanoelectrode ensembles followed by hybridization with glucose oxidase-labeled target DNA and the success of the immobilization determined by detection of an electrocatalytic current after addition of the glucose substrate and a redox mediator [34].

In addition to the immobilization of proteins, several applications were published using DDI to capture cells on functionalized surfaces. Niemeyer et al. [35] compared the labeling efficacy, specificity, and effects on cell viability of different methods and found that there is a strong influence of the employed cell line. Besides previously published methods based on direct coupling of *N*-hydroxysuccinimide or strain-promoted alkyne-azide cycloaddition, they utilized a new method based on the oxime ligation. Cell-surface glycans were modified by periodate cleavage to form aldehyde groups, which reacted with amino-biotin before STV–DNA conjugates were coupled to the cell surfaces leading to the high labeling densities while retaining a high percentage of living cells. Another approach to capture specific cells on a surface was developed by DDI of the antibodies, which selectively bind kidney or lymphoma cells [36]. Using this modular setup of biotinylated antibodies in combination with STV–DNA conjugates, the method could be applied to immobilize modified cells and investigate the protein interactions within (Fig. 4) [37].

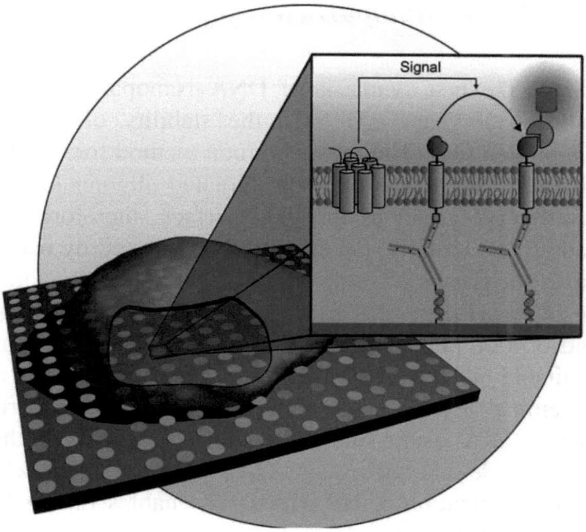

Fig. 4 Measurement of protein interactions inside living cells by immobilization of cells via DNA-modified antibodies on a solid support. Reprinted with permission from Ref. [37]

DNA modification was also applied to prepare glycoconjugates in order to design carbohydrate microarrays for the analysis of their interactions with the galactose-specific lectin PA-IL in search of a suitable inhibitor. PA-IL is suspected to play an important role in the adhesion and biofilm formation of *Pseudomonas aeruginosa* (PA), the bacterium involved in various infections. Use of the DDI, in comparison to other direct immobilization methods for attachment of glycoconjugates, enabled the specific recognition of binding events in a complex mixture [38].

In another application, Homola and coworkers developed a system to detect protein biomarkers relevant to cancer diagnostics in buffer solution as well as in the complex mixture of blood serum, utilizing surface plasmon resonance (SPR) imaging [39]. Antibody–DNA conjugates were immobilized using DDI on the sensing area of an SPR sensor previously spotted with complementary thiol–DNA enabling a high-throughput screening of biomarkers with low limits of detection.

Since the first use of complementary DNA as immobilization tool for proteins two decades ago, the method has been employed for a vast variety of applications including development of biosensors and diagnostic methods. This methodology evolved even further and enabled the immobilization of complex systems such as the whole living cells. Overcoming the limitation of the DNA conjugation to a range of molecular species will open up a whole range of new exciting applications of DDI as immobilization tool.

Besides being used in immobilization of biomolecules onto planar surfaces, DNA can also be employed for functionalization of NPs to design functional devices such as biosensors or plasmonic materials. Next paragraph will present some of the developments in the field of NP–DNA conjugates.

2.2 DNA–Nanoparticle Conjugates

Since the first synthesis of DNA–Nanoparticle (DNA–NP) (AuNPs) conjugates almost 20 years ago [40], the stability of these conjugates was continually improved [41]. The most common method for the synthesis of DNA–AuNP today is the ligand exchange with thiolated oligonucleotides, which are bound through sulfhydryl group to the gold surface, therefore enabling the coupling without additives. Usually, the stability is improved by using salt aging of the conjugates over the course of some days [42], but recently, a faster method has been reported using a pH-depended attachment of DNA to gold or silver NP to yield quantitative adsorption [43]. Other techniques involve first the functionalization of NPs with a suitable bifunctional linker followed by coupling of an oligonucleotide using mild chemical approaches. For example, silver nanoparticles (AgNP) were modified with DNA using one pot synthesis of NP in the presence of benzotriazole maleimide and subsequent Diels–Alder reaction with furan-modified DNA [44]. The conjugation of NP with DNA enables further hybridization, therefore facilitating arrangement of DNA-modified NP into secondary structures with different geometries using complementary-modified NP in order to develop new materials [41]. For a controlled assembly, several methods were developed to immobilize DNA–NPs at specific position and in fixed angles to each other. An overview over these methods is shown in Fig. 5.

To achieve a linear geometry of several DNA–AuNP, a template-assisted approach can be used, fixing two NP-binding groups in a 180° angle to each other. Following the binding of AuNP to these groups, the template strand is removed enabling the successive hybridization with similarly modified DNA–AuNP conjugates (Fig. 5b) [43]. Because of the limits of single-stranded DNA as template, branched DNA structures can be employed covalently coupling two DNA strands in a fixed angle to one internal cyclic dithiol for binding of AuNP (Fig. 5c) [45]. Depending on the choice of the DNA, trapezoid or triangle geometries could also be obtained. Combining the programmable hybridization of DNA–AuNP with layer-by-layer thin-film fabrication, it is also possible, as shown recently, to synthesize free-standing NP films that are capable of programmable transformation [46].

All of those different forms of DNA–NP conjugates can be used as solid support for the immobilization of several DNA-modified molecules, such as biomarkers or proteins using the DDI approach [47, 48]. The advantage of using DNA is in controllable hybridization and dehybridization, leading to programmable biosensing nanomaterials [49, 50].

The specific recognition between two complementary DNA strands led to the development of colorimetric sensors for the detection of mutations in DNA sequences using DNA-modified AuNPs [51, 52]. This molecular recognition can be analyzed by utilizing the spectrometric properties of AuNPs, which change color upon aggregation caused by hybridization. Simple methods of exploring this phenomenon employ UV–Vis spectroscopy or size monitoring by DLS [53]. A

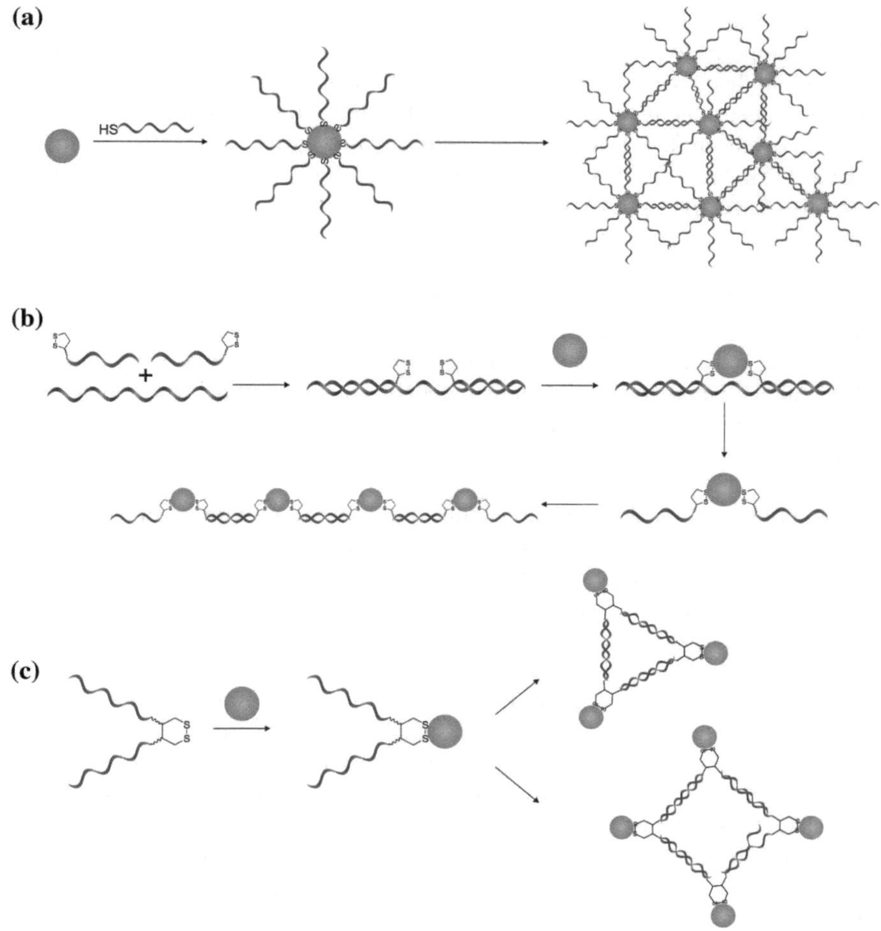

Fig. 5 Schematic illustration of the immobilization of DNA onto AuNP using monothiol DNA (**a**) [40], template-assisted immobilization (**b**) [43], and branched DNA structures for flexible assembly of different geometries (**c**) [45]

more sensitive method was developed by combining DNA-modified NPs with surface-enhanced Raman scattering (SERS). Using SERS-sensitive dye and two differently modified NP containing complementary oligonucleotide strands, SERS can be exploited in identification of target oligonucleotides [54].

To identify single-base mismatches with high specificity, a method was developed using hairpin oligonucleotide-modified AuNP [55]. The hairpin oligonucleotide was designed in a way that only the hybridization of two single-base mismatches triggers the conformation change, which enables recognition of an immobilized oligonucleotide with the stem region of the hairpin. By using this double-target binding approach, it was possible to accumulate AuNPs in a testing zone of a lateral flow strip, therefore enabling a visual detection of red AuNP bands. Compared to perfectly matched DNA, the single-base mismatches yielded a 28 times higher signal. To demonstrate its possible application as nucleic acid-based biosensor, the method was applied to the detection of two different mutation sites involved in the skin disorder autosomal-recessive congenital ichthyosis (ARCI) [55]. Apart from single-base mismatches in genes, DNA-modified AuNPs were also used for miRNA profiling of prostate cancer markers in fM concentrations [56]. A scanometric assay was developed utilizing a universal DNA sequences for the modification of AuNP. The complementary sequence was used as universal cloning linker, which was ligated to synthetic miRNA by T4 ligase. Following this modification, miRNA was captured by a low-density DNA array, leaving the universal cloning linker free for hybridization with complementary DNA–AuNP. Upon immobilization of AuNP onto the array, where binding of miRNA occurred, binding could be detected by gold enhancement. Therefore, this method enables the read out of hybridization events of AuNP on the surface with a light scanner [56].

Gao et al. combined the electrochemical enhancement with immobilized hairpin oligonucleotides for efficient detection of target DNA. Addition of target DNA causes a structural change in the hairpin, which enables binding of the reporter DNA–AuNP. So-called circular strand-replacement polymerization (CSRP) causes the release of target DNA by elongation of the reporter strand through addition of polymerase. This target DNA can be bound again by another hairpin oligonucleotide, therefore increasing the sensitivity of this assay. Bound AuNP catalyzed the deposition of silver, which was analyzed by electrochemical stripping, allowing the differentiation of single-base mismatched DNA (Fig. 6) [57].

These examples and those covered in a range of excellent reviews [47, 58–64] show that DNA–NP offer a broad range of possible applications due to their unique optical and electrochemical properties, which enable immobilization of molecules of interest in a programmable manner as well as DNA mismatch detection and design of protein biosensing platforms. The next step is to use both biomolecule–DNA and NP–DNA conjugates on a solid support to design sensing chips, new plasmonic materials, or even programmable catalytic platforms. Recent introduction of DNA origami, a folding methodology to enable structuring of DNA in an easily programmable way, opened new avenues to the use of different conjugates.

3 DNA Origami: Combination of Natural and Chemically Synthesized DNA

DNA origami technique is one of the fastest-growing fields in structural DNA nanotechnology, and nanostructures prepared in this way are characterized by higher yields and higher mechanical flexibility of 2D and 3D structure assemblies.

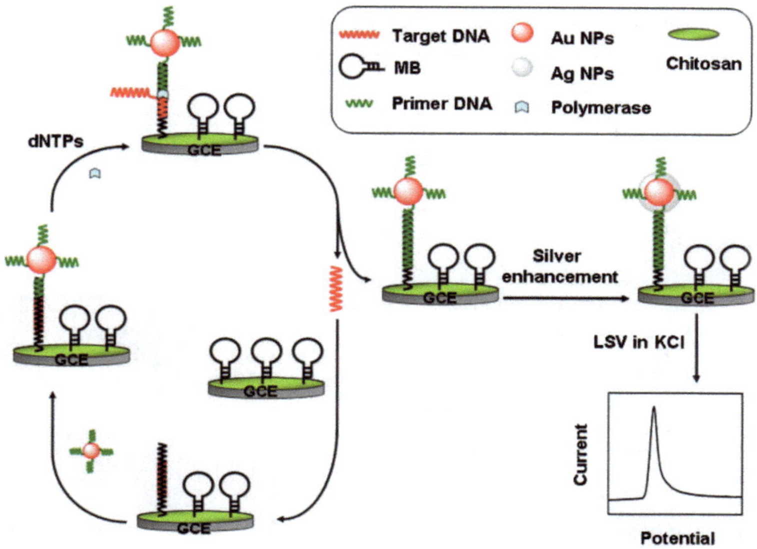

Fig. 6 Schematic illustration of dual-signal amplification strategy for DNA detection by electrochemical enhancement and circular strand-replacement polymerization. Reprinted with permission from [57]

This method of DNA folding, named after Japanese traditional technique of origami (From *ori*: folding and *kami*: paper), was first reported by Rothemund, who used a long viral circular ssDNA strand from bacteriophage (7.25 kilobases) to prepare customized shapes [4]. This can be achieved using short synthetic DNA strands, called the staple strands (mostly between 20 and 50 bases), which can fold long viral DNA by a simple thermal annealing process (Fig. 7a). While the first publication was limited to 2D structures such as rectangles, triangles, and stars, more complex planar geometrical forms as well as 3D structures such as tetrahedral shapes [65], DNA boxes with a controllable lid [66] or honeycomb [67] structures followed quickly few years after, indicating the applicability of presented approach.

The scaffold assembling strategy involves special staple strand design, whereby the customized folding process of the natural scaffold is controlled by base pairing with the staple strands following geometrical rules of helical structure formation and crossover design (Fig. 7b for 2-dimensional design). In the case of 2D objects, the helix row formation is stabilized by a regular progression of crossovers, which are replicated every 1,5 helical turns (16 bases in natural B-DNA) between two neighboring helices (Fig. 7b), which means that the alternated crossovers take place every 180° resulting in a folding process of the helices in a planar shape. By changing the crossover angles between two helices inside of the structure, folding process of 3D structures can be achieved. A thorough explanation on crossover

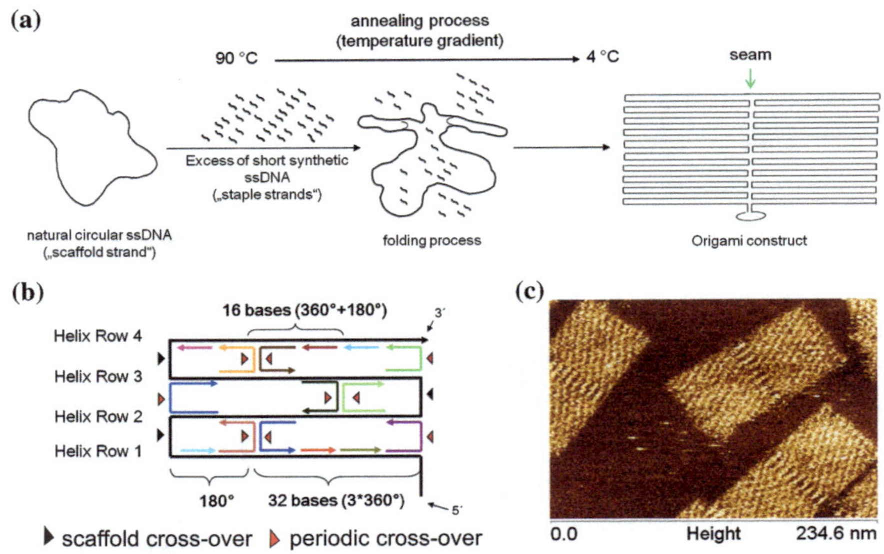

Fig. 7 **a** Scheme for the folding process of 2D origami construct in a shape of a *rectangle*. The scaffold strand is heated with an excess of hundreds of staple strands and cooled down for the folding process to proceed. **b** Scheme of the scaffold strand pathway (*black arrow*) and staple strand pathways (*colored arrows*) in design of 2D. The triangles represent the distances for periodic crossovers of the staple strands (*red triangles*) and the crossovers of the scaffold strand (*black arrows*). **c** Characterization of a rectangle-shaped DNA origami by atomic force microscopy (*AFM*) (authors unpublished work)

patterns of 2D and 3D origami shapes is given in a mini-review by Sacca et al. [68].

It is important to note that DNA itself is limited in terms of chemical, optical, or electronic properties and that the functionalization of DNA origami for future applications has been, and it will continue to be one of the research challenges. To enable further modification of DNA origami, short staple strands containing various functional groups can be employed as already described previously in the case of DNA conjugate preparation. In origami, there are more than 200 unique staple strands, which can individually be modified at defined positions up to a resolution of 6 nm [69]. Nowadays, the automated synthesis of oligonucleotides in combination with different orthogonal chemical reactions has led to development of diverse strategies to functionalize DNA nanostructures. One of the first strategies employed was STV–biotin interaction, with origami surface being decorated with biotinylated staple strands at nanometer-defined positions. For example, Kuzuya et al. [70] described a planar punched DNA origami structure with periodic wells in nanometer scale (6.8 × 12 × 2.0 nm). These wells were functionalized with biotinylated triethylene glycol (TEG) staple strands, which act as capture points for single STV protein. In a subsequent study, it was also shown that a small modification of the staple strands with toehold anchor extensions inside the wells could be used to produce a stepwise and reversible nanoplatform for the selective binding and removal of STV by single-strand displacement, which can be followed by AFM imaging [71].

To illustrate how the staple strand modifications can be used as a versatile method for the imaging of orthogonal chemical reactions at single-molecular level, one could use a recent example by Voigt et al. [72]. In this work, staple strands on customized positions were modified with biotin groups containing selectively cleavable linkers such as disulfide bridges, which can be reduced by DDT, or an electron-rich 1,2-bis(alkylthio)ethene moiety, for the cleavage by singlet oxygen generated with light in the presence of a singlet oxygen photosensitizer (PS). The orthogonal addressing of different functional biotin groups on the origami platform surface was achieved by the introduction of alkyne- and azide-groups for click chemistry or primary amines for active ester amidation, whereby the orthogonal cleavage process was easily characterized by STV binding and AFM imaging. Orthogonal decoration of origami with proteins can be also achieved using protein tag strategies such as SnapTag and HaloTag as recently demonstrated by Niemeyer and coworkers (Fig. 8a) [73].

On the other side, direct immobilization of DNA-tagged molecules can be employed to introduce NPs or proteins directly to the staple strands. For example, origami can be decorated with complementary protruding staple strands, which act as a capture oligonucleotides. In this way, different origami platforms were designed, which organized DNA-tagged molecules in predefined 2D or 3D structures [75, 77–81].

As described previously, DNA modification of metallic NP is now a well-established strategy. However, designing devices that make use of plasmon-distance effects of different NPs for new applications in nanophotonics and nanoelectronics still poses a challenge [77, 82–85], and DNA origami might be just the right solution. The immobilization of NPs of different sizes was recently described by Ding et al. [85] who used a triangle-shaped DNA origami structure for the immobilization of AuNPs with different sizes to form a linear structure with well-controlled orientation and <10-nm spacing. This spacing and orientation could be used to generate an extremely high field enhancement for the construction of nanolenses.

The DNA origami nanostructures could also be used for the preparation of new hybrid materials with different metallic NPs as in the case of rectangle-shaped origami decorated with AuNP (5 nm)–AgNP (20 nm) heterodimer structures [74] (Fig. 8b). Besides NPs, DNA-modified nanorods (NRs) were also successfully immobilized onto origami structures. Yan and coworkers [75] used the origami platform for the bottom-up construction of well-ordered AuNR–AuNR and AuNR–AuNP hybrid nanoarchitectures with precise inter-rod angles and distances (Fig. 8c). These new hybrid materials showed unique plasmonic properties, which enhanced the possibility to characterize the effect of interparticle distances and geometrically dependent photonic interactions leading to a new class of plasmonic materials.

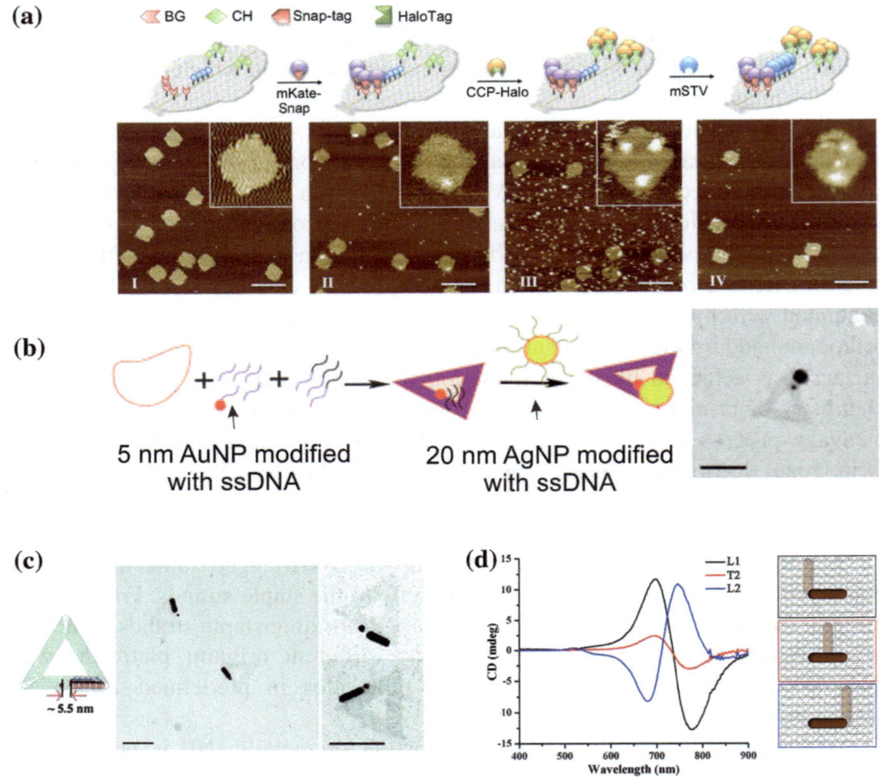

Fig. 8 **a** Orthogonal decoration of DNA origami with proteins by using protein tags and biotin–STV binding strategies [73]. **b** Schematic representation of a 5-nm AuNP and 20-nm AgNP dimer on a triangle-shaped origami structure and characterization with TEM [74]. **c** Scheme diagram and TEM images of an end on AuNR–AuNP heterodimer structure on an origami triangle platform [75]. **d** Scheme and CD spectra of discrete 3D plasmonic AuNR dimer structures on a rectangle-shaped origami nanostructure [76]. All scale bars are 100 nm. All figures are adopted from named publications with kind permission

One interesting model system, to use plasmonic effects for the design of new materials with useful biosensing properties, was described by Fan and Govorov [86]. In this model system, the authors described that neighboring NPs can create a circular dichroism (CD) effect when they are organized in asymmetrical geometries such as pyramides, tetramers, or helices with the strongest signal being obtained from the helical structures. The effect is produced by a collective chiral plasmon resonance, which is released by Coulomb and electromagnetic interactions [87]. Moreover, the response of CD activity could be controlled by the metal compositions and different particle sizes [78, 80]. Also NRs showed tailored optical chirality in well-defined spatial configuration, which was described the first time by Lan et al. [76] (Fig. 8d). These systems demonstrate that shifting the location of single AuNR on an origami platform at the nanometer scale has a strong influence of the plasmonic CD effect and could be use for new material design named by authors the "chiral plasmonic ruler." Recently, the metallization of DNA was utilized to grow arbitrarily shaped AuNPs on the origami structure [88]. Au clusters with a size of 1.4 nm were coated with positively charged amines and were used to bind to negatively charged origami templates with different shapes, in which they act as seeding sites for the cluster growth. The pre-seeded origami shapes were metallized by electroless deposition of gold ions from solution and could grow in predefined formations such as gold nanocuboids, polymerized nanorods, or nanodonuts. In this case, it was shown that DNA origami could be used as an addressable platform for the controlled metallization of seeding sites in customized shapes with the nanometer precision, which could find new applications in optical and electronical devices.

All of above examples used original viral DNA scaffold strand. However, the customized changes of the scaffold strands could lead to new designs and new modification capabilities. For this purpose, several different techniques were developed for the scaffold strand production, keeping in mind that the scaffold extraction from natural sources or the use of natural DNA as templates for the production seemed to be the most promising route. Several methods have already been described for the scaffold production. The one introduced by Pound et al. [89] involves the use of a polymerase chain reaction (PCR)-based techniques and two different viral DNA sources as templates, the λ-DNA (48.5 kp) and the commonly used M13Mp18 DNA (7.25 kb). Resulting dsDNA PCR products with a length of 765–4,808 bp are flanked with 3′ biotinylated primer, whereby one of the complementary strands could be bound to streptavidin-coated magnetic beads and used for purification. Nevertheless, the production of the scaffold using this method is limited by time-consuming purification steps, which decrease the amount of obtained DNA that can be further used. The viral λ-DNA was also used by Zhang et al. [90] as a template to produce a 26-kb-long dsDNA strand, which can then selectively be digested into ssDNA by using enzyme λ-exonuclease. This enzyme catalyzes the digestion of dsDNA into mononucleotides from 5′ to 3′ direction in a highly progressive manner and if a primer is designed in such a way that one strand of dsDNA is protected, this strand can be used further, while the other is digested by the enzyme. In another variation of PCR methodology, Said et al. [91] used the M13Mp18 DNA as a template for the production of a 1.4-kb-long origami scaffold with the help of restriction and ligation processes for the cyclization of the ssDNA scaffold product. Recently, a new methodology was employed based on an isothermal nucleic acid amplification technique called rolling circle amplification (RCA) [92]. In this technique, a long ssDNA is directly produced in the presence of a specific Γ29-Polymerase by using deoxynucleotide triphosphates (dNTPs) and short DNA primer strands [93]. A significant advance in origami design was made by enabling direct folding of dsDNA by using a derivative of M13 and plasmids [94] or digested fragments of λ-DNA [95]. Use of dsDNA as a scaffold could have a huge potential as there is an abundant supply of natural dsDNA and that would also remove the need of ssDNA production prior to folding. However, the staple strand design for such scaffolds is more complex as it needs to take into account the proper folding paths from both ssDNA strands and it will take time to optimize.

Alternative scaffolds can also be obtained by generating circular ssDNA from plasmid sources. In such a way, Douglas et al. [96] established a convenient methodology, which employs the ability of Nb.BsrDI enzyme, a nicking endonuclease, to site specifically cleave the phosphodiester bond of one strand of dsDNA plasmids. The nicked strand could be selectively digested by treatment with T7- and λ-exonuclease and the constructed ssDNA scaffolds then used to fold 2- and 3-dimensional origami structures. This technique was recently further extended by Erkelenz et al. [97], who explored a highly efficient method for a large-scale production to prepare scaffold strands with customized length and sequences from plasmids. To achieve this, molecular cloning techniques of site-directed mutagenesis [98] and SLIC cloning [99] were combined to produce tailored plasmids, which also contain genetic regulatory elements and tailored nicking endonuclease restriction site for the production of circular scaffold strands in high amounts. Besides viral ssDNA, bacterial ssDNA and dsDNA, eukaryotic ssDNA could also be used as a scaffold for future origami design. The application of the origami technology could further be extended by the customized scaffold strand production from natural sources, to which genetic modification is introduced to enable additional modification or preparation of custom length DNA.

In such a way, the combination of a nanoscale control over design and tailored genetic information could be used for design and decoration of origami with signal molecules to control different biological process or enable controlled transport of genetic information to the targets within cell. Origami methodology is based on the use of both natural scaffolding DNA strand and numerous short, chemically prepared DNA strand. However, when larger quantities of DNA are desirable, cheap sources of natural DNA are needed. This is the case with DNA-based optoelectronic devices, which move away from nano- to microscale. Next paragraph will give a short overview of recent developments in DNA thin-film design and their uses in preparation of optoelectronic devices.

4 DNA in Design of Thin Films and Optoelectronic Devices: Use of DNA from Natural Sources

Some optoelectronic devices, which will be discussed in this section, have been realized with enhanced efficiency due to the special properties of dsDNA. Namely, DNA molecule has a π–π stacking structure basically forming a tunnel ready for electron transfer, it has high affinity for metal ions acting as a good template for metallic NP preparation, DNA films exhibit high optical transparency in a broad spectrum range, and finally, the fluorescence efficiency of numerous fluorescence dyes can be enhanced due to the intercalation or insertion between adjacent base pairs or minor grooves. Such intercalation or groove binding of dyes to DNA helix is particularly interesting as it has been shown that in such a way the aggregation of dyes at high concentration is reduced, and therefore, DNA is superior to conventional polymer hosts. However, it should be taken into account that DNA is also more sensitive than polymers to the environmental conditions, such as moisture and temperature, and does not possess sufficient mechanical strength for practical microdevice implementation. Additional processing is therefore required to make natural DNA biopolymer suitable for application in thin-film production. In 2001, Ogata et al. described the large-scale preparation and characterization of a series of natural DNA–cationic surfactant complexes, demonstrating a simple method to prepare optical-quality natural DNA biopolymer film from raw DNA derived from salmon testes [100]. Since then, several research groups worldwide have been devoted to advance the material processing and broaden the device implementation.

Experiments have shown that the use of DNA biopolymers as host materials can dramatically enhance light emission compared to conventional polymer hosts [101]. It is also reported that organic light-emitting diodes (OLEDs) that incorporate DNA thin films as electron-blocking layers show improvements in one to two orders of magnitude in terms of efficiency and brightness [102]. We can recommend several excellent reviews and books that summarize these developments in the past few years [103–109], but in the following paragraphs, we have chosen to focus onto the most recent developments in DNA thin-film preparation and application.

One of the most widely used DNA in optoelectronics is DNA from salmon testes. The marine-derived raw DNA normally has a high molecular weight, which makes it hard to process. Furthermore, such high molecular weight DNA increases the viscosity when dissolved in solvents, which, in turn, affects the film morphology and uniformity. Therefore, it is necessary to reduce the molecular weight by sonification to afford uniform thin films with desired thicknesses [110]. Following the reduction of the molecular weight, formation of DNA–cationic surfactant complexes is performed via an ionic exchange reaction, where sodium ions within natural DNA are substituted by cationic surfactants. The most commonly used surfactants are long alkyl quaternary ammonium compounds, such as cetyl trimethylammonium (CTMA) chloride [100, 107]. The advantage of such protocol is that all of the used materials are commercially available in large amounts. The use of different surfactants for preparation of DNA complexes can extend the range of available solvents, and it has also been reported that they can affect the material performance and optical switching and lasing properties.

The most common techniques to form thin films of DNA–surfactant complexes are by spin-coating or casting. Depending on the coating protocols, nanometers to micrometers of films can be prepared with high optical quality [107, 111]. Detailed characterization of the films, including thermal, optical, and electrical properties, have been carried out, and the results indicate that the films exhibit high transparency within a broad spectrum range from visible to infrared.

Great progress has been made in use of DNA–surfactant complexes for optoelectronic applications. Previously, DNA–surfactant complexes have been utilized

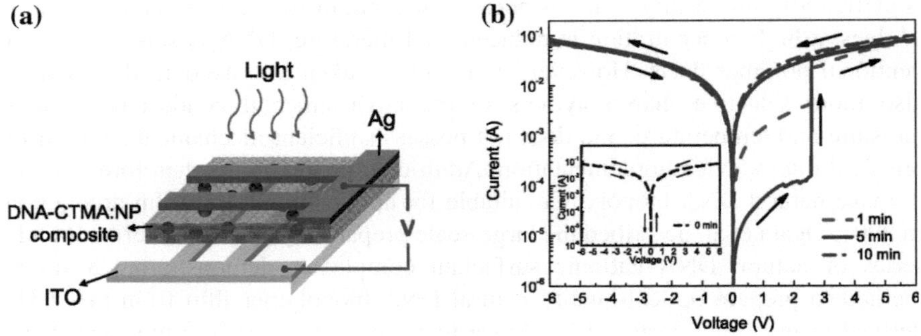

Fig. 9 **a** An illustration of the memory device based on natural DNA films and **b** a current–voltage (I–V) curve of device under different periods of UV exposure time. *Inset* I–V curve of device without UV irradiation. Reprinted with permission from Ref. [114]

as an effective electron-blocking material in organic light-emitting devices (OLEDs) and as an insulating layer in organic field-effect transistors (OFETs). Due to the high transparency in the communication wavelengths, they were also employed in preparation of electro-optic modulators.

Owing to the high dielectric constant and insulating characteristic, DNA complex films have been incorporated in field-effect transistors (FETs) as gate dielectrics [112, 113]. The device can also be operated as a memory device owing to the hysteresis property of the transfer function [113]. Recently, it was demonstrated that an optical memory device based on photoinduced DNA complex–NP nanocomposite can be prepared [114]. This is achieved by phototriggered growth of AgNP on a DNA film using Ag salt precursor and photoinitiator [107]. Memory device consists of a DNA complex thin film sandwiched between two electrodes, and the electrical bistability is activated by in situ formation of silver NPs embedded in biopolymer upon light irradiation (Fig. 9).

This facile technique takes the advantage of DNA's affinity for metal ions and solution processing and can optically manipulate the properties of DNA nanocomposite thin films and therefore holds a promise for optical storage and plasmonic applications.

Many researchers working in the field of optoelectronics have at first considered DNA to be an exotic material with limited use. However, recent results have shown that not only DNA can be an excellent and versatile device platform but in combination with the new advances in nanotechnology can also lead to new DNA-based materials with enhanced properties useful for engineering of new generation of biohybrid devices. One interesting generation of these materials combines DNA with polymers and will be discussed in our closing paragraph.

5 DNA-Based Materials

5.1 DNA-Based Hydrogels

Hydrogels are water-swollen cross-linked polymeric networks with distinct three-dimensional structure [115–117], which can adapt to their environment. This stimulus responsiveness is primarily associated with swelling and de-swelling of the polymeric materials, which is mediated by change in temperature, pH, ionic strength, solvent type, or photon flux. Hydrogels have been made out of pure DNA [117, 118], but they are expensive and do not have good sensitivity. However, sensing gels were prepared by combining polymers and DNA by first attaching DNA to the polymer backbone and using different strategies to transform it into a cross-linked architecture. The first strategy based on gel-to-sol transition was reported by Nagahara and Matsuda [119] in 1996, and it involves the hybridization of two conjugates (two different polymers, each with different DNA sequence attached). In second strategy, hydrogel was built up by chemically cross-linking dsDNA strands [120] using ethylene glycol diglycidyl ether as a cross-linking agent. Recently, Tan and coworkers prepared fast-responding gels by using adenosine-binding DNA aptamer as a cross-linker [121]. Without adenosine, the aptamer acted as a stable cross-linker to hold the gel together. When adenosine is added, it induces aptamer folding, which breaks the cross-link causing the change in materials' properties. In another application, pH-triggered, fast-responding DNA hydrogel was prepared by Liu and coworkers by cross-linking the gel using a pH-sensitive DNA I-motif, which causes the disruption of the gel within a minute by adding appropriate base [122].

Hydrogels do not only have interesting properties, but can be employed as effective sensing platforms. For example, amylase/iodine was entrapped within the hydrogel cross-linked with cocaine aptamer [123]. Such hydrogel changes color from blue to colorless in the presence of cocaine due to conversion of amylase (amylase/iodine is blue) into glucose (colorless). Willner and coworkers designed an ion sensor using a switchable sol–gel transition in the presence of K^+ by using G-quadruplex DNA [124]. Addition of free K^+ to G-rich DNA attached to the polymer chains induced the formation of G-quadruplex, which cross-linked the gel in a reversible way as the sol state could be achieved by removal of K^+ ions by addition of K^+-binding 18-crown-6-ether. Another interesting approach was recently presented by Yang group, and it involves use of DNAzyme cross-linker, which cleaves DNA substrate in the presence of Cu^{2+} and allows detection by a color change in gel induced by the presence of pre-trapped AuNPs [125].

Recently, hydrogels were used for sensing of target DNA [126] and, in combination with quartz crystal microbalance (QCM) and appropriate aptamer, for detection of avian influenza virus [127]. Furthermore, DNA hydrogels are also interesting as they can be an excellent platform for preparation of thermoreversible materials [119], drug release from nanosized [128] and macroscopic networks

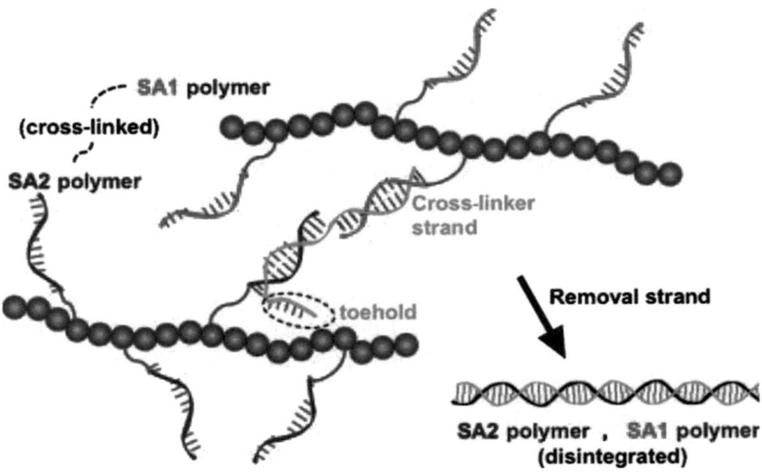

Fig. 10 Scheme of reversible DNA cross-linked polyacrylamide hydrogel. The overhang of the cross-linker DNA enables the reversible disintegration of the network by the hybridization of removal strand, which is complementary to the cross-linker strand in full length [136]

[129, 130], and detection of target DNA by use of stem-loop cross-linking structures [131].

DNA hydrogels with convertible mechanical states induced by hybridization have also been prepared by making use of the switching process previously employed for design of DNA-fueled machines, where changes can be induced by addition of short DNA [132, 133]. A further advance to this concept was achieved by use of toehold, overhang DNA, which can be used as a zipper to open up the gels (Fig. 10) [134]. Using this strategy, Simmel et al. [135] presented the controlled trapping and release of quantum dots in DNA hydrogel by employing single-molecule fluorescence microscopy and fluorescence correlation spectroscopy. They employed the DNA cross-linker strands equipped with an additional unhybridized "toehold" section that acts as a "recognition tag" for DNA release strands. When release strands fully complementary to the cross-linker strands are added to the gel, they attach to the toeholds and remove the cross-linker strands via branch migration causing the gel to dissolve into a solution and in such a way liberate the trapped particles.

DNA hydrogel networks containing only DNA have also been produced by chemical cross-linking of salmon testes DNA (st-DNA) with ethylene glycol diglycidyl ether [120]. A discontinuous volume transition upon addition of acetone to the aqueous DNA hydrogel was observed with 15-fold shrinking of the material relative to its initial volume. Such cross-linked DNA networks can be generated by PCR where two Y-shaped trisoligonucleotides in combination with a double-stranded template can be transformed into hydrogels with segment lengths between the branching points ranging from 70 to 1062 nucleotides [137]. DNA hydrogels were also prepared by enzyme-catalyzed assembly [117], in vitro transcription/translation system for functional proteins with DNA hydrogels [138] and responsive DNA-based hydrogels from Y-shaped building blocks [122]. Interesting photoresponsive platform was presented by Kang et al. in which an azo molecule was incorporated into the backbone of the cross-linker DNA sequences. Under visible light, the trans-form of azo molecule allowed cross-linker DNA to hybridize with DNAs on the polymer side chains forming three-dimensional hydrogel networks. When the gel was irradiated with UV light, the azo molecule photoisomerized to the cis-form, which prevented hybridization, thus reverting gel to sol state [139].

The DNA components in gel bare some level of mechanical roles to support the gel–sol transition or volume expansion. One drawback of using DNA as a cross-link to observe gel mechanical property is the high concentration of DNA required, which not only increases the cost but also reduces the sensitivity of the system. To overcome this problem, the gel matrix was used for sensor immobilization (in particular heavy-metal detection) and the DNA, instead of being used as building material, was used to generate optical signals. In one such example, Hg^{2+}-binding DNA, rich on thymine, was embedded in the gel [140]. In the presence of Hg^{2+}, the aptamer adopts a hairpin configuration that produces strong green fluorescence with SYBR Green I [141], which can further be improved by tuning the gel charge [142]. Similar gel–DNA sensor for lead detection using a guanine-rich DNA, which allowed visual detection of Pb^{2+} at the concentration as low as 20 nM, was also designed [143]. Recently, Ag nanoclusters (AgNC) in combination with DNA-based hydrogels have also been employed for detection of Hg^{2+} [144]. Namely, DNA of particular sequence can bind Ag^{+} and induce the formation of two types of AgNCs emitting red and green fluorescence, respectively. Interestingly, some of the red emitter can be converted to the green due to the intrinsic properties of AgNC in the presence of Hg^{2+} [145–148], leading to development of Hg^{2+} ion sensor.

However, one of the most exciting applications of DNA hydrogels is probably their potential use as a biofriendly platform either for cell growth (for example stem cells) or for production of proteins. Latter was already demonstrated by Park et al. who prepared a DNA hydrogel that produces functional proteins without any living cells with efficiency about 300 times higher than current, solution-based systems. This protein-producing gel termed as "the P-gel system" or "P-gel" contains genes as gel scaffolding [138]. The P-gel can be fabricated by ligating X-DNA and linear plasmids within a polydimethylsiloxane (PDMS) micromould. Subsequently, proteins can be expressed simply by incubating the P-gel micropads with cell lysates for a specific time period (Fig. 11).

Besides above examples, DNA hydrogels have also been shown to be useful for preparation of high surface area materials [149], DNA hydrogel-based supercapacitors [150], hydrogels for controlled release of plasmid DNA [151–156], and DNA hydrogel bead as selective adsorbent of dioxins [157]. Taking into account vast possibilities for the use of such hybrid material in both materials science and bionanotechnology, it could be expected that more work will be done in controlling their architecture, mechanical properties, and function in the near future.

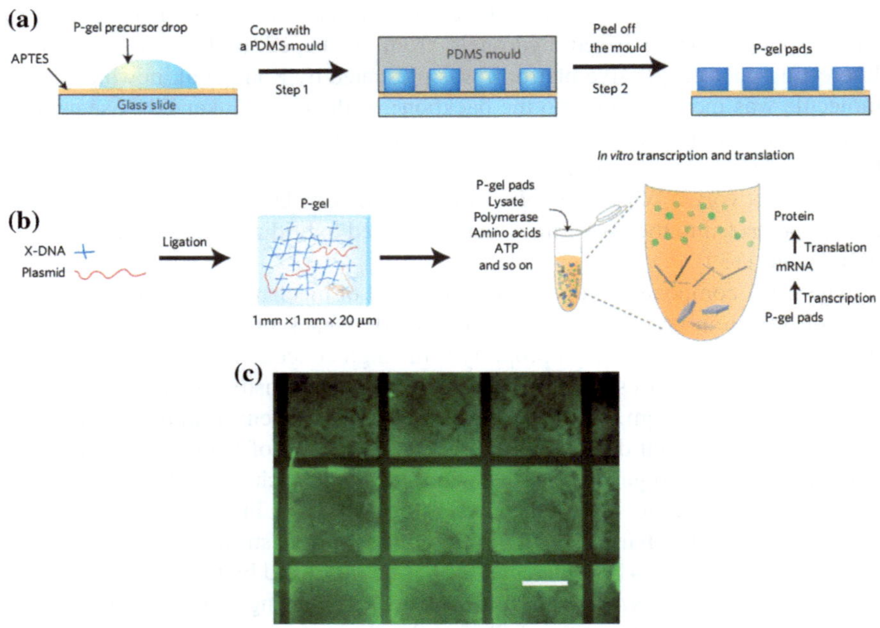

Fig. 11 Fabrication of protein-making gel (P-gel) matrix. **a** schematic diagram illustrating the formation of P-gel micropads (*side view*). The P-gel precursor drop, which contains X-DNA, genes, and T4 DNA ligase, was confined within a PDMS mold with precisely defined dimensions. Extra solution was able to flow out owing to the surface modification (*Step 1*). After gelation, the PDMS mold was peeled off from the substrate (*Step 2*), and P-gel micropads were formed. **b** A scheme of the gelation process through enzymatic cross-linking and cell-free expression with P-gel pads. **c** A fluorescent image of the P-gel pads after staining with SYBR I (scale bar: 500 μm). Reprinted with permission from Ref. [138]

5.2 DNA-Polymeric Hybrid Materials

Over the past two decades, one of the most exciting and emerging goals in chemical synthesis was the macromolecule generation with a well-defined monomer sequence, which performs complex functions. The combination of synthetic macromolecules with biopolymers is an interesting and powerful opportunity for merging the distinctive properties of each class of material and at the same time, overcoming specific limitations such as decreased stability of biopolymers and poor structural control of synthetic systems [158, 159]. The combination of proteins and peptides with synthetic macromolecules has been explored in depth. Extensive efforts have been made for the development of covalently bonded hybrid structures consisting of biomacromolecules and organic polymers. Different architectures are observed depending on how the biological moieties are connected to the synthetic macromolecules. For example, the connecting building blocks can be arranged in an alternating fashion leading to linear

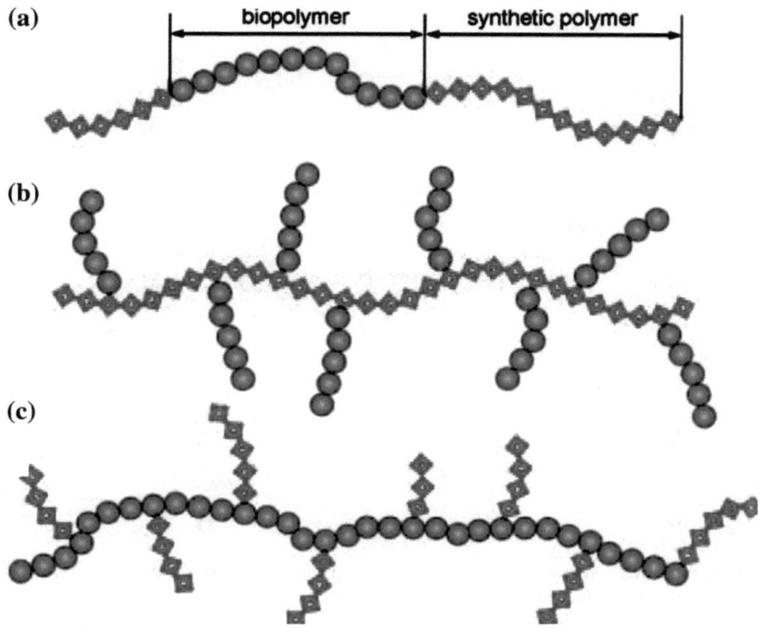

Fig. 12 Illustrations of covalently bonded biopolymers (*green circles*) and synthetic (*yellow diamonds*) polymers. **a** Linear block copolymer composed of biopolymer and synthetic polymer blocks. **b** Biomacromolecules grafted onto a synthetic polymer backbone. **c** Synthetic polymer side chains on a biomacromolecule backbone. Reprinted with permission from Ref. [160]

block copolymers (Fig. 12a) [160] and side-chain polymer structures can be obtained by arranging several peptides or nucleic acid moieties along a polymer backbone (Fig. 12b). The opposite case in which several synthetic polymers are attached to a biomacromolecular backbone is rare (Fig. 12c) [161].

Combination of peptides and the organic polymers was already explored [162, 163], but much less attention went to an important class of bioorganic hybrids consisting of nucleic acids and synthetic polymers [164]. The very first reports of DNA–polymer conjugates were presented in 1980s where antisense oligonucleotides were grafted onto a poly(L-lysine) (PLL) backbone. These conjugates were employed to inhibit the synthesis of vesicular stomatitis virus proteins, therefore basically acting as antiviral agents [165–167]. There are numerous applications of these materials, besides the gene or oligonucleotide delivery, which range from the purification of biomaterials to the detection of DNA. Recently, hydrophobic polymers instead of hydrophilic polymers such as PLL and poly(ethylene glycol) (PEG) have been connected to oligonucleotides to generate macromolecular amphiphiles, which tend to self-assemble into micellar systems with nanometer dimensions. Such materials have promising potential for applications in the fields of bionanotechnology and nanomedicine, i.e., as drug delivery systems.

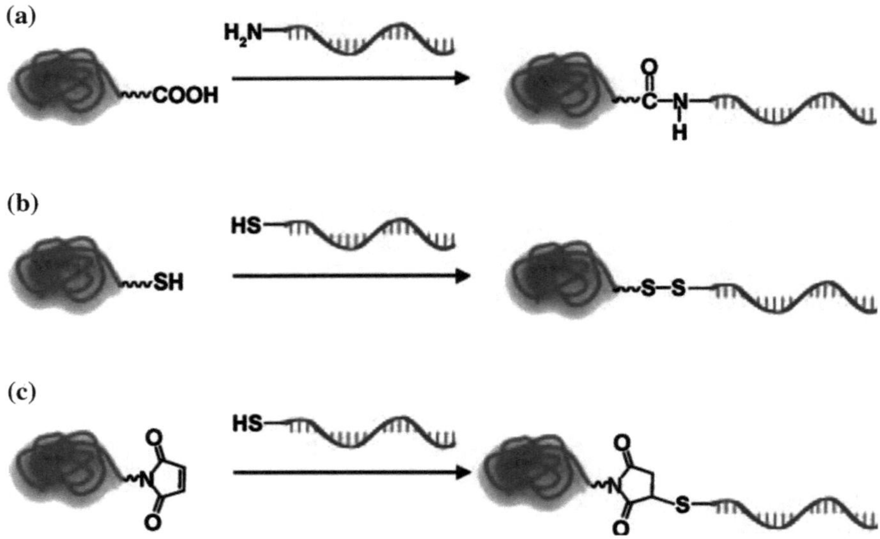

Fig. 13 Coupling methods for the synthesis of DNA block copolymers in solution. **a** Carboxylic acid-terminated polymer coupled with an amino-DNA. **b** Disulfide bond formation between polymer and thiolated DNA thiols. **c** Michael addition of a polymer containing a terminal maleimide and thiolated DNA. Reprinted with permission from Ref. [160]

The coupling of the nucleic acid blocks and the organic polymers can be done in solution or on solid support. For example, amino-functionalized short DNA can be reacted with water-soluble polymer having terminal carboxylic acid group such as PEG [168, 169], poly(*N*-isopropylacrylamide) (PNIPAM) [170, 171], poly(D,L-lactic-*co*-glycolic acid) [172], and poly(*p*-phenyleneethynyl-ene) [173] to form an amide bond (Fig. 13a). In another approach, disulfide bonds are formed between thiolated DNA and PEG-containing 2-pyridyl disulfide (Fig. 13b) or maleimide (Fig. 13c) group. Besides purely chemical strategies, Marx and coworkers used DNA polymerases to catalyze the reaction for the synthesis of DNA polymers that can be grafted with short polymer blocks leading to entities of high molecular weight, modification density, and defined structure [174]. Herrmann and coworkers used PCR successfully to prepare DNA–polymer hybrids with ultrahigh molecular weight and high structural accuracy [175].

When hybrid materials such as amphiphilic DNA coblock polymers are desired, it is difficult to prepare them in solution due to different solubility properties of each component. In that case, the strategy involves grafting polymers on DNA on a solid support using phosporamidite strategy (Fig. 14a). After deprotection with ammonia from the solid support, ssDNA diblock copolymers could be obtained, for example, DNA-*b*-polystyrene (PS) [176] and DNA-*b*-poly(propylene oxide) (PPO) [177], and triblock architectures of the type DNA-*b*-PEG-*b*-DNA can be made [178] when both termini of the organic polymer were activated as

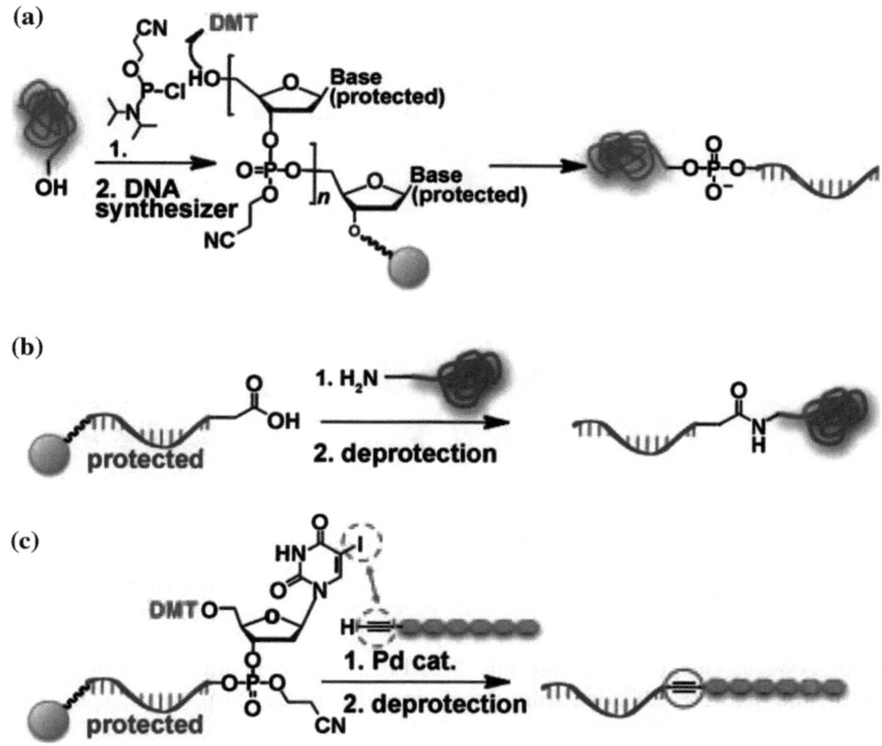

Fig. 14 Coupling reactions that can be used to prepare DNA block copolymers on a solid support. **a** A hydroxylated polymer was converted into the phosphoramidite and coupled with 5′-OH group of DNA in the last step of the DNA synthesis. DMT = dimethoxytrityl. **b** An amino-modified polymer was added to a carboxylic acid-modified DNA on a solid support. **c** Pd-catalyzed Sonogashira–Hagihara coupling of an acetylene-terminated p-conjugated polymer and a 5-iodouracil nucleobase on the solid phase. Reprinted with permission from Ref. [136]

phosphoramidites. Besides using phosphoramidite strategy, terminal amino group could be grafted onto immobilized carboxyl-modified DNA through amide bond formation (Fig. 14b) [179] or DNA coupled by Sonogashira–Hagihara coupling (Fig. 14c) [180].

The advantages of DNA are specific recognition and self-assembly of the oligonucleotides of appropriate sequences, which can be employed in design of programmable nanoreactors based on DNA block copolymers (DBC) [177, 181]. The ssDBC aggregates were hybridized with reactant DNA-bearing functional groups, and due to the close proximity of these groups within the micellar confinement, several organic reactions such as the Michael addition and amide bond formation could be carried out sequence-specifically within the programmable nanoreactors. It is even possible to spatially define the transformations to take

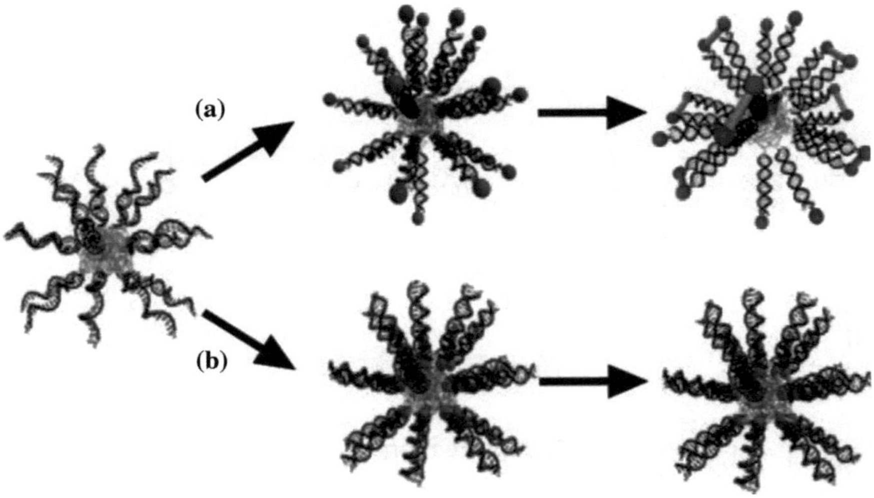

Fig. 15 DNA-templated synthesis on the surface of a PPO-b-DNA micelle (**a**) and in the core of the micelle (**b**). *Red and green spheres* correspond to reacting functional groups, and *yellow sticks* represent the resulting covalent bonds. Reprinted with permission from Ref. [177]

place either at the surface of the micelles (Fig. 15a) or within the interior of the bioorganic particles shielded from the environment (Fig. 15b).

DBC micelles can be used as a scaffold for the enzymatic conversions [182] or to generate higher-ordered organic/inorganic particle networks through sequence specific hybridization with other DNA-modified nanoparticles [176]. Exciting application is also a design of microcapsules, which can be obtained through alternating layers of DBCs assembled on silicon templates and then used for the delivery of small interfering RNA (siRNA) and anticancer drugs [183]. Recent examples have also shown that such materials could be used in therapeutic purposes by being effective drug delivery systems [169, 184, 185].

5.3 DNA Side-Chain Polymers

The side-chain polymer structures can be obtained by arranging several peptides or nucleic acid moieties along a polymer backbone or less frequently by attaching several synthetic polymers to a biomacromolecular backbone [161, 186]. In one of the first examples, polyacrylamide or PNIPAM was attached to bacteriophage T4 (166 kbp) or λ-phage DNA (48.5 kbp) [187, 188]. In addition to direct attachment, intercalators such as phenzinium or psoralen can be used to combine polymer and DNA in a non-covalent fashion [189–191]. Such DNA polymers have found and will continue to find application in DNA detection assays [192–194] and in design of novel drug delivery systems.

After the potential of DNA was recognized in particular for design of programmable platforms and hybrids, search for new DNA-based materials continues. One recent example is preparation of spherical nucleic acids (SNA), densely packed highly oriented nucleic acids in a spherical geometry-termed SNA [40, 60]. This kind of unique arrangement and orientation of one-dimensional linear nucleic acids within this three-dimensional framework offers new chemical, biological, and physical properties. For example, such SNA can be used for intracellular gene regulation [195–198], in vitro biodetection [51, 199–203], intracellular assays [204–207] and as a potent-cell transfection tool [195, 208–210]. Initially, SNA consisted of 3′-SH-terminated oligonucleotides covalently attached to the surface of spherical AuNPs by salt aging in order to obtain dense nucleic acid loading [40]. These nanostructures exhibit properties (physical and chemical) that are distinct from those of both the NPs and DNA from which they derive. Although different core materials such as inorganic nanoparticles [211–213] and semiconductor materials [214] were used to provide physical and chemical stability as well as to act as a scaffold, many of the properties of these nanostructures were shown to be core independent but rather originated from the dense layer of oriented nucleic acids. This idea was further exploited by Mirkin and coworkers to prepare polyvalent nucleic acids (PNANs) [215] from alkyne-modified 3′-thiolated DNA which comprised of only cross-linked and oriented nucleic acids, which were capable of effecting high cellular uptake and gene regulation without the need of a cationic polymer cocarrier. These examples show that there is more to short DNA than it meets the eye—new combinations of different nanoelements and DNA might lead to powerful tools, which can be used in drug delivery or in vivo biosensing.

6 Conclusion

Being a natural polymer with highly programmable sequence, DNA emerged as an exciting material for both molecular structuring and design of novel hybrids with applications ranging from molecular biology to catalysis and optoelectronic device design. Advances in chemical modification of short synthetic oligonucleotides and the folding or film casting of the natural DNA showed that the field of DNA research is far from exhausted. Future will probably focus on merging DNA and nanomaterials to produce new therapeutic (clever drug delivery) systems or robust but biodegradable materials for applications in energy material design and electronic devices. DNA has indeed shown that it is more than a genetic carrier and it will continue to be one of the pillars of interdisciplinary research crossing the fields of molecular biology, chemistry, physics, and engineering.

References

1. Seeman NC (1982) Nucleic acid junctions and lattices. J Theor Biol 99:237–247
2. Seeman NC (2007) An overview of structural DNA nanotechnology. Mol Biotechnol 37:246–257
3. Niemeyer CM, Sano T, Smith CL, Cantor CR (1994) Oligonucleotide-directed self-assembly of proteins: semisynthetic DNA–streptavidin hybrid molecules as connectors for the generation of macroscopic arrays and the construction of supramolecular bioconjugates. Nucleic Acids Res 22:5530–5539
4. Rothemund PW (2006) Folding DNA to create nanoscale shapes and patterns. Nature 440:297–302
5. Gilham PT, Khorana HG (1958) Studies on Polynucleotides. 1. A new and general method for the chemical synthesis of the C5′–C3′ internucleotidic linkage—Syntheses of Deoxyribo-Dinucleotides. J Am Chem Soc 80:6212–6222
6. Beaucage SL, Caruthers MH (1981) Deoxynucleoside Phosphoramidites—a new class of key intermediates for Deoxypolynucleotide synthesis. Tetrahedron Lett 22:1859–1862
7. Matteucci MD, Caruthers MH (1981) Nucleotide Chemistry. 4. Synthesis of Deoxyoligonucleotides on a polymer support. J Am Chem Soc 103:3185–3191
8. Letsinger RL, Lunsford WB (1976) Synthesis of thymidine oligonucleotides by phosphite triester intermediates. J Am Chem Soc 98:3655–3661
9. Sinha ND, Grossbruchhaus V, Koster H (1983) A new synthesis of Oligodeoxynucleoside Methylphosphonates on control pore glass polymer support using phosphite approach. Tetrahedron Lett 24:877–880
10. Sacca B, Niemeyer CM (2011) Functionalization of DNA nanostructures with proteins. Chem Soc Rev 40:5910–5921
11. Niemeyer CM (2010) Semisynthetic DNA-protein conjugates for biosensing and nanofabrication. Angew Chem Int Ed 49:1200–1216
12. Braun G, Diechtierow M, Wilkinson S, Schmidt F, Huben M, Weinhold E, Reich NO (2008) Enzyme-directed positioning of nanoparticles on large DNA templates. Bioconjug Chem 19:476–479
13. Goodman RP, Erben CM, Malo J, Ho WM, McKee ML, Kapanidis AN, Turberfield AJ (2009) A facile method for reversibly linking a recombinant protein to DNA. ChemBioChem 10:1551–1557
14. Fruk L, Kuo CH, Torres E, Niemeyer CM (2009) Apoenzyme reconstitution as a chemical tool for structural enzymology and biotechnology. Angew Chem Int Ed 48:1550–1574
15. Fruk L, Niemeyer CM (2005) Covalent hemin-DNA adducts for generating a novel class of artificial heme enzymes. Angew Chem Int Ed 44:2603–2606
16. Fruk L, Muller J, Weber G, Narvaez A, Dominguez E, Niemeyer CM (2007) DNA-directed immobilization of horseradish peroxidase-DNA conjugates on microelectrode arrays: towards electrochemical screening of enzyme libraries. Chemistry 13:5223–5231
17. Kuo CH, Fruk L, Niemeyer CM (2009) Addressable DNA-myoglobin photocatalysis. Chem Asian J 4:1064–1069
18. Simon P, Dueymes C, Fontecave M, Decout JL (2005) DNA detection through signal amplification by using NADH: flavin oxidoreductase and oligonucleotide-flavin conjugates as cofactors. Angew Chem Int Ed 44:2764–2767
19. Lapiene V, Kukolka F, Kiko K, Arndt A, Niemeyer CM (2010) Conjugation of fluorescent proteins with DNA oligonucleotides. Bioconjug Chem 21:921–927
20. Kukolka F, Schoeps O, Woggon U, Niemeyer CM (2007) DNA-directed assembly of supramolecular fluorescent protein energy transfer systems. Bioconjug Chem 18:621–627
21. Fruk L, Muller J, Niemeyer CM (2006) Kinetic analysis of semisynthetic peroxidase enzymes containing a covalent DNA-heme adduct as the cofactor. Chemistry 12:7448–7457
22. Bauer DM, Ahmed I, Vigovskaya A, Fruk L (2013) Clickable tyrosine binding bifunctional linkers for preparation of DNA-protein conjugates. Bioconjug Chem 24:1094–1101

23. Bauer DM, Rogge A, Stolzer L, Barner-Kowollik C, Fruk L (2013) Light induced DNA-protein conjugation. Chem Commun (Camb, UK) 49:8626–8628
24. Lovrinovic M, Spengler M, Deutsch C, Niemeyer CM (2005) Synthesis of covalent DNA-protein conjugates by expressed protein ligation. Mol BioSyst 1:64–69
25. Duckworth BP, Chen Y, Wollack JW, Sham Y, Mueller JD, Taton TA, Distefano MD (2007) A universal method for the preparation of covalent protein-DNA conjugates for use in creating protein nanostructures. Angew Chem Int Ed 46:8819–8822
26. Chandra RA, Douglas ES, Mathies RA, Bertozzi CR, Francis MB (2006) Programmable cell adhesion encoded by DNA hybridization. Angew Chem Int Ed 45:896–901
27. Dey S, Sheppard TL (2001) Ketone-DNA: a versatile postsynthetic DNA decoration platform. Org Lett 3:3983–3986
28. Tominaga J, Kemori Y, Tanaka Y, Maruyama T, Kamiya N, Goto M (2007) An enzymatic method for site-specific labeling of recombinant proteins with oligonucleotides. Chem Commun (Camb, UK) 401–403
29. Jongsma MA, Litjens RH (2006) Self-assembling protein arrays on DNA chips by auto-labeling fusion proteins with a single DNA address. Proteomics 6:2650–2655
30. Lee HS, Dimla RD, Schultz PG (2009) Protein-DNA photo-crosslinking with a genetically encoded benzophenone-containing amino acid. Bioorg Med Chem Lett 19:5222–5224
31. Bano F, Fruk L, Sanavio B, Glettenberg M, Casalis L, Niemeyer CM, Scoles G (2009) Toward multiprotein nanoarrays using nanografting and DNA directed immobilization of proteins. Nano Lett 9:2614–2618
32. Shtenberg G, Massad-Ivanir N, Moscovitz O, Engin S, Sharon M, Fruk L, Segal E (2013) Picking up the pieces: a generic porous Si biosensor for probing the proteolytic products of enzymes. Anal Chem 85:1951–1956
33. Gogolin L, Schroeder H, Itzen A, Goody RS, Niemeyer CM, Becker CF (2013) Protein-DNA arrays as tools for detection of protein-protein interactions by mass spectrometry. ChemBioChem 14:92–99
34. Silvestrini M, Fruk L, Ugo P (2013) Functionalized ensembles of nanoelectrodes as affinity biosensors for DNA hybridization detection. Biosens Bioelectron 40:265–270
35. Vogel K, Glettenberg M, Schroeder H, Niemeyer CM (2013) DNA-modification of eukaryotic cells. Small 9:255–262
36. Reisewitz S, Schroeder H, Tort N, Edwards KA, Baeumner AJ, Niemeyer CM (2010) Capture and culturing of living cells on microstructured DNA substrates. Small 6:2162–2168
37. Gandor S, Reisewitz S, Venkatachalapathy M, Arrabito G, Reibner M, Schroder H, Ruf K, Niemeyer CM, Bastiaens PI, Dehmelt L (2013) A protein-interaction array inside a living cell. Angew Chem Int Ed 52:4790–4794
38. Goudot A, Pourceau G, Meyer A, Gehin T, Vidal S, Vasseur JJ, Morvan F, Souteyrand E, Chevolot Y (2013) Quantitative analysis (K(d) and IC(50)) of glycoconjugates interactions with a bacterial lectin on a carbohydrate microarray with DNA direct immobilization (DDI). Biosens Bioelectron 40:153–160
39. Piliarik M, Bockova M, Homola J (2010) Surface plasmon resonance biosensor for parallelized detection of protein biomarkers in diluted blood plasma. Biosens Bioelectron 26:1656–1661
40. Mirkin CA, Letsinger RL, Mucic RC, Storhoff JJ (1996) A DNA-based method for rationally assembling nanoparticles into macroscopic materials. Nature 382:607–609
41. Mucic RC, Storhoff JJ, Mirkin CA, Letsinger RL (1998) DNA-directed synthesis of binary nanoparticle network materials. J Am Chem Soc 120:12674–12675
42. Elghanian R, Storhoff JJ, Mucic RC, Letsinger RL, Mirkin CA (1997) Selective colorimetric detection of polynucleotides based on the distance-dependent optical properties of gold nanoparticles. Science 277:1078–1081
43. Zhang T, Dong Y, Sun Y, Chen P, Yang Y, Zhou C, Xu L, Yang Z, Liu D (2012) DNA bimodified gold nanoparticles. Langmuir 28:1966–1970

44. Chen C, Fruk L (2013) Functionalization of maleimide-coated silver nanoparticles through Diels–Alder cycloaddition. RSC Adv 3:1709–1713
45. Wen Y, Chen L, Wang W, Xu L, Du H, Zhang Z, Zhang X, Song Y (2012) A flexible DNA modification approach towards construction of gold nanoparticle assemblies. Chem Commun (Camb, UK) 48:3963–3965
46. Estephan ZG, Qian Z, Lee D, Crocker JC, Park SJ (2013) Responsive multidomain free-standing films of gold nanoparticles assembled by DNA-Directed Layer-by-Layer approach. Nano Lett 13:4449–4455
47. Peng HI, Miller BL (2011) Recent advancements in optical DNA biosensors: exploiting the plasmonic effects of metal nanoparticles. Analyst 136:436–447
48. Obliosca JM, Wang PC, Tseng FG (2012) Probing quenched dye fluorescence of Cy3-DNA-Au-nanoparticle hybrid conjugates using solution and array platforms. J Colloid Interface Sci 371:34–41
49. Hazarika P, Kukolka F, Niemeyer CM (2006) Reversible binding of fluorescent proteins at DNA-gold nanoparticles. Angew Chem Int Ed 45:6827–6830
50. Niemeyer CM, Ceyhan B (2001) DNA-directed functionalization of colloidal gold with proteins. Angew Chem Int Ed 40:3685–3688 This work was supported by Deutsche Forschungsgemeinschaft and Fonds der Chemischen Industrie. We thank Prof. D. Blohm for helpful discussions and generous support
51. Storhoff JJ, Elghanian R, Mucic RC, Mirkin CA, Letsinger RL (1998) One-pot colorimetric differentiation of polynucleotides with single base imperfections using gold nanoparticle probes. J Am Chem Soc 120:1959–1964
52. Oh JH, Lee JS (2011) Designed hybridization properties of DNA-gold nanoparticle conjugates for the ultraselective detection of a single-base mutation in the breast cancer gene BRCA1. Anal Chem 83:7364–7370
53. Yin H, Huang X, Ma W, Xu L, Zhu S, Kuang H, Xu C (2014) Ligation chain reaction based gold nanoparticle assembly for ultrasensitive DNA detection. Biosens Bioelectron 52:8–12
54. Graham D, Stevenson R, Thompson DG, Barrett L, Dalton C, Faulds K (2011) Combining functionalised nanoparticles and SERS for the detection of DNA relating to disease. Faraday Discuss 149:291–299; discussion 333–256
55. He Y, Zhang X, Zhang S, Kris MK, Man FC, Kawde AN, Liu G (2012) Visual detection of single-base mismatches in DNA using hairpin oligonucleotide with double-target DNA binding sequences and gold nanoparticles. Biosens Bioelectron 34:37–43
56. Alhasan AH, Kim DY, Daniel WL, Watson E, Meeks JJ, Thaxton CS, Mirkin CA (2012) Scanometric microRNA array profiling of prostate cancer markers using spherical nucleic acid-gold nanoparticle conjugates. Anal Chem 84:4153–4160
57. Gao F, Zhu Z, Lei J, Geng Y, Ju H (2013) Sub-femtomolar electrochemical detection of DNA using surface circular strand-replacement polymerization and gold nanoparticle catalyzed silver deposition for signal amplification. Biosens Bioelectron 39:199–203
58. Fischler M, Simon U (2009) Metal nanoparticle-DNA hybrids—from assembly towards functional conjugates. J Mater Chem 19:1518–1523
59. Hu R, Zhang X-B, Kong R-M, Zhao X-H, Jiang J, Tan W (2011) Nucleic acid-functionalized nanomaterials for bioimaging applications. J Mater Chem 21:16323–16334
60. Cutler JI, Auyeung E, Mirkin CA (2012) Spherical nucleic acids. J Am Chem Soc 134:1376–1391
61. Freeman R, Girsh J, Willner I (2013) Nucleic Acid/Quantum Dots (QDs) hybrid systems for optical and Photoelectrochemical sensing. ACS Appl Mater Interfaces 5:2815–2834
62. Heuer-Jungemann A, Harimech PK, Brown T, Kanaras AG (2013) Gold nanoparticles and fluorescently-labelled DNA as a platform for biological sensing. Nanoscale 5:9503–9510
63. Lee J-H, Hwang J-H, Nam J-M (2013) DNA-tailored plasmonic nanoparticles for biosensing applications. Wiley Interdiscip Rev Nanomed Nanobiotechnol 5:96–109
64. Su S, Zuo X, Pan D, Pei H, Wang L, Fan C, Huang W (2013) Design and applications of gold nanoparticle conjugates by exploiting biomolecule-gold nanoparticle interactions. Nanoscale 5:2589–2599

65. Ke Y, Sharma J, Liu M, Jahn K, Liu Y, Yan H (2009) Scaffolded DNA origami of a DNA tetrahedron molecular container. Nano Lett 9:2445–2447
66. Andersen ES, Dong M, Nielsen MM, Jahn K, Subramani R, Mamdouh W, Golas MM, Sander B, Stark H, Oliveira CL, Pedersen JS, Birkedal V, Besenbacher F, Gothelf KV, Kjems J (2009) Self-assembly of a nanoscale DNA box with a controllable lid. Nature 459:73–76
67. Ke Y, Douglas SM, Liu M, Sharma J, Cheng A, Leung A, Liu Y, Shih WM, Yan H (2009) Multilayer DNA origami packed on a square lattice. J Am Chem Soc 131:15903–15908
68. Sacca B, Niemeyer CM (2012) DNA origami: the art of folding DNA. Angew Chem Int Ed 51:58–66
69. Hung AM, Noh H, Cha JN (2010) Recent advances in DNA-based directed assembly on surfaces. Nanoscale 2:2530–2537
70. Kuzuya A, Kimura M, Numajiri K, Koshi N, Ohnishi T, Okada F, Komiyama M (2009) Precisely programmed and robust 2D streptavidin nanoarrays by using periodical nanometer-scale wells embedded in DNA origami assembly. ChemBioChem 10:1811–1815
71. Numajiri K, Kimura M, Kuzuya A, Komiyama M (2010) Stepwise and reversible nanopatterning of proteins on a DNA origami scaffold. Chem Commun (Camb, UK) 46:5127–5129
72. Voigt NV, Torring T, Rotaru A, Jacobsen MF, Ravnsbaek JB, Subramani R, Mamdouh W, Kjems J, Mokhir A, Besenbacher F, Gothelf KV (2010) Single-molecule chemical reactions on DNA origami. Nat Nanotechnol 5:200–203
73. Sacca B, Meyer R, Erkelenz M, Kiko K, Arndt A, Schroeder H, Rabe KS, Niemeyer CM (2010) Orthogonal protein decoration of DNA origami. Angew Chem Int Ed 49:9378–9383
74. Pal S, Deng Z, Ding B, Yan H, Liu Y (2010) DNA-origami-directed self-assembly of discrete silver-nanoparticle architectures. Angew Chem Int Ed Engl 49:2700–2704
75. Pal S, Deng Z, Wang H, Zou S, Liu Y, Yan H (2011) DNA directed self-assembly of anisotropic plasmonic nanostructures. J Am Chem Soc 133:17606–17609
76. Lan X, Chen Z, Dai G, Lu X, Ni W, Wang Q (2013) Bifacial DNA origami-directed discrete, three-dimensional, anisotropic plasmonic nanoarchitectures with tailored optical chirality. J Am Chem Soc 135:11441–11444
77. Zhao Z, Jacovetty EL, Liu Y, Yan H (2011) Encapsulation of gold nanoparticles in a DNA origami cage. Angew Chem Int Ed 50:2041–2044
78. Shen X, Song C, Wang J, Shi D, Wang Z, Liu N, Ding B (2012) Rolling up gold nanoparticle-dressed DNA origami into three-dimensional plasmonic chiral nanostructures. J Am Chem Soc 134:146–149
79. Shen X, Asenjo-Garcia A, Liu Q, Jiang Q, Garcia de Abajo FJ, Liu N, Ding B (2013) Three-dimensional plasmonic chiral tetramers assembled by DNA origami. Nano Lett 13:2128–2133
80. Kuzyk A, Schreiber R, Fan Z, Pardatscher G, Roller EM, Hogele A, Simmel FC, Govorov AO, Liedl T (2012) DNA-based self-assembly of chiral plasmonic nanostructures with tailored optical response. Nature 483:311–314
81. Fu Y, Zeng D, Chao J, Jin Y, Zhang Z, Liu H, Li D, Ma H, Huang Q, Gothelf KV, Fan C (2013) Single-step rapid assembly of DNA origami nanostructures for addressable nanoscale bioreactors. J Am Chem Soc 135:696–702
82. Prodan E, Radloff C, Halas NJ, Nordlander P (2003) A hybridization model for the plasmon response of complex nanostructures. Science 302:419–422
83. Jain PK, Huang WY, El-Sayed MA (2007) On the universal scaling behavior of the distance decay of plasmon coupling in metal nanoparticle pairs: a plasmon ruler equation. Nano Lett 7:2080–2088
84. Sharma J, Chhabra R, Andersen CS, Gothelf KV, Yan H, Liu Y (2008) Toward reliable gold nanoparticle patterning on self-assembled DNA nanoscaffold. J Am Chem Soc 130:7820–7821
85. Ding B, Deng Z, Yan H, Cabrini S, Zuckermann RN, Bokor J (2010) Gold nanoparticle self-similar chain structure organized by DNA origami. J Am Chem Soc 132:3248–3249

86. Fan Z, Govorov AO (2010) Plasmonic circular dichroism of chiral metal nanoparticle assemblies. Nano Lett 10:2580–2587
87. Wang ZG, Song C, Ding B (2013) Functional DNA nanostructures for photonic and biomedical applications. Small 9:2210–2222
88. Schreiber R, Kempter S, Holler S, Schuller V, Schiffels D, Simmel SS, Nickels PC, Liedl T (2011) DNA origami-templated growth of arbitrarily shaped metal nanoparticles. Small 7:1795–1799
89. Pound E, Ashton JR, Becerril HA, Woolley AT (2009) Polymerase chain reaction based scaffold preparation for the production of thin, branched DNA origami nanostructures of arbitrary sizes. Nano Lett 9:4302–4305
90. Zhang H, Chao J, Pan D, Liu H, Huang Q, Fan C (2012) Folding super-sized DNA origami with scaffold strands from long-range PCR. Chem Commun (Camb, UK)
91. Said H, Schuller VJ, Eber FJ, Wege C, Liedl T, Richert C (2013) M1.3–a small scaffold for DNA origami. Nanoscale 5:284–290
92. Ouyang X, Li J, Liu H, Zhao B, Yan J, Ma Y, Xiao S, Song S, Huang Q, Chao J, Fan C (2013) Rolling circle amplification-based DNA origami nanostructrures for intracellular delivery of immunostimulatory drugs. Small 9:3082–3087
93. Lizardi PM, Huang XH, Zhu ZR, Bray-Ward P, Thomas DC, Ward DC (1998) Mutation detection and single-molecule counting using isothermal rolling-circle amplification. Nat Genet 19:225–232
94. Hogberg B, Liedl T, Shih WM (2009) Folding DNA origami from a double-stranded source of scaffold. J Am Chem Soc 131:9154–9155
95. Yang Y, Han D, Nangreave J, Liu Y, Yan H (2012) DNA origami with double-stranded DNA as a unified scaffold. ACS Nano 6:8209–8215
96. Douglas SM, Dietz H, Liedl T, Hogberg B, Graf F, Shih WM (2009) Self-assembly of DNA into nanoscale three-dimensional shapes. Nature 459:414–418
97. Erkelenz M, Bauer DM, Meyer R, Gatsogiannis C, Raunser S, Sacca B, Niemeyer CM (2013) A facile method for preparation of tailored scaffolds for DNA-Origami. Small
98. Zheng L, Baumann U, Reymond JL (2004) An efficient one-step site-directed and site-saturation mutagenesis protocol. Nucleic Acids Res 32:e115
99. Li MZ, Elledge SJ (2007) Harnessing homologous recombination in vitro to generate recombinant DNA via SLIC. Nat Methods 4:251–256
100. Wang LL, Yoshida J, Ogata N (2001) Self-assembled supramolecular films derived from marine deoxyribonucleic acid (DNA)-cationic surfactant complexes: large-scale preparation and optical and thermal properties. Chem Mater 13:1273–1281
101. Yu Z, Li W, Hagen JA, Zhou Y, Klotzkin D, Grote JG, Steckl AJ (2007) Photoluminescence and lasing from deoxyribonucleic acid (DNA) thin films doped with sulforhodamine. Appl Opt 46:1507–1513
102. Hagen JA, Li W, Steckl J, Grote JG (2006) Enhanced emission efficiency in organic light-emitting diodes using deoxyribonucleic acid complex as an electron blocking layer. Appl Phys Lett 88
103. Liu X, Diao H, Nishi N (2008) Applied chemistry of natural DNA. Chem Soc Rev 37:2745–2757
104. Singh TB, Sariciftci NS, Grote JG (2010) Bio-organic optoelectronic devices using DNA. Org Electron 223:189–212
105. Kwon YW, Choi DH, Jin JI (2012) Optical, electro-optic and optoelectronic properties of natural and chemically modified DNAs. Polym J (Tokyo, Jpn) 44:1191–1208
106. Chen IC, Chiu YW, Hung YC (2012) Efficient biopolymer blue organic light-emitting devices with low driving voltage. Jpn J Appl Phys 51
107. Hung YC, Lin TY, Hsu WT, Chiu YW, Wang YS, Fruk L (2012) Functional DNA biopolymers and nanocomposite for optoelectronic applications. Opt Mater 34:1208–1213
108. Hung YC, Mueller P, Wang YS, Fruk L (2012) Phototriggered growth of crystalline Au structures in the presence of a DNA-surfactant complex. Nanoscale 4:5585–5587

109. Hung YC, Su CH, Huang HW (2012) Low threshold amplified spontaneous emission from dye-doped DNA biopolymer. J Appl Phys 111
110. Heckman EM, Hagen JA, Yaney PP, Grote JG, Hopkins FK (2005) Processing techniques for deoxyribonucleic acid: biopolymer for photonics applications. Appl Phys Lett 87
111. Steckl AJ (2007) DNA—a new material for photonics? Nat Photonics 1:3–5
112. Singh B, Sariciftci NS, Grote JG, Hopkins FK (2006) Bio-organic-semiconductor-field-effect-transistor based on deoxyribonucleic acid gate dielectric. J Appl Phys 100
113. Stadler P, Oppelt K, Singh TB, Grote JG, Schwodiauer R, Bauer S, Piglmayer-Brezina H, Bauerle D, Sariciftci NS (2007) Organic field-effect transistors and memory elements using deoxyribonucleic acid (DNA) gate dielectric. Org Electron 8:648–654
114. Hung YC, Hsu WT, Lin TY, Fruk L (2011) Photoinduced write-once read-many-times memory device based on DNA biopolymer nanocomposite. Appl Phys Lett 99
115. Wichterle O, Lim D (1960) Hydrophilic gels for biological use. Nature 185:117–118
116. Kopecek J, Yang JY (2007) Review—Hydrogels as smart biomaterials. Polym Int 56:1078–1098
117. O'Neill P, Rothemund PW, Kumar A, Fygenson DK (2006) Sturdier DNA nanotubes via ligation. Nano Lett 6:1379–1383
118. Xing Y, Cheng E, Yang Y, Chen P, Zhang T, Sun Y, Yang Z, Liu D (2011) Self-assembled DNA hydrogels with designable thermal and enzymatic responsiveness. Adv Mater 23:1117–1121
119. Nagahara S, Matsuda T (1996) Hydrogel formation via hybridization of oligonucleotides derivatized in water-soluble vinyl polymers. Polym Gels Netw 4:111–127
120. Amiya T, Tanaka T (1987) Phase-transitions in cross-linked gels of natural polymers. Macromolecules 20:1162–1164
121. Yang HH, Liu HP, Kang HZ, Tan WH (2008) Engineering target-responsive hydrogels based on aptamer—Target interactions. J Am Chem Soc 130:6320+
122. Cheng EJ, Xing YZ, Chen P, Yang Y, Sun YW, Zhou DJ, Xu LJ, Fan QH, Liu DS (2009) A pH-triggered, fast-responding DNA hydrogel. Angew Chem Int Ed 48:7660–7663
123. Zhu Z, Wu C, Liu H, Zou Y, Zhang X, Kang H, Yang CJ, Tan W (2010) An aptamer cross-linked hydrogel as a colorimetric platform for visual detection. Angew Chem Int Ed 49:1052–1056
124. Lu CH, Qi XJ, Orbach R, Yang HH, Mironi-Harpaz I, Seliktar D, Willner I (2013) Switchable catalytic acrylamide hydrogels cross-linked by Hemin/G-Quadruplexes. Nano Lett 13:1298–1302
125. Lin H, Zou Y, Huang Y, Chen J, Zhang WY, Zhuang Z, Jenkins G, Yang CJ (2011) DNAzyme crosslinked hydrogel: a new platform for visual detection of metal ions. Chem Commun (Camb, UK) 47:9312–9314
126. Gao M, Gawel K, Stokke BT (2011) Toehold of dsDNA exchange affects the hydrogel swelling kinetics of a polymer-dsDNA hybrid hydrogel. Soft Matter 7:1741–1746
127. Wang RH, Li YB (2013) Hydrogel based QCM aptasensor for detection of avian influenza virus. Biosens Bioelectron 42:148–155
128. Nayak S, Lyon LA (2005) Soft nanotechnology with soft nanoparticles. Angew Chem Int Ed 44:7686–7708
129. Peppas NA, Hilt JZ, Khademhosseini A, Langer R (2006) Hydrogels in biology and medicine: from molecular principles to bionanotechnology. Adv Mater (Weinheim, Ger) 18:1345–1360
130. Miyata T, Asami N, Uragami T (1999) A reversibly antigen-responsive hydrogel. Nature 399:766–769
131. Murakami Y, Maeda M (2005) DNA-responsive hydrogels that can shrink or swell. Biomacromolecules 6:2927–2929
132. Yurke B, Turberfield AJ, Mills AP Jr, Simmel FC, Neumann JL (2000) A DNA-fuelled molecular machine made of DNA. Nature 406:605–608
133. Feng LP, Park SH, Reif JH, Yan H (2003) A two-state DNA lattice switched by DNA nanoactuator. Angew Chem Int Ed 42:4342–4346

134. Lin DC, Yurke B, Langrana NA (2004) Mechanical properties of a reversible, DNA-crosslinked polyacrylamide hydrogel. J Biomech Eng 126:104–110
135. Liedl T, Dietz H, Yurke B, Simmel FC (2007) Controlled trapping and release of quantum dots in a DNA-switchable hydrogel. Small 3:1688–1693
136. Kwak M, Herrmann A (2010) Nucleic acid/organic polymer hybrid materials: synthesis, superstructures, and applications. Angew Chem Int Ed 49:8574–8587
137. Keller S, Wang J, Chandra M, Berger R, Marx A (2008) DNA polymerase-catalyzed DNA network growth. J Am Chem Soc 130:13188–13189
138. Park N, Um SH, Funabashi H, Xu J, Luo D (2009) A cell-free protein-producing gel. Nat Mater 8:432–437
139. Kang H, Liu H, Zhang X, Yan J, Zhu Z, Peng L, Yang H, Kim Y, Tan W (2011) Photoresponsive DNA-cross-linked hydrogels for controllable release and cancer therapy. Langmuir 27:399–408
140. Dave N, Chan MY, Huang PJJ, Smith BD, Liu JW (2010) Regenerable DNA-functionalized hydrogels for ultrasensitive, instrument-free Mercury(II) detection and removal in water. J Am Chem Soc 132:12668–12673
141. Wang J, Liu B (2008) Highly sensitive and selective detection of Hg(2+) in aqueous solution with mercury-specific DNA and Sybr Green I. Chem Commun (Camb) 4759–4761
142. Baeissa A, Dave N, Smith BD, Liu J (2010) DNA-functionalized monolithic hydrogels and gold nanoparticles for colorimetric DNA detection. ACS Appl Mater Interfaces 2:3594–3600
143. Jacobi ZE, Li L, Liu J (2012) Visual detection of lead(II) using a label-free DNA-based sensor and its immobilization within a monolithic hydrogel. Analyst 137:704–709
144. Xie J, Zheng Y, Ying JY (2010) Highly selective and ultrasensitive detection of Hg(2+) based on fluorescence quenching of Au nanoclusters by Hg(2+)-Au(+) interactions. Chem Commun (Camb) 46:961–963
145. MacLean JL, Morishita K, Liu J (2013) DNA stabilized silver nanoclusters for ratiometric and visual detection of Hg(2)(+) and its immobilization in hydrogels. Biosens Bioelectron 48:82–86
146. Morishita K, MacLean JL, Liu B, Jiang H, Liu J (2013) Correlation of photobleaching, oxidation and metal induced fluorescence quenching of DNA-templated silver nanoclusters. Nanoscale 5:2840–2849
147. Helwa Y, Dave N, Froidevaux R, Samadi A, Liu J (2012) Aptamer-functionalized hydrogel microparticles for fast visual detection of mercury(II) and adenosine. ACS Appl Mater Interfaces 4:2228–2233
148. Joseph KA, Dave N, Liu J (2011) Electrostatically directed visual fluorescence response of DNA-functionalized monolithic hydrogels for highly sensitive Hg(2)+ detection. ACS Appl Mater Interfaces 3:733–739
149. Lee SH, Lee CK, Shin SR, Gu BK, Kim SI, Kang TM, Kim SJ (2010) Enhanced actuation of PPy/CNT hybrid fibers using porous structured DNA hydrogel. Sens Actuators B Chem 145:89–92
150. Hur J, Im K, Hwang S, Choi B, Kim S, Hwang S, Park N, Kim K (2013) DNA hydrogel-based supercapacitors operating in physiological fluids. Sci Rep 3
151. Kushibiki T, Tomoshige R, Iwanaga K, Kakemi M, Tabata Y (2006) Controlled release of plasmid DNA from hydrogels prepared from gelatin cationized by different amine compounds. J Controlled Release 112:249–256
152. Megeed Z, Cappello J, Ghandehari H (2002) Controlled release of plasmid DNA from a genetically engineered silk-elastinlike hydrogel. Pharm Res 19:954–959
153. Gupta M, Gupta AK (2004) Hydrogel pullulan nanoparticles encapsulating pBUDLacZ plasmid as an efficient gene delivery carrier. J Controlled Release 99:157–166
154. Ma D, Zhang HB, Chen DH, Zhang LM (2011) Novel supramolecular gelation route to in situ entrapment and sustained delivery of plasmid DNA. J Colloid Interface Sci 364:566–573

155. Kimura T, Iwai S, Moritan T, Nam K, Mutsuo S, Yoshizawa H, Okada M, Furuzono T, Fujisato T, Kishida A (2007) Preparation of poly(vinyl alcohol)/DNA hydrogels via hydrogen bonds formed on ultra-high pressurization and controlled release of DNA from the hydrogels for gene delivery. J Artif Organs 10:104–108
156. Nishikawa M, Mizuno Y, Mohri K, Matsuoka N, Rattanakiat S, Takahashi Y, Funabashi H, Luo D, Takakura Y (2011) Biodegradable CpG DNA hydrogels for sustained delivery of doxorubicin and immunostimulatory signals in tumor-bearing mice. Biomaterials 32:488–494
157. Liu XD, Murayama Y, Matsunaga M, Nomizu M, Nishi N (2005) Preparation and characterization of DNA hydrogel bead as selective adsorbent of dioxins. Int J Biol Macromol 35:193–199
158. Klok HA (2005) Biological-synthetic hybrid block copolymers: Combining the best from two worlds. J Polym Sci Part A: Polym Chem 43:1–17
159. Hawker CJ, Wooley KL (2005) The convergence of synthetic organic and polymer chemistries. Science 309:1200–1205
160. Kwak M, Herrmann A (2010) Nucleic acid/organic polymer hybrid materials: synthesis, superstructures, and applications. Angew Chem Int Ed Engl 49:8574–8587
161. Soh N, Umeno D, Tang Z, Murata M, Maeda M (2002) Affinity precipitation separation of DNA binding protein using block conjugate composed of poly(N-isopropylacrylamide) grafted double-stranded DNA and double-stranded DNA containing a target sequence. Anal Sci 18:1295–1299
162. Perly B, Douy A, Gallot B (1974) Synthesis of Polybutadiene-Benzylpoly-L-Glutamate sequential copolymers and structural study of Mesophases. C R Acad Sci 279:1109–1111
163. Yamashita Y, Iwaya Y, Ito K (1975) Block Copolymerization. 9. Polymerization of Nca of Methyl D-Glutamate by Telechelic Polystyrene Having Glycyl groups as active chain ends. Macromol Chem Phys 176:1207–1216
164. Alemdaroglu FE, Herrmann A (2007) DNA meets synthetic polymers–highly versatile hybrid materials. Org Biomol Chem 5:1311–1320
165. Lemaitre M, Bayard B, Lebleu B (1987) Specific antiviral activity of a poly(L-lysine)-conjugated oligodeoxyribonucleotide sequence complementary to vesicular stomatitis virus N protein mRNA initiation site. Proc Natl Acad Sci USA 84:648–652
166. Leonetti JP, Degols G, Milhaud P, Gagnor C, Lemaitre M, Lebleu B (1989) Antiviral activity of antisense oligonucleotides linked to Poly(L-Lysine)—Targets on genomic RNA and or messenger-RNA of Vesicular Stomatitis-Virus. Nucleos Nucleot 8:825–828
167. Leonetti JP, Rayner B, Lemaitre M, Gagnor C, Milhaud PG, Imbach JL, Lebleu B (1988) Antiviral activity of conjugates between poly(L-lysine) and synthetic oligodeoxyribonucleotides. Gene 72:323–332
168. Jeong JH, Kim SH, Kim SW, Park TG (2005) Polyelectrolyte complex micelles composed of c-raf antisense oligodeoxynucleotide-poly(ethylene glycol) conjugate and poly(ethylenimine): effect of systemic administration on tumor growth. Bioconjug Chem 16:1034–1037
169. Oishi M, Nagasaki Y, Itaka K, Nishiyama N, Kataoka K (2005) Lactosylated poly(ethylene glycol)-siRNA conjugate through acid-labile beta-thiopropionate linkage to construct pH-sensitive polyion complex micelles achieving enhanced gene silencing in hepatoma cells. J Am Chem Soc 127:1624–1625
170. Takei YG, Aoki T, Sanui K, Ogata N, Okano T, Sakurai Y (1993) Temperature-responsive bioconjugates. 1. Synthesis of temperature-responsive oligomers with reactive end groups and their coupling to biomolecules. Bioconjug Chem 4:42–46
171. Fong RB, Ding Z, Long CJ, Hoffman AS, Stayton PS (1999) Thermoprecipitation of streptavidin via oligonucleotide-mediated self-assembly with poly(N-isopropylacrylamide). Bioconjug Chem 10:720–725
172. Jeong JH, Park TG (2001) Novel polymer-DNA hybrid polymeric micelles composed of hydrophobic poly(D, L-lactic-co-glycolic acid) and hydrophilic oligonucleotides. Bioconjug Chem 12:917–923

173. Lee K, Povlich LK, Kim J (2007) Label-free and self-signal amplifying molecular DNA sensors based on bioconjugated polyelectrolytes. Adv Funct Mater 17:2580–2587
174. Baccaro A, Marx A (2010) Enzymatic synthesis of organic-polymer-grafted DNA. Chemistry 16:218–226
175. Alemdaroglu FE, Zhuang W, Zophel L, Wang J, Berger R, Rabe JP, Herrmann A (2009) Generation of multiblock copolymers by PCR: synthesis, visualization and nanomechanical properties. Nano Lett 9:3658–3662
176. Li Z, Zhang Y, Fullhart P, Mirkin CA (2004) Reversible and chemically programmable micelle assembly with DNA block-copolymer amphiphiles. Nano Lett 4:1055–1058
177. Alemdaroglu FE, Ding K, Berger R, Herrmann A (2006) DNA-templated synthesis in three dimensions: introducing a micellar scaffold for organic reactions. Angew Chem Int Ed 45:4206–4210
178. Alemdaroglu FE, Safak M, Wang J, Berger R, Herrmann A (2007) DNA multiblock copolymers. Chem Commun (Cambridge, UK) 1358–1359
179. Teixeira F Jr, Rigler P, Vebert-Nardin C (2007) Nucleo-copolymers: oligonucleotide-based amphiphilic diblock copolymers. Chem Commun (Cambridge, UK) 1130–1132
180. Yang CYJ, Pinto M, Schanze K, Tan WH (2005) Direct synthesis of an oligonucleotide-poly(phenylene ethynylene) conjugate with a precise one-to-one molecular ratio. Angew Chem Int Ed 44:2572–2576
181. Li XY, Liu DR (2004) DNA-Templated organic synthesis: Nature's strategy for controlling chemical reactivity applied to synthetic molecules. Angew Chem Int Ed 43:4848–4870
182. Alemdaroglu FE, Wang J, Borsch M, Berger R, Herrmann A (2008) Enzymatic control of the size of DNA block copolymer nanoparticles. Angew Chem Int Ed 47:974–976
183. Cavalieri F, Postma A, Lee L, Caruso F (2009) Assembly and functionalization of DNA-polymer microcapsules. ACS Nano 3:234–240
184. Kim SH, Jeong JH, Lee SH, Kim SW, Park TG (2008) Local and systemic delivery of VEGF siRNA using polyelectrolyte complex micelles for effective treatment of cancer. J Controlled Release 129:107–116
185. Alemdaroglu FE, Alemdaroglu NC, Langguth P, Herrmann A (2008) DNA block copolymer micelles—A combinatorial tool for cancer Nanotechnology. Adv Mater (Weinheim, Ger) 20:899+
186. Kleiner RE, Brudno Y, Birnbaum ME, Liu DR (2008) DNA-templated polymerization of side-chain-functionalized peptide nucleic acid aldehydes. J Am Chem Soc 130:4646–4659
187. Minagawa K, Matsuzawa Y, Yoshikawa K, Masubuchi Y, Matsumoto M, Doi M, Nishimura C, Maeda M (1993) Change of the higher-order structure of DNA induced by the complexation with intercalating synthetic-polymer, as is visualized by fluorescence microscopy. Nucl Acids Res 21:37–40
188. Maeda M, Nishimura C, Inenaga A, Takagi M (1993) Modification of DNA with Poly(N-Isopropylacrylamide) for thermally-induced affinity separation. React Polym 21:27–35
189. Maeda M, Nishimura C, Umeno D, Takagi M (1994) Psoralen-containing vinyl monomer for conjugation of double-helical DNA with vinyl polymers. Bioconjug Chem 5:527–531
190. Umeno D, Maeda M (1997) Poly(N-isopropylacrylamide) carrying double-stranded DNA for affinity separation of genotoxins. Anal Sci 13:553–556
191. Umeno D, Kawasaki M, Maeda M (1998) Photolithographic coating of polymers on DNA as an approach for construction of nano-structures. Supramol Sci 5:427–431
192. Minard-Basquin C, Chaix C, Pichot C, Mandrand B (2000) Oligonucleotide-polymer conjugates: effect of the method of synthesis on their structure and performance in diagnostic assays. Bioconjug Chem 11:795–804
193. Minard-Basquin C, Chaix C, D'Agosto F, Charreyre MT, Pichot C (2004) Oligonucleotide synthesis onto poly(N-acryloylmorpholine-co-N-acryloxysuccinimide): assessment of the resulting conjugates in a DNA sandwich hybridization test. J Appl Polym Sci 92:3784–3795
194. Lee K, Rouillard JM, Pham T, Gulari E, Kim J (2007) Signal-amplifying conjugated polymer-DNA hybrid chips. Angew Chem Int Ed 46:4667–4670

195. Rosi NL, Giljohann DA, Thaxton CS, Lytton-Jean AK, Han MS, Mirkin CA (2006) Oligonucleotide-modified gold nanoparticles for intracellular gene regulation. Science 312:1027–1030
196. Giljohann DA, Seferos DS, Prigodich AE, Patel PC, Mirkin CA (2009) Gene regulation with polyvalent siRNA-nanoparticle conjugates. J Am Chem Soc 131:2072–2073
197. Davis ME, Chen ZG, Shin DM (2008) Nanoparticle therapeutics: an emerging treatment modality for cancer. Nat Rev Drug Discov 7:771–782
198. Dobson J (2006) Gene therapy progress and prospects: magnetic nanoparticle-based gene delivery. Gene Ther 13:283–287
199. Jain KK (2003) Nanodiagnostics: application of nanotechnology in molecular diagnostics. Expert Rev Mol Diagn 3:153–161
200. Farokhzad OC, Langer R (2006) Nanomedicine: developing smarter therapeutic and diagnostic modalities. Adv Drug Deliv Rev 58:1456–1459
201. Taton TA, Mirkin CA, Letsinger RL (2000) Scanometric DNA array detection with nanoparticle probes. Science 289:1757–1760
202. Rosi NL, Mirkin CA (2005) Nanostructures in biodiagnostics. Chem Rev (Washington, DC, USA) 105:1547–1562
203. Bruchez M Jr, Moronne M, Gin P, Weiss S, Alivisatos AP (1998) Semiconductor nanocrystals as fluorescent biological labels. Science 281:2013–2016
204. Wilson SR, Sabatine MS, Braunwald E, Sloan S, Murphy SA, Morrow DA (2009) Detection of myocardial injury in patients with unstable angina using a novel nanoparticle cardiac troponin I assay: observations from the PROTECT-TIMI 30 Trial. Am Heart J 158:386–391
205. Verma A, Simard JM, Worrall JW, Rotello VM (2004) Tunable reactivation of nanoparticle-inhibited beta-galactosidase by glutathione at intracellular concentrations. J Am Chem Soc 126:13987–13991
206. Kyriacou SV, Brownlow WJ, Xu XH (2004) Using nanoparticle optics assay for direct observation of the function of antimicrobial agents in single live bacterial cells. Biochemistry 43:140–147
207. Seferos DS, Giljohann DA, Hill HD, Prigodich AE, Mirkin CA (2007) Nano-flares: probes for transfection and mRNA detection in living cells. J Am Chem Soc 129:15477–15479
208. Giljohann DA, Seferos DS, Daniel WL, Massich MD, Patel PC, Mirkin CA (2010) Gold nanoparticles for biology and medicine. Angew Chem Int Ed 49:3280–3294
209. Sandhu KK, McIntosh CM, Simard JM, Smith SW, Rotello VM (2002) Gold nanoparticle-mediated transfection of mammalian cells. Bioconjug Chem 13:3–6
210. Ravi Kumar MN, Sameti M, Mohapatra SS, Kong X, Lockey RF, Bakowsky U, Lindenblatt G, Schmidt H, Lehr CM (2004) Cationic silica nanoparticles as gene carriers: synthesis, characterization and transfection efficiency in vitro and in vivo. J Nanosci Nanotechnol 4:876–881
211. Lee JS, Lytton-Jean AK, Hurst SJ, Mirkin CA (2007) Silver nanoparticle-oligonucleotide conjugates based on DNA with triple cyclic disulfide moieties. Nano Lett 7:2112–2115
212. Cutler JI, Zheng D, Xu X, Giljohann DA, Mirkin CA (2010) Polyvalent oligonucleotide iron oxide nanoparticle "click" conjugates. Nano Lett 10:1477–1480
213. Xue C, Chen X, Hurst SJ, Mirkin CA (2007) Self-assembled monolayer mediated silica coating of silver triangular nanoprisms. Adv Mater (Weinheim, Ger) 19:4071+
214. Mitchell GP, Mirkin CA, Letsinger RL (1999) Programmed assembly of DNA functionalized quantum dots. J Am Chem Soc 121:8122–8123
215. Cutler JI, Zhang K, Zheng D, Auyeung E, Prigodich AE, Mirkin CA (2011) Polyvalent nucleic acid nanostructures. J Am Chem Soc 133:9254–9257

Fast Force Clamp in Optical Tweezers: A Tool to Study the Kinetics of Molecular Reactions

Pasquale Bianco, Lorenzo Bongini, Luca Melli, Giulia Falorsi, Luca Salvi, Dan Cojoc and Vincenzo Lombardi

Abstract A dual-laser optical tweezers has been developed to study the mechanics of motor proteins or DNA filaments. A bead attached to one end of the specimen is trapped in the confocal point of the two lasers, while the other end is connected to a three-dimensional piezo-stage. The instrument can be operated under computer control either as a length clamp, applying length steps or ramps, or as a force clamp, applying abrupt changes in load of fixed magnitude and direction. The dynamic range of the instrument (0.5–75,000 nm in length and 0.5–200 pN in force) and the speed of the force feedback permit recording the kinetics of molecular and intermolecular phenomena such as the overstretching transition in double-stranded DNA (ds-DNA) or the generation of force and shortening by an ensemble of myosin motors pulling on an actin filament. We demonstrate the performance of the system

P. Bianco · L. Bongini · L. Melli · G. Falorsi · L. Salvi · V. Lombardi (✉)
Laboratory of Physiology, Department of Biology, University of Florence, Florence, Italy
e-mail: vincenzo.lombardi@unifi.it

P. Bianco
e-mail: pasquale.bianco@unifi.it

L. Bongini
e-mail: lnzbng@gmail.com

L. Melli
e-mail: luca.melli@unifi.it

G. Falorsi
e-mail: giuliafalorsi@gmail.com

L. Salvi
e-mail: luca.salvi1@stud.unifi.it

D. Cojoc
IOM—National Research Council, Trieste, Italy
e-mail: cojoc@iom.cnr.it

F. Benfenati et al. (eds.), *Novel Approaches for Single Molecule Activation and Detection*, Advances in Atom and Single Molecule Machines,
DOI: 10.1007/978-3-662-43367-6_7,

in recording for the first time the transient kinetics of the ds-DNA overstretching transition, which allows the determination of the underlying reaction parameters, such as rate constants and distance to the transitions state.

1 Single-Molecule Mechanics with Optical Tweezers

Optical traps (or tweezers) are an efficient method for studying the properties of biological systems such as motor proteins [18, 25, 37], nucleic acid structures [24, 34], and processive enzymes [28, 48], which work in the nanometer-piconewton scale. The method allows recording molecular or intermolecular events one at a time, which is otherwise impossible from molecule ensemble experiments in cell or tissue. In single-molecule mechanical experiments, the light from a focused laser beam is used to trap a microscopic dielectric bead (made of silica or polystyrene) in a three-dimensional potential well centered near the focal point [3]. For a few hundreds of nanometers displacement from the equilibrium position, the trapping potential is harmonic, the restoring force varies linearly with the displacement, and the optical trap can be approximated to a linear spring. The spring constant, or stiffness, depends on the steepness of the optical gradient, i.e., how tightly the laser is focused and the laser power. In addition to the gradient force, there is a scattering force, due to the reflection on the bead surface, directed along the beam direction, which results in a shift of the equilibrium trapping position slightly past the focus. The high intensity of the trapping laser near the focus produces local heating, which is minimized using transparent dielectric beads and laser power not larger than 100 mW. To minimize photodamage of the biological specimen, laser wavelengths in the near infrared (800–1,100 nm) are preferred. With a laser power of 100 mW, the maximum force attained before exiting the linear region is around 50 pN and force resolution is approximately 0.1 pN.

The properties of the biological system under study are investigated by attaching the molecules of interest to the microscopic bead to bring it into contact with a fixed partner and measure the force and the displacement generated by the interaction. This type of assay has been first applied to the mechanical characterization of optically trapped motor protein kinesin moving along fixed microtubules (Fig. 1a; [6, 38, 42]). Processive motors, such as RNA polymerase, kinesin, and myosin V, spend a large part of the ATPase cycle time attached to their track, undergoing several steps from the start of the interaction, so that it is easy to measure the movement produced by a single motor and its speed in relation to the load carried by the motor. Muscle myosin II is not a processive motor and spends only a small fraction of the ATPase cycle time attached to the actin filament. Motor systems based on myosin II are organized in arrays (as it occurs for the myosin filament in the muscle sarcomere), and by this collective organization, they cooperatively maintain continuous interactions with the actin filament, producing steady force and filament sliding at high velocity. However, in single-molecule experiments, the

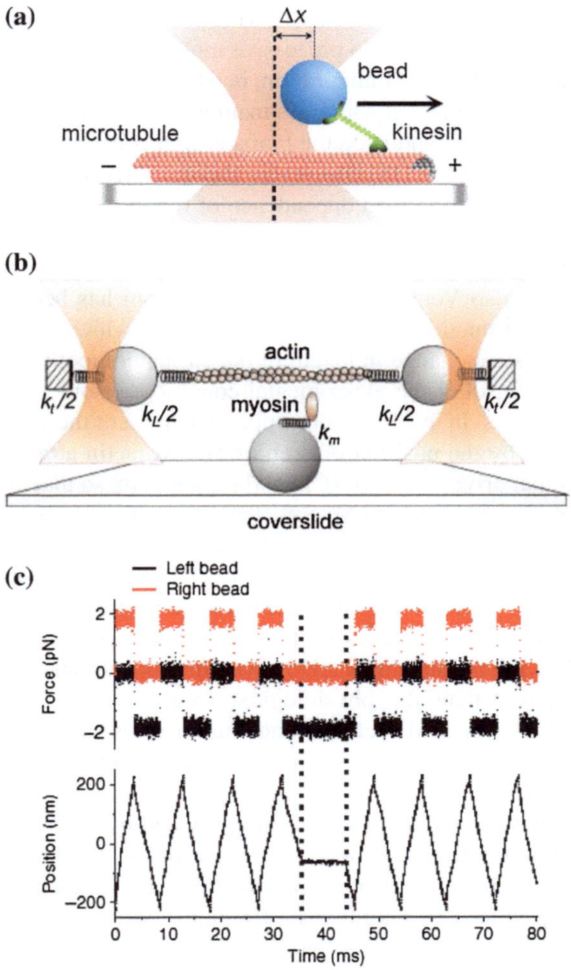

Fig. 1 **a** Motility assay used to record the activity of the processive motor kinesin (*green*) stepping along a microtubule (*red*). The distance between bead (*blue*) and trap (*pink*), Δx, was fixed at 175 nm, corresponding to a load of 6.5 pN, by moving the trap with AOD during the stepping. From Visscher et al. [42]. **b** Three-bead assay (*TBA*) to record the interaction of a myosin II motor with an actin filament. Both traps are static, and the interaction is recorded by the reduction of noise in the bead position due to the increase in stiffness of the system for the addition in parallel to the link stiffness (k_L) and trap stiffness (k_t) of the myosin stiffness (k_m). Modified from Capitanio et al. [10]. **c** Records of force on the left and the right beads (*top*) and position of one trap (*bottom*) in a TBA with fast force clamp. The force command that drives the AODs is switched periodically between positive and negative to keep the dumbbell within a confined spatial interval (±200 nm). The dumbbell stops when a myosin motor binds to the actin filament (between the *dotted lines*). From Capitanio et al. [11]

fast intermittent interactions of myosin II with actin cannot be resolved with the same trapping technique applied to processive motors. This limitation is overcame in the three-bead assay (TBA) configuration, in which the laser is split in two to trap the two ends of an actin filament which are manipulated by means of Acousto-Optic Deflectors (AOD) to bring and keep the filament into contact with a myosin motor attached to a third fixed bead [18]. However, also the TBA proved to be inadequate to measure the force and the load dependence of the movement generated by one myosin motor because the rise of force during a single myosin–actin interaction is largely affected by the low stiffness of attachments of the actin filament to the beads (k_L in Fig. 1b; see also Veigel et al. [41]). The system has been recently implemented with the introduction of the force clamp mode, which fixes the load on the bead and eliminates the change in length of the elastic components of the systems. This enables the control of the experimental conditions (isometric/isotonic) under which the motor works. In this ingenious implementation of the TBA [11], the force signal (determined by the position of each of the two actin attached beads) is the feedback signal that drives the two AOD in the laser path, so that the actin filament moves at the velocity necessary to generate the viscous force equal to the command force. When the myosin attaches to the actin filament, the filament stops moving as the force is now exerted by the myosin (Fig. 1c). The analysis of filament velocity reveals the duration of actin–myosin interactions and the associated length step. The system allows the mechanics of the event to be resolved within the first millisecond following the attachment. However, the concentration of ATP is kept at least 40 times lower than the physiological value to increase the lifetime of the events and make them detectable, and this in turns limits the power of the method for the determination of the load-dependent kinetics of the myosin–actin interaction.

Steady force and shortening in muscle are provided by cyclic actin attachment/detachment of myosin II motors working in parallel in the half-sarcomere. The collective nature of this motor can be studied at the nanometer scale in a synthetic machine made by a linear array of myosin II motors carried by a structured surface and brought to interact with an actin filament. In this way, the myosin motors work in parallel like in the half-sarcomere and provide the condition for cyclic interactions with the actin filament. The mechanical outputs of this collective motor would be in the range 0.5–200 pN force and 1–10,000 nm movement. To measure and control the mechanical performance of such machines is beyond the possibility of the single-laser optical tweezers considered so far. The range of the needed movement is above what can be attained with AODs, but this limit can overcame keeping fixed the trap position and using a piezo-nanopositioner to move the trapping chamber. The other major limit is the reduced force range of the single laser. This is a general problem that prevents the application of the technique for studying all molecular and intermolecular functions that occur at forces >50 pN, such as unfolding of proteins or structural transitions in nucleic acids, and has given way to the double-laser optical tweezers and to AFM-based mechanics.

AFM has the advantage of high-resolution imaging, but, as far as the applications of interest in this report, has limits in the lower force range sensitivity ($\sim$10 pN) and, because of the large dimension of the probe, in the possibility to discriminate between specific and non-specific interactions.

2 The Dual-Laser Optical Tweezers

The dual-laser optical tweezers technique enhances the upper force limit utilizing two counter-propagating beams [2, 34, 35]. The instrument described here, which operates in a dynamic force range 0.5–200 pN, has been developed in our laboratory to study the structure of DNA and the mechanical performance of a motility system made of an ensemble of myosin motors pulling on an actin filament. In this design, two microscope objectives face each other and focus two separate laser beams to the same spot. Since the scattering force on the bead is approximately the same for each laser, these forces cancel out each other and the axial trap stability is greatly enhanced. The dual-laser optical tweezers can therefore generate higher trapping forces for a given laser power. In this case, the upper force limit reaches 200 pN with 100 mW power, while the beams size can be reduced to fill only partially the back focal plane of the objectives, so that all the photons leaving the trapped bead are efficiently collected with the opposite objective lens and are projected onto a position-sensing photodiode. Under these conditions, the force can be measured by the change in light momentum measured by the deflection of the laser beams (Fig. 2a), a method that has the advantage to depend only on the shift of the refracting object from the equilibrium position and not on the size, shape, and, to a give extent, refractive index of the object [35].

While a single-laser optical tweezers can be constructed using a commercial inverted microscope, a dual-laser instrument is typically custom built from opto-mechanical components. The two laser beams must be aligned to within less than a bead diameter, and the resulting measurements must be corrected for errors due to the drift in the relative beam alignment. Because of these alignment issues, a dual-laser tweezers can hardly be used with beads of diameters less than 1 μm. The bead trapped in the focus of the two laser beams is in a flow-through micro-chamber, the position of which is controlled by means of a low-profile three-axes piezo-stage with capacitive position sensing (Mad City Lab, Madison WI, USA). With this device, the position of the specimen is digitally controlled with the precision of 0.5 nm over a volume range 75 $\times$ 75 $\times$ 50 μm.

Data acquisition and control of the piezoelectric stage are done by means of a NI PCI-6251 board driven by LabVIEW software (National Instruments, Austin TX, USA). Joystick control of the stage allows the instrument to be remotely operated. The electronics of the servo loop is implemented to provide position steps complete in 1 ms.

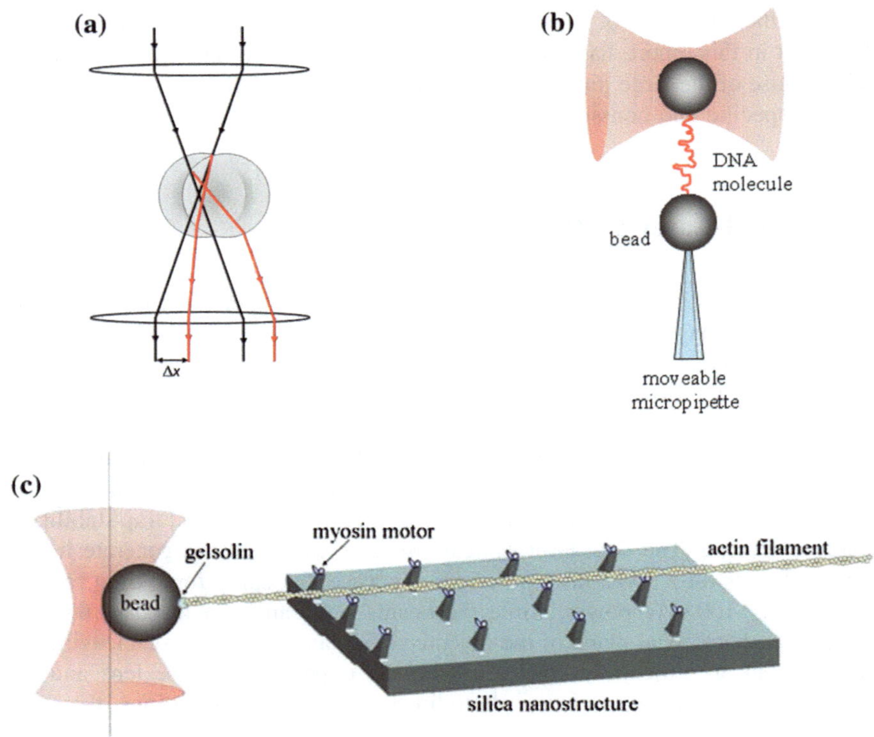

Fig. 2 **a** Change in light momentum (from *black* to *red*) produced by change in the position of the trapped bead and measured by the change in position of the laser beam (Δx) in the back focal plane of the objective. **b** Application of the dual-laser tweezers to mechanics of a DNA molecule (*red*). The ends of the molecule are attached to polystyrene beads one of which is held by suction on the tip of a glass micropipette, while the other is held in the optical trap. **c** Application of the dual-laser tweezers to mechanics of a biomachine made by an array of myosin II motors disposed on a structured silica surface and an actin filament attached with the right polarity (via gelsolin, [36]) to the trapped bead

2.1 Mounting the Specimen in the Experimental Chamber Between the Force and Length Transducers

The flow chamber is assembled by placing two layers of parafilm cut in the chamber shape between two microscope coverslips, to have a chamber depth of 150 μm. The design of the flow chamber is optimized for manipulation and dynamic force and extension measurements of a single double-stranded DNA (ds-DNA) molecule, to determine the kinetics of the overstretching transition from the basic conformation to the 1.7 longer conformation [13, 34]. The chamber is clamped on a holder carried by the piezoelectric stage with two aluminum brackets with threaded holes to connect tubing to the chamber. The flow chamber allows maintaining a laminar flow that, during the experiment, can be used to provide jumps in the composition of the solution. A micropipette is placed between the two parafilm layers prior to sealing and thus is integral with the flow chamber and the piezo-stage and its position is controlled with sub-nanometer precision. Two streptavidin-coated beads are held: one by the optical trap and the other by the tip of the micropipette through suction (Fig. 2b). A diluted solution of DNA with biotinylated 3' ends is flown through the cell, until one end of a DNA molecule is attached to the trapped bead; then, the bead on the pipette is moved toward the trapped bead until the opposite end of the molecule is bound. Changes in the extension of the molecule are measured by the movements of the two beads. The absolute extension of the molecule is estimated at the end of the experiment by moving the pipette toward the trapped bead and measuring the position of the pipette, with 5-nm precision, when the two beads start to separate as judged from the force signal.

The temperature in the chamber is controlled in the range 4–40 °C by pumping fluid at the desired temperature through copper jackets placed on the objectives [26]. The time to reset the temperature of the solution after solution exchange is 3–5 s. The vibrations due to the fluid circulation pump, tested with power spectrum measurements, are minimized by tightly fitting the jackets around the objectives and by mechanical insulation of the water circulator from the optical table.

For the application of the dual-laser tweezers to the study of the mechanical performance of a motility system made by an ensemble of myosin motors pulling on an actin filament, the design of the chamber and the connections of the specimen to the transducers have to be specifically adapted.

The actin filament is attached with the correct polarity (the barbed or + end) to the bead acting as the force transducer and the structured surface with the array of myosin motors is brought by a support fixed to the chamber carried by the piezo-stage (Fig. 2c).

2.2 Calibration of Force

The force is recorded by the position of the trapped bead, measured by the change in the light momentum (Fig. 2a) with a precision of ~0.2 pN. The force calibration is made by determining the trap stiffness in response to known forces of viscous or thermal nature. Actually, to increase the accuracy, the calibration must be done using both viscous drag (1) and thermal noise (2), as follows:

(1) The force due to viscous drag on a sphere of known radius can be calculated according to Stokes' law. If a liquid with viscosity η flows past a sphere of radius R with velocity v, the force due to viscous drag, F_v, is given by:

$$F_v = \gamma v = 6\pi\eta Rv, \qquad (1)$$

where γ is the viscous drag coefficient. Viscous drag to the trapped bead is applied by moving at preset velocities the flow chamber with the piezo-stage. Because the movement of the flow chamber involves the movement of the entire block of fluid surrounding the trapped bead, the bead is exposed to preset viscous drag forces.

(2) The second technique to calibrate trap stiffness is by measurement of the Brownian motion of the trapped bead. A bead captured in an optical trap experiences random forces due to thermal fluctuations. By Fourier analysis of the positional fluctuations of the trapped bead, the trap stiffness (k) can be directly obtained as

$$k = 2\pi\gamma f_c, \qquad (2)$$

where f_c is the corner frequency, the frequency above which the system fails to respond and the position fluctuation is attenuated, as determined by the power spectrum. The viscous drag method involving the movement of the piezo-stage has the advantage of providing a measure of the trap stiffness at large displacements.

2.3 Mechanical Protocols: Length Versus Force Clamp

The instrument can be operated under computer control either as a length clamp, applying length steps or ramps with the three-dimensional piezo-stage connected to one end of the specimen, or as a force clamp, applying abrupt changes in load of fixed magnitude and direction to the trapped bead connected to the other end of the specimen (Fig. 3). In a length clamp experiment, the feedback signal is the position of the piezo-stage and the state of the specimen can be perturbed by lengthening or shortening steps with a risetime ($t_{r,}$ the time to 95 % of the step) that is limited only by the frequency response of the piezo-stage (2.5 kHz in our improved version). In this case, a stepwise perturbation can be much faster than the elicited molecular reaction, allowing the force response to be analyzed in terms of the elastic response, simultaneous with the step, and the following molecular relaxation to the new equilibrium. However, the interpretation of the reaction elicited by the step is complicated by the changes of the potential energy landscape generated by the molecular motion against a changing load. In a force clamp experiment, the feedback signal is the instantaneous position of the trapped bead connected to the other end of the specimen and the feedback system moves the piezo-stage to achieve and maintain a set position of the trapped bead. The force clamp is a very powerful tool and is the protocol of election in the experiments described here, for two important reasons. The first reason is that the molecular or intermolecular process elicited following a step to a given force level occurs in isotonic conditions and thus without the effect of any series compliance between

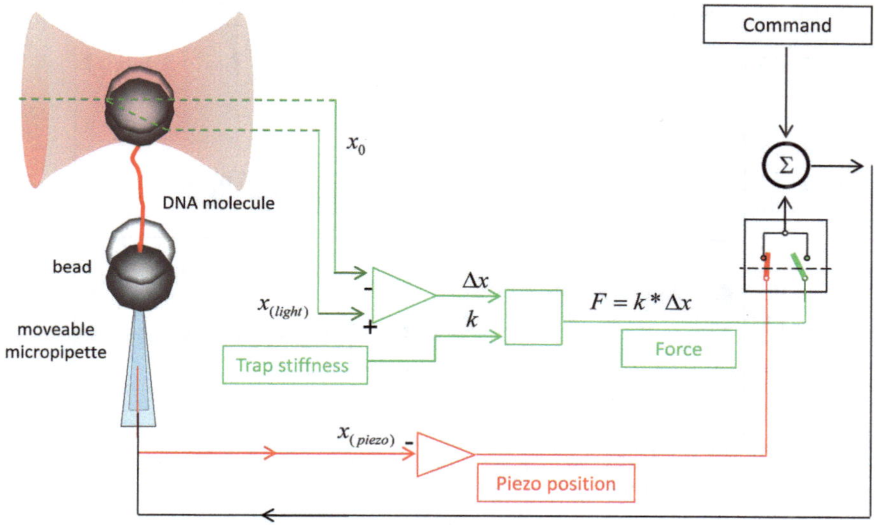

Fig. 3 Block diagram of the electronic circuit that controls the position of the piezo-stage (*black*) in the dual-laser tweezers for DNA mechanics. The summing amplifier (Σ) compares the command with either the force (*green*, calculated by the product of the change in light momentum Δx and the stiffness of the trap k) or the length of the molecule (*red*, measured by the position of the micropipette integral to the piezo-stage, x_{piezo}). The switch selects either the length clamp (position shown) or the force clamp

the specimen and the force transducer (the trap) and the length transducer (the support connected to the piezo-stage), which otherwise would require corrections for the actual movement of the specimen. The second reason is that, following the step, the force and thus the potential energy landscape remain constant and the reaction is followed through the change in length, which serves at the same time as the reaction coordinate. The two-state reaction nature of a given event can be defined by measuring the force dependence of the lifetimes (or rate constants) of the conformational change along the reaction coordinate, and consequently, the relative energy profile and reaction distance can be described.

However, the force clamp suffers of the main limit that the finite response time of the feedback loop (τ_f) reduces the frequency response of the system to <1 kHz. Beyond the frequency response of the piezo-stage, τ_f depends on the trap stiffness, bead size, compliance of tethers connecting the specimen to the transducers, and eventually the specimen compliance. In a recent paper, the problem is described in relation to the folding/unfolding of DNA hairpin (Fig. 4, from Elms et al. [16]). The reduced frequency response in force clamp, with respect to length clamp, causes the failure to record short-lived dwell times in the equilibrium behavior of the two-state reaction. As a consequence, the load dependence of the rate constants obtained from the analysis of the respective lifetimes, interpreted in terms of Bell theory, provides estimates of the distance to the transition state that are

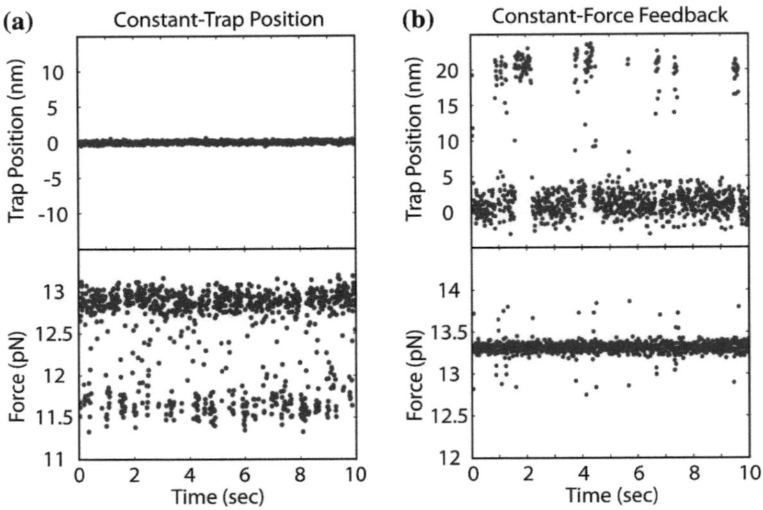

Fig. 4 **a** Record of force fluctuations of a DNA hairpin undergoing folding/unfolding in length clamp. **b** Record of length fluctuations in force clamp. Data averaged down to 100 Hz. From Elms et al. [16]

significantly larger (33 %) than the observed change in the molecular extension. Thus, missing to record the faster events affects the definition of the structure of the reaction producing an overestimate of the reaction distance, and the error increases with the fraction (f_m) of fast events missed. f_m depends on the relative timescales of the molecular reaction under study and the response time of the feedback system τ_f, as defined by:

$$f_m = 1 - \exp\left(-\tau_f k_A \exp\left(\frac{\Delta F_{A-B}\,\Delta x^{\ddagger}}{k_B T}\right)\right) \quad (3)$$

where k_A is the rate constant in state A, ΔF is the change in force between the states A and B, $\Delta x^{\ddagger}$ is the distance to the transition state, k_B is the Boltzmann constant, T is the absolute temperature, and τ_f is the response time of the system. As mentioned above, τ_f depends on the sum of the compliances of the series of elements that constitute the feedback loop, included the compliance of the specimen under study that may change in the different states.

In addition to the limit described above, occurring with force clamp at constant force, there is another major source of error that shows up in force feedback protocols with force steps.

Due to the finite-time limitations of the force feedback, the system can only follow the stepwise command signal on a timescale greater than the timescale of the feedback. Force deviations from the command due to the reaction of the specimen that occurs at shorter timescales would modify the length response in the

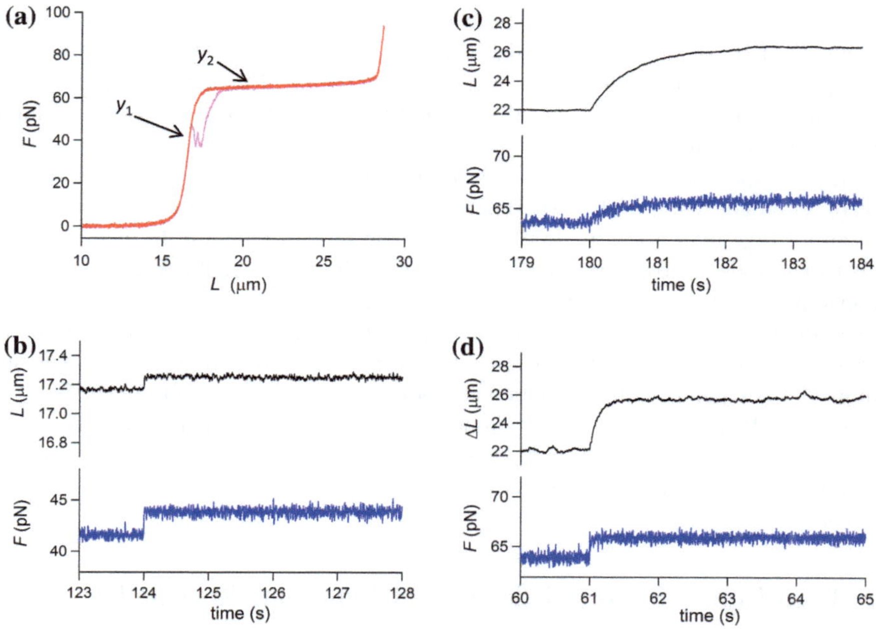

Fig. 5 **a** Force–extension curve of a ds-DNA molecule in physiological solution (150 mM NaCl) and 25 °C. Lengthening (*red*) and shortening (*pink*) are imposed at velocity of 1.4 μm/s. This molecule shows hysteresis under shortening. **b** Length response (*black*) to a 2 pN step (*blue*) imposed on a steady force of 42 pN corresponding to y_1 in **a**. **c** Length response (*black*) to the 2 pN step (*blue*) imposed on a steady force of 63 pN (corresponding to y_2 in a) with the same feedback gain as in **b**. **d** Length response (*black*) to the same step as in **c** (*blue*), but with the feedback gain adjusted to account for the drop in stiffness of the molecule in the overstretching transition region

direction that produces an underestimate of the rate of the length change expected by the commanded force.

To illustrate this point, consider a molecule of ds-DNA on which a force step of 2 pN is superimposed on a steady force level in correspondence of two different regions of the characteristic force–extension (*F–L*) curve (Fig. 5a): y_1(42 pN), in the region of intrinsic elasticity of the molecule, and y_2(63 pN), in the region of the overstretching transition. The command signal has a t_r of 1 ms, and the feedback gain is set for the optimization of the force step (t_r ~ 2 ms) in the elastic region. In this case (Fig. 5b), the length response is an almost simultaneous length change of ~60 nm revealing the elastic properties of the molecule in the basic conformation, with a compliance of ~30 nm/pN. When the 2 pN step is superimposed on force y_2 (Fig. 5c), the feedback signal is deteriorated, with a t_r that is increased to ~1 s. The amplitude of the length response is increased by one order of magnitude in comparison with that expected from the compliance in b and has a time course almost parallel to that of the force perturbation. In Fig. 5d, the same force step command is imposed with the feedback gain, and thus the complex stiffness of the feedback loop, adequately increased to counteract the increased compliance of the molecule in the overstretching region. The consequent reduction in τ_f allows the stepwise shape of the force perturbation to be recovered, which reveals the exponential nature of the lengthening response.

In the experiment described in Fig. 4, fluctuations of a single two-state reaction were recorded, while in the experiment of Fig. 5, we recorded the time evolution of an average over a multitude of copies of the unitary reaction. In this case, the artifact introduced by finite-time limitations of the force feedback consists in an apparent slowing of the transient kinetics (Fig. 5c). Like in the single reaction experiment, in order to check the absence of any artifact in the definition of transient kinetics, the structural information obtained from the transient analysis might be compared with that obtained from a standard equilibrium analysis.

The application of a fast force clamp to the study of the kinetics and energetics of the overstretching transition of ds-DNA is described in the next section, which summarizes the work of Bianco et al. [5].

3 Transition Kinetics of the DNA Elongation Revealed by Force Steps

To describe the coupling between structure and force in ds-DNA is fundamental for understanding the mechanism of the molecular machines used by cells to duplicate and repair their genome and modulate the accessibility of the genetic information [1, 9]. When stretched under forces lower than 60 pN, ds-DNA maintains its basic conformation (B-form) and displays the elastic response of an extensible worm-like chain [8, 27, 33, 43]. At forces higher than 60 pN, the molecule abruptly becomes much more extensible (overstretching transition) and within a few piconewtons acquires an extended conformation (S-form) 1.7 times longer than the B-form [13, 34].

Although the overstretching phenomenon has long been known, a detailed description of its mechanism is still missing. There are contributions [29, 30, 32, 40, 44, 46, 47] in favor of the idea that the elongation could merely be due to a force-induced melting with strand separation, as indicated by the hysteretic behavior in consecutive stretching and shortening cycles. However, other experiments support the view that overstretched DNA is characterized by a reduced helicity [7, 22, 31] with a stiffness higher than expected for single stranded (ss-) DNA [14, 45]. In this case, melting should occur at nicks in the phosphate backbone [34, 43] or at the free DNA ends, by an extent that depends on the bp sequence, solution conditions and manipulation of the molecule [12, 15, 19, 20, 40, 47]. In particular, it has been shown in force clamp experiments [19, 20] that DNA overstretching under melting-preventing conditions (such as high salt concentration, GC-rich sequences) involves only a rapid non-hysteretic transition to the elongated double-stranded S-form, while removing these conditions induces also a slow hysteretic strand separation.

All the above studies have a limit in defining the nature of the overstretching transition, because they use the equilibrium force–extension relation and do not have the time resolution for determining the non-equilibrium kinetics. The development of our dual-laser tweezers with a sufficiently fast force clamp has provided for the first time the opportunity for recording the length transient elicited by force steps and its dependence on the load, providing the constraints for the definition of the structure of the unitary reaction.

In the experiment of Fig. 6a, a staircase of force steps of amplitude 2 pN is imposed on a molecule of ds-DNA with ends opened and kept in a solution with physiological salt composition and room temperature. In the region of the overstretching transition (between 60 and 72 pN, Fig. 6b), the length response to a 2 pN step is characterized by an exponential time course (Fig. 6c), which implies per se that the main transition mechanism cannot be force-induced melting starting at the free ends and then propagating inward [40], since this process would be characterized by linear kinetics.

Five to eight steps of 2 pN are necessary to complete the B–S transition. Superposing the elongations elicited by a series of six 2-pN steps, identified in Fig. 6c by the different colors, it can be seen how the amplitude and the speed of the response depend on the force attained at the end of each step. Going from the first (purple) to the sixth (green) step, the amplitude of the elongation (ΔL_e) increases abruptly up to a maximum of ~3.3 μm (third step, black) and then reduces again. After the elastic response has been subtracted, the time course of the elongation (ΔL) is fitted with the exponential equation $\Delta L = \Delta L_e \cdot (1 - \exp(-rt))$, where t is the time elapsed after the step, r is the rate constant, and ΔL_e is the asymptotic value of the elongation. ΔL_e is smaller, and r is larger at the beginning and at the end of the B–S transition. The dependence of r on the force attained after the 2 pN step, shown by blue symbols in Fig. 6d, is U-shaped, showing maxima of ~50 s^{-1} at the beginning ($F \sim 62$ pN) and at the end ($F \sim 72$ pN) of the overstretching transition and a minimum of $3.82 \pm 0.68\ s^{-1}$ (mean and SD, calculated in the range 65.5–66.5 pN) in the plateau region. If the size of the step is reduced to 0.5 pN, the amplitude of the elongation response reduces and the number of steps necessary to complete the B–S transition increases, but the r–F points (green circles) superpose on those obtained with 2 pN steps (blue symbols). Thus, the r–F relation depends uniquely on the final force attained by the step and is independent of the size of the step and, moreover, of the occurrence of some degree of melting, as shown by the hysteresis upon relaxation (compare blue squares, from 4 molecules without hysteresis, and triangles, from the 13 molecules with hysteresis).

In force clamp experiments with the dual-laser tweezers apparatus, the length change elicited by a force step is realized through movement of the piezo-stage and thus of the fluid surrounding the bead. Consequently, the change in position of the trapped bead reliably measures the tension on the molecule only if it is not influenced by a significant drag due to the movement of the solution accompanying

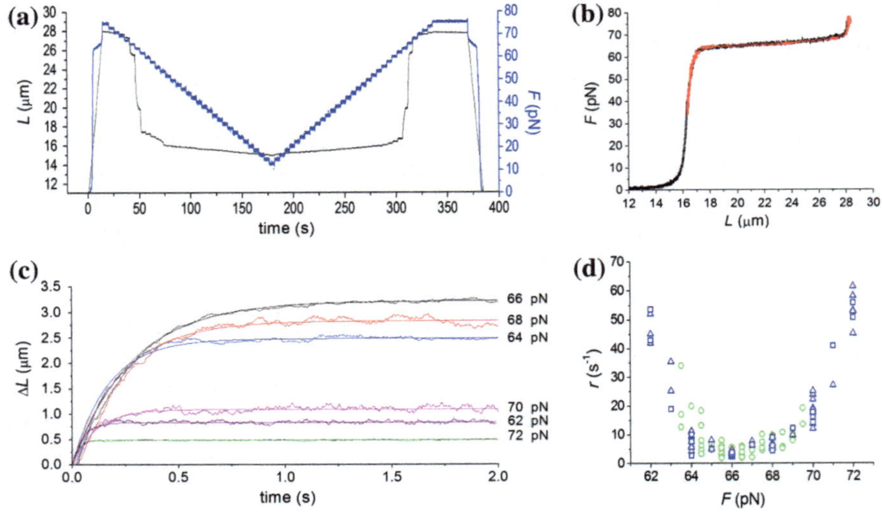

Fig. 6 **a** Time course of the length (*black*) and force (*blue*) of ds-DNA first stretched in length clamp (velocity 1.4 μm/s) then shortened in force clamp with a staircase of 2 pN steps at 5 s interval, then stretched again with a similar staircase of 2 pN steps and eventually shortened in length clamp at velocity 1.4 μm/s. **b** Force–extension curves in length clamp (*black*) and in force clamp (*red*) drawn from a protocol like that in **a** for a molecule without hysteresis. **c** Superimposed time courses of elongation (ΔL) following a series of 2 pN force steps starting at 60 pN during the staircase. The level of force attained by the step is reported next to the trace. The lines are single exponential fits to the traces. **d** Relationship between the rate of elongation (r) and force attained at the end of the step (F) following 0.5 pN (*green circles*) and 2 pN (*blue symbols*) steps. 2 pN data are from 13 molecules showing hysteresis in relaxation (*triangles*) and 4 molecules without hysteresis (*squares*). From Bianco et al. [5]

the movement of the stage. Moreover, there may be a significant rotational drag while ds-DNA elongates/untwists during the overstretching transition, which could affect the rate of elongation. All these problems are discussed in detail in the Appendix, showing the absence of any significant influence of viscosity. Under these conditions, the r–F relation is an intrinsic property of the molecule, related to its transition kinetics.

4 The Structure of the DNA Overstretching Transition

The results of non-equilibrium kinetics just described are the basis for the definition of a two-state reaction model that predicts the structure of the overstretching transition. The model assumes that ds-DNA is composed of an ensemble of units which can attain two different conformational states, a compact B state (with a

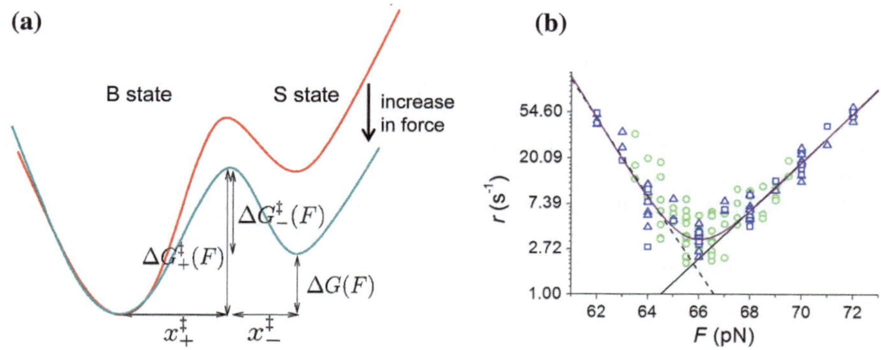

Fig. 7 **a** Free-energy profile of a two-state unit and the effect of an external force (from *red* to *blue*). The force *tilts* the energy landscape and thus lowers the energy barrier for the transition to the extended state. $\Delta G^{\ddagger}_{+}$, energy barrier for the forward transition; $\Delta G^{\ddagger}_{-}$, energy barrier for the backward transition; ΔG, free-energy difference between the two states. The transition state distance is $x^{\ddagger}_{+}$ from the B state and $x^{\ddagger}_{-}$ from the S state. **b** Relation between ln r and F. (*Black straight lines*): Fits of Eq. 7 (*continuous line*) to data in the force range 67–72 pN and Eq. 8 (*dashed line*) to data in the range 61–65 pN. (*Purple line*) four-parameter fit with Eq. 9 to all data. From Bianco et al. [5]

molecular extension of 0.33 nm per bp) and an extended S state (with a molecular extension of 0.56 nm per bp). The molecular extension of each of these units provides a convenient reaction coordinate to study the overstretching transition. The free-energy profile of each unit along this reaction coordinate, schematically represented in Fig. 7a, is dictated by four fundamental parameters: the free-energy difference between B and S states ΔG, the forward energy barrier $\Delta G^{\ddagger}_{+}$ (the backward barrier $\Delta G^{\ddagger}_{-}$ being $= \Delta G^{\ddagger}_{+} - \Delta G$), and the distances of the transition state from both the B state $x^{\ddagger}_{+}$and the S state $x^{\ddagger}_{-}$.

When a force F is applied to the molecule, the free-energy profile tilts (from red to blue in Fig. 7a), leading to a decrease in the free-energy difference ΔG and a decrease of the free-energy barrier for stretching and a corresponding increase of the free-energy barrier for shortening:

$$\Delta G^{\ddagger}_{\pm} = \Delta G^{\ddagger}_{\pm} \mp F x^{\ddagger}_{\pm}. \tag{4}$$

The rate constants for the elongation (k_{+}) and the shortening (k_{-}) of the molecule depend exponentially on the force according to Kramers–Bell theory [4, 17, 21]:

$$k_{+} = A_{+} \cdot \exp\left(\frac{F \cdot x^{\ddagger}_{+}}{k_{\mathrm{B}} T}\right) \tag{5}$$

$$k_- = A_- \cdot \exp\left(-\frac{F \cdot x_-^{\ddagger}}{k_\mathrm{B}T}\right), \tag{6}$$

where k_B is the Boltzmann constant, T is the temperature in Kelvin, and A_+ and $A_-(\mathrm{s}^{-1})$ are the rate constants at zero force.

The unidirectional rate constants expressed in logarithmic units have a linear dependence on force, and the slope of the relation is the distance from the starting state to the transition state divided by $k_\mathrm{B}T$:

$$\ln k_+ = \ln A_+ + \left(\frac{F \cdot x_+^{\ddagger}}{k_\mathrm{B}T}\right) \tag{7}$$

$$\ln k_- = \ln A_- - \left(\frac{F \cdot x_-^{\ddagger}}{k_\mathrm{B}T}\right). \tag{8}$$

In our experiments, the observed rate constant r (Fig. 6d) is the sum of the forward and backward rate constants. However, since the equilibrium is dominated by k_- in the low force side of the overstretching transition and by k_+ in the high force side, we used the logarithmic relations in these two regions to separately estimate the slopes of the respective unidirectional reactions and thus of the distances to the transition state (Fig. 7b). The parameters of the linear regression equations fitted to ln r with Eq. 7 in the force range 67–72 pN and Eq. 8 in the range 61–65 pN are reported in Table 1 either for the 0.5 pN steps (green circles) or for the 2 pN steps (blue symbols). It can be seen that all the kinetic parameters are not significantly influenced by the force step size ($P > 0.1$). The linear fits on the pooled 0.5 and 2 pN data within the same force ranges give similar values.

Alternatively, a more complex, four-parameter fit with the expression

$$r = k_+ + k_- = A_+ \exp\left(\frac{Fx_+^{\ddagger}}{k_\mathrm{B}T}\right) + A_- \exp\left(\frac{Fx_-^{\ddagger}}{k_\mathrm{B}T}\right) \tag{9}$$

can be performed on the entire force range (purple line in Fig. 7b) and gives values of the four parameters A_+, A_-, $x_+^{\ddagger}$, and $x_-^{\ddagger}$ (Table 1) that are almost identical to those determined by fitting separately the low and high force branch of the relation.

The distances to the transition state, $x_+^{\ddagger}$ and $x_-^{\ddagger}$, are similar showing that the transition occurs roughly midway between the B state and the S state of DNA, slightly shifted toward the B state. The sum of the two distances, $\left(x_-^{\ddagger} + x_-^{\ddagger}\right)$, gives a total length change of 5.85 ± 0.20 nm for the unitary reaction Δx. The ratio of Δx over the elongation undergone by each bp ($0.33 \times 1.7 - 0.33 = 0.231$ nm)

Table 1 Kinetic parameters of the two-state reaction estimated with Eqs. 7 and 8 (first three rows) and with Eq. 9 (fourth row)

	$\ln A_+$	$x_+^\ddagger$ (nm)	$\ln A_-$	$x_-^\ddagger$ (nm)
0.5 pN step	-32 ± 7	2.1 ± 0.4	64 ± 12	3.9 ± 0.8
2 pN step	-34.4 ± 1.3	2.18 ± 0.07	61 ± 3	3.7 ± 0.2
Pooled	-34.1 ± 1.1	2.18 ± 0.07	56 ± 3	3.7 ± 0.2
Pooled, four parameters	-34.1 ± 1.7	2.2 ± 0.1	59 ± 4	3.7 ± 0.2

measures the number of bp involved in the unitary reaction, or cooperativity coefficient, and is 25.32 ± 0.86.

The finding that the transition state is almost midway between the compact and extended states implies that the force (F_m) at which the relaxation rate is minimum also corresponds, within our resolution limit, to the force (F_e) at which the work done for the elongation (W_e) equals the free-energy difference between the compact and the extended states: $W_e = F_e \times \Delta x$. With $\Delta x = 5.85$ nm, W_e is (66 × 5.85 =) 386 zJ per molecule (or 236 kJ per mol), that at 25 °C corresponds to 94 k_BT. The average binding free energy per bp obtained from these mechanical measurements is ($F_e \times 0.231$ nm =) 15.25 zJ, corresponding to only 3.71 k_BT per molecule. Thus, the cooperative mechanism for DNA elongation increases the stability of the DNA structure in the B state by increasing the minimal free-energy change necessary for the elongation reaction to 94 k_BT.

5 Assessing the Force Clamp Description of DNA Overstretching Kinetics: Comparison Between Non-equilibrium and Equilibrium Results

As demonstrated by Elms et al. [16] for a single two-state reaction (Fig. 4), also for the time course of a multitude of copies of the unitary reaction finite-time limitations of the force feedback would produce an apparent slowing of the length transient. To check the absence of any artifact in the definition of transient kinetics, here the structural information obtained from non-equilibrium analysis is compared with that obtained from equilibrium analysis.

The unidirectional transition rates between the two conformations k_+ and k_- depend on an additional parameter Ω, a kinetic pre-factor which is related to the viscous drag experienced by the molecule in its motion along the reaction coordinate and to the shape of the molecular potential energy near the transition state. According to Kramers-Bells theory, the A_+ and A_- parameters are:

$$A_+ = \Omega \exp\left(-\frac{\Delta G_+^\ddagger}{k_B T}\right) \tag{10}$$

and

$$A_{-} = \Omega \exp\left(-\frac{\Delta G_{+}^{\ddagger} - \Delta G}{k_{\mathrm{B}}T}\right) \tag{11}$$

The force (F_{m}) corresponding to the minimal relaxation rate $r(F)$ is

$$F_{\mathrm{m}} = \frac{\Delta G - k_{\mathrm{B}}T \ln\left(x_{+}^{\ddagger}/x_{-}^{\ddagger}\right)}{x_{+}^{\ddagger} + x_{-}^{\ddagger}} = F_{\mathrm{e}} - \frac{k_{\mathrm{B}}T \ln\left(x_{+}^{\ddagger}/x_{-}^{\ddagger}\right)}{x_{+}^{\ddagger} + x_{-}^{\ddagger}} \tag{12}$$

where F_{e} is the coexistence force, the force at which the probabilities to reside in the B and S state (p_{B} and p_{S}) are equal.

Also, the equilibrium probabilities at each force are known:

$$p_{\mathrm{B}}(F) = \frac{1}{1 + \exp\left(-\frac{\Delta G - F\left(x_{+}^{\ddagger} + x_{-}^{\ddagger}\right)}{k_{\mathrm{B}}T}\right)} \tag{13}$$

and

$$p_{\mathrm{S}}(F) = \frac{\exp\left(-\frac{\Delta G - F\left(x_{+}^{\ddagger} + x_{-}^{\ddagger}\right)}{k_{\mathrm{B}}T}\right)}{1 + \exp\left(-\frac{\Delta G - F\left(x_{+}^{\ddagger} + x_{-}^{\ddagger}\right)}{k_{\mathrm{B}}T}\right)} \tag{14}$$

$p_{\mathrm{S}}(F)$ grows monotonically with F, having maximal first derivative at the coexistence force, $F_e = \Delta G/\left(x_{+}^{\ddagger} + x_{-}^{\ddagger}\right)$. At this force $\Delta G = F \cdot \left(x_{+}^{\ddagger} + x_{-}^{\ddagger}\right)$, that is, the work done for the elongation equals the free-energy difference between the B and the S state. We note that only for symmetric landscapes, $x_{+}^{\ddagger} = x_{-}^{\ddagger}$, F_{e} coincides with F_{m}, the force of minimal relaxation rate.

During a positive staircase of force steps, when the force is changed from F to $F + \Delta F$, the probability to be in the extended state grows and, after equilibration, $N(p_S(F + \Delta F) - p_S(F))$units have extended, where N is the total number of extensible units. It is therefore straightforward to compute the equilibrium force–extension profile which, according to the expressions for p_S and p_B, depends only on the two parameters ΔG and $x_{+} + x_{-}$. In Fig. 8, the experimental force–extension relation (black trace) from Fig. 6b is compared to the theoretical relation (red dashed line) obtained with the ΔG and $x_{+} + x_{-}$ chosen by fitting the elongation rates with the procedure described in the test. The agreement shows that the model developed to fit the observed relaxation kinetics is capable of reproducing the equilibrium force–extension relation, proving that our force clamp has the time resolution for correctly recording the transient elicited by force steps.

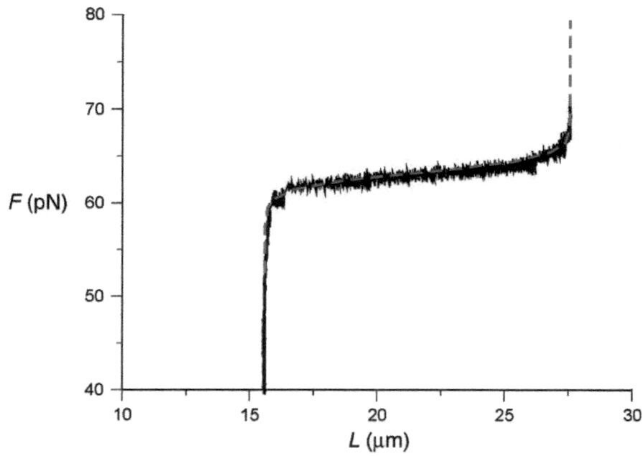

Fig. 8 Comparison between the experimental equilibrium force–extension relation (*black traces*) and the simulated relation (*red dashed trace*), calculated with the parameters extracted from the transient kinetics analysis of the responses to force steps. The base pair separations in states B and S were set to 0.325 and 0.57 nm, respectively. From Supplementary material in Bianco et al. [5]

Eventually, it is worth assessing whether the exponential kinetics observed for λ-DNA contrasts with the results of other recent overstretching experiments under force clamp. Fu et al. [19] report on constant force DNA overstretching experiments performed by means of magnetic tweezers. Under melting-preventing conditions (high salt concentration or moderately GC-rich sequences), DNA constructs of approximately 600 bp show, in response to force increase, discrete lengthening steps followed by length fluctuations, instead of a smooth exponential elongation (Fig. 9a). The predictions of our two-state model for such conditions are tested by performing a series of Monte Carlo simulations. A set of 23 two-state units, each 25 bp long, was defined in order to simulate a molecule 575 bp long. All units were initially assigned a compact conformation, and force was gradually increased with 1 pN steps. For each force value, the system was integrated for 20 s according to a stochastic dynamics where each unit could change its state from compact to extended and from extended to compact according to their transition probabilities computed as $k_{\pm}$ dt with dt the integration time. $k_{\pm}$ were computed according to Eqs. 5 and 6, using the parameters resulting from the fit of the rate–force relation as in Fig. 6d. Figure 9b confirms that, for a molecular size as low as 575 bp, random length fluctuations are so prevalent to mask almost completely the shape of the relaxation. On the contrary, as shown in Fig. 9c, that reports the result of a simulation with 1,936 two-state units, or (1,936 $\times$ 25 =) 48,400 bp, the entire DNA molecule exhibits a much smoother lengthening, that with an adequately expanded timescale (Fig. 9d), reveals its exponential kinetics. This is due to the

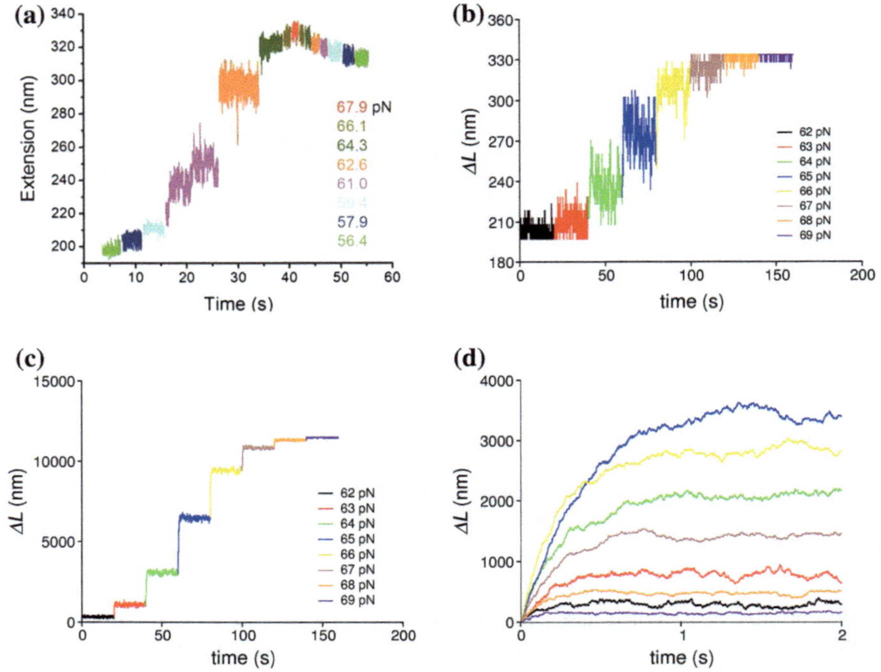

Fig. 9 **a** Length response to a staircase of force increments of ~1.5 pN imposed on a DNA construct of ~600 bp under melting-preventing conditions throughout the overstretching transition region (corresponding force attained by the step identified by the colors, pH 7.5, 24 °C). From Fu et al. [20]. **b** Simulated length responses to eight consecutive force steps of 1 pN (force attained by the step identified by the colors) imposed on 23 identical two-state units of 25 bp (total number of bp 575). **c** Simulated length responses to eight consecutive force steps of 1 pN as in **b**, imposed on 1,936 two-state units of 25 bp (total number of bp 48,400). **d** Superposed length responses as in **c**, on a faster time base, adequate to resolve the exponential time course of the early phase of the response, similar to the experimental responses in Fig. 6c. This indicates that the stepwise length changes in **a** are due to the absence of time resolution in those experiments. **b–d** From Bianco et al. [5]

self-averaging effect of length fluctuations in the units in a linear chain when the number of units is sufficiently high: for each extending unit another one contracts compensating the effect of the first. A striking result of this simulation is that the number of bp involved in the elementary reaction (25, the cooperativity coefficient), selected for fitting the responses of a molecule 48 kbp long with the two-state model, is able to predict to a very good approximation the length fluctuations measured by Fu and coworkers for molecules two orders of magnitude shorter.

Appendix: Influence of Viscosity on the Kinetics of the Elongation–Untwisting of the ds-DNA During the Overstretching Transition

Drag Produced on the Trapped Bead by the Viscosity of the Medium

A general problem with force clamp experiments made using the dual-laser tweezers apparatus is that the length change elicited by a force step is realized through movement of the piezo-stage and thus of the fluid surrounding the bead. Consequently, the change in the position of the trapped bead reliably measures the tension on the molecule only if it is not influenced by the drag due to the movement of the solution accompanying the movement of the stage. The drag on the bead is $F_v = 6\pi\eta Rv$ (Eq. 1), where η, the viscosity of the solution, is 10^{-3} Pa s, at 25 °C, R, the radius of the bead, is 1.64 or 1.09 μm, and v is the translational velocity of the bead. Considering that the stiffness of the molecule is ~60 pN μm^{-1} and the stiffness of the trap is 150 pN μm^{-1}, a step of 2 pN complete in 2 ms implies a bead movement of ~40 nm at a velocity of ~20 μm s^{-1}. Consequently, for the bead with $R = 1.09$ μm, F_v attains a value of 0.5 pN and decays with a time constant of 0.5 ms (1/4 the risetime of the step). This analysis indicates that the viscous drag on the bead does not significantly influence the position of the bead during the step nor the observed elongation kinetics.

A direct test of this conclusion is obtained by comparing the *r*–*F* relations obtained with different bead diameters. In Fig. 10a, the blue points data from Fig. 7c are unpooled to identify those obtained with beads of 3.28 μm diameter (black symbols, 12 molecules) and 2.18 μm diameter (red symbols, 5 molecules). Moreover, in Fig. 10b, *r* is plotted on a log–log scale against the final length change (ΔL_e) induced by a step, for the same data as in a. In both cases, it is evident the absence of any effect of the bead diameter on the overstretching kinetics.

Rotational Drag of the Molecule While Untwisting

A rotational drag of the ds-DNA while untwisting in response to a rise in torque has been directly measured by attaching a bead near a nick and determining the angular velocity [7]. In this way, it has been shown that the untwisting takes several minutes and the drag dominates the elongation–untwisting velocity. However, in that experiment, the drag should be several times larger than in our experiment, as it is generated by the revolutions of a large bead accompanying the untwisting of the molecule.

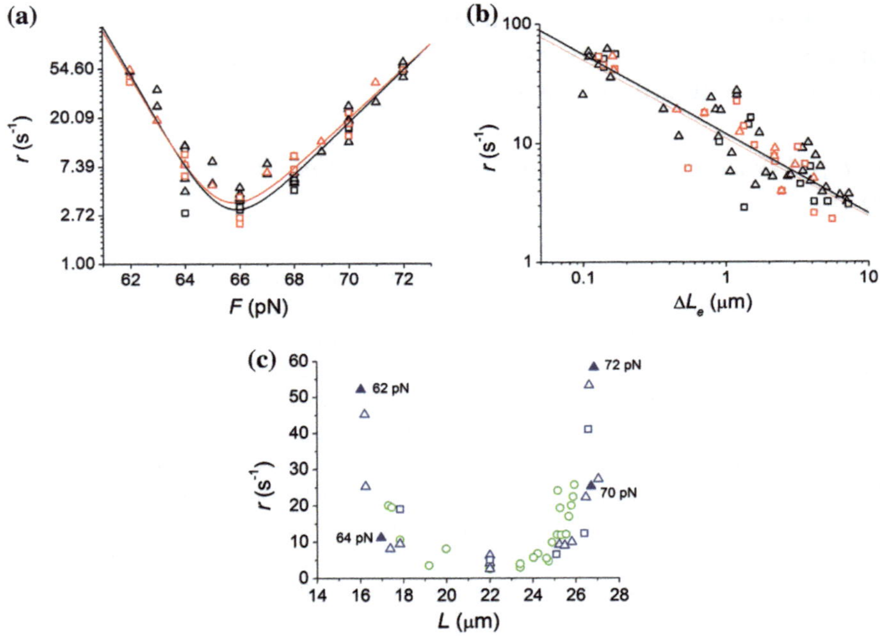

Fig. 10 **a** Relation of ln *r* versus *F* for the 2 pN steps (*blue symbols*) from Fig. 7b. *Red symbols* refer to data obtained with 2.18 μm bead diameter (5 molecules); *black symbols* to data obtained with 3.28 μm bead diameter (12 molecules). **b** Log–log relation between *r* and amount of lengthening (ΔL_e) for the same 2 pN steps as in **a**. **c** Relation of *r* versus the total molecular length (*L*) attained following 0.5 pN (*green circles*) and 2 pN (*blue symbols*) steps. 2 pN data pooled from the same molecules as in Fig. 6d. Figures close to *filled symbols* indicate the respective forces. From Supplementary Material in Bianco et al. [5]

During the overstretching transition under our conditions, the molecule of DNA elongates by 11 μm, while it reduces the number of turns from 4,500 to 1,450, that is, it untwists by 278 turns μm^{-1}. The largest elongation in response to a 2 pN step is ~5 μm attained within 0.5 s (Fig. 6). This implies a rotational speed ω of (278 × 5/0.5 =) 2,780 turns s^{-1}, so that each of the two ends of the molecule should counter-rotate at 1,390 turns s^{-1}. According to measurements of rotational drag in the experiment of Thomen et al. [39], where the two strands of DNA are attached to two independent beads and separated at different velocities, a torque of 0.6 k_BT would be necessary for the rotation at the speed of our overstretching transition. However, in that experiment, the rotating stretch is at longitudinal force zero (and therefore, the molecule is not straight), while during the overstretching transition, the longitudinal force is ~65 pN and the molecule can be assimilated to a rigid rod. Modeling a rigid rod [23] leads to a torque $\vartheta = 4\pi\eta R_H^2 L_{\mathrm{eff}}\,\omega$, where η is the viscosity of the solution as above, R_H the hydrodynamic radius of DNA (1.05 nm, [39]), and L_{eff} the extension of the portion of DNA which rotates in order to release the torsional stress. The molecular extension in the middle of the plateau of the overstretching transition is approximately 22 μm which, under the assumption of torsional stress accumulating in the middle of the molecule, gives a L_{eff} of 11 μm. With $\omega = 1{,}390$ turns s^{-1} (= 8,730 rad s^{-1}), the maximal frictional torque results to be 0.3 k_BT. If, however, the torsional stress is distributed uniformly along the whole length of the molecule, the resulting frictional torque should drop to 0.15 k_BT. A frictional reaction of 0.15–0.3 k_BT is expected to have a negligible effect on the kinetics of the B–S transition as demonstrated below.

The presence of an external torque (in this case of frictional origin) opposing the B–S transition not only modifies the free-energy difference between S and B states but also the free-energy barriers that the system must overcome for the transition (Fig. 7a). The B–S barrier will be increased (more difficult transition), while the S–B barrier will be decreased (easier transition). Let us define x_B and x_S the distances between consecutive bps in the two states and θ_B and θ_S the twist angles between consecutive bps, with numerical values: $x_B = 0.34$ nm, $x_S = 0.58$ nm $\theta_B = 1/10$ turns $= 0.63$ rad, $\theta_S = 1/30$ turns $= 0.21$ rad. Assuming for simplicity that the transition state is in the middle between both x_S and x_B and θ_S and θ_B, the change in barrier height due to the presence of a viscous torque can be estimated and compared with the analogous change due to the presence of an external force. The maximum torque-induced barrier change is $\Delta E_\tau = \tau(\theta_S - \theta_B)/2 = 0.3\ k_BT \times 0.4/2 = 0.25$ pN nm or 0.06 k_BT, while the barrier change caused by the external force F (~65 pN at the transition) is $\Delta E_F = F(x_S - x_B)/2 = 65 \times 0.24/2 = 7.8$ pN nm or 1.9 k_BT. Thus, since the barrier change introduced by the viscous torque is smaller by a factor of 32 relative to the barrier change caused by the external force, the torsional viscosity contribution can be disregarded.

A simple direct test of the influence of the rotational drag of the molecule on the kinetics of elongation is obtained by plotting the elongation rate r as a function of the length of the molecule L during the overstretching transition (Fig. 10c). If the elongation kinetics was dominated by the viscous friction, one would expect the relaxation rates to depend on L. Actually, the r–L relation shows a U-shaped dependence that excludes the hypothesis of a significant effect of the rotational drag on the elongation rate.

A further test is provided by the comparison of the responses to 2 and 0.5 pN force steps. For the same force, the extent of elongation (and thus ω) is smaller with the smaller step . Since the rotational drag depends linearly on ω, if it was dominating the kinetics of the process, it would have generated a reduction of the rate constant of elongation for the larger step. However, the observed rate constant is the same at the same force for either step size (Fig. 6d, blue 2 pN, green 0.5 pN).

In conclusion, both the theoretical treatment and the experimental evidences given above support the conclusion that the rate of DNA elongation following force steps applied in the overstretching transition region is not affected by viscosity and is mainly determined by the kinetics of the two-state reaction.

References

1. Allemand JF, Bensimon D, Croquette V (2003) Stretching DNA and RNA to probe their interactions with proteins. Curr Opin Struct Biol 13:266
2. Ashkin A (1970) Acceleration and trapping of particles by radiation pressure. Phys Rev Lett 24:156
3. Ashkin A, Dziedzic JM, Bjorkholm JE, Chu S (1986) Observation of a single-beam gradient force optical trap for dielectric particles. Opt Lett 11:288
4. Bell GI (1978) Models for the specific adhesion of cells to cells. Science 200:618
5. Bianco P, Bongini L, Melli L, Dolfi M, Lombardi V (2011) Piconewton-millisecond force steps reveal the transition kinetics and mechanism of the double-stranded DNA elongation. Biophys J 101:866
6. Block SM, Goldstein LS, Schnapp BJ (1990) Bead movement by single kinesin molecules studied with optical tweezers. Nature 348:348
7. Bryant Z, Stone MD, Gore J, Smith SB, Cozzarelli NR, Bustamante C (2003) Structural transitions and elasticity from torque measurements on DNA. Nature 424:338
8. Bustamante C, Marko JF, Siggia ED, Smith S (1994) Entropic elasticity of lambda-phage DNA. Science 265:1599
9. Bustamante C, Bryant Z, Smith SB (2003) Ten years of tension: single-molecule DNA mechanics. Nature 421:423
10. Capitanio M, Canepari M, Cacciafesta P, Lombardi V, Cicchi R, Maffei M, Pavone FS, Bottinelli R (2006) Two independent mechanical events in the interaction cycle of skeletal muscle myosin with actin. Proc Natl Acad Sci USA, vol 103
11. Capitanio M, Canepari M, Maffei M, Beneventi D, Monico C, Vanzi F, Bottinelli R, Pavone FS (2012) Ultrafast force-clamp spectroscopy of single molecules reveals load dependence of myosin working stroke. Nat Methods 9:1013
12. Clausen-Schaumann H, Rief M, Tolksdorf C, Gaub HE (2000) Mechanical stability of single DNA molecules. Biophys J 78:1997
13. Cluzel P, Lebrun A, Heller C, Lavery R, Viovy JL, Chatenay D, Caron F (1996) DNA: an extensible molecule. Science 271:792
14. Cocco S, Yan J, Leger JF, Chatenay D, Marko JF (2004) Overstretching and force-driven strand separation of double-helix DNA. Phys Rev E Stat Nonlin Soft Matter Phys 70:011910-1
15. Danilowicz C, Limouse C, Hatch K, Conover A, Coljee VW, Kleckner N, Prentiss M (2009) The structure of DNA overstretched from the 5'5' ends differs from the structure of DNA overstretched from the 3'3' ends. Proc Natl Acad Sci USA 106:13196
16. Elms PJ, Chodera JD, Bustamante CJ, Marqusee S (2012) Limitations of constant-force-feedback experiments. Biophys J 103:1490
17. Evans E (2001) Probing the relation between force - lifetime - and chemistry in single molecular bonds. Annu Rev Biophys Biomol Struct 30:105
18. Finer JT, Simmons RM, Spudich JA (1994) Single myosin molecule mechanics: piconewton forces and nanometre steps. Nature 368:113
19. Fu H, Chen H, Marko JF, Yan J (2010) Two distinct overstretched DNA states. Nucleic Acids Res 38:5594
20. Fu H, Chen H, Zhang X, Qu Y, Marko JF, Yan J (2011) Transition dynamics and selection of the distinct S-DNA and strand unpeeling modes of double helix overstretching. Nucleic Acids Res 39:3473
21. Kramers HA (1940) Brownian motion in a field of force and the diffusion model of chemical reactions. Physica 7:284
22. Léger JF, Romano G, Sarkar A, Robert J, Bourdieu L, Chatenay D, Marko JF (1999) Structural Transitions of a Twisted and Stretched DNA Molecule. Phys Rev Lett 83:1066
23. Levinthal C, Crane HR (1956) On the Unwinding of DNA. Proc Natl Acad Sci USA 42:436
24. Liphardt J, Onoa B, Smith SB, Tinoco I Jr, Bustamante C (2001) Reversible unfolding of single RNA molecules by mechanical force. Science 292:733

25. Mallik R, Carter BC, Lex SA, King SJ, Gross SP (2004) Cytoplasmic dynein functions as a gear in response to load. Nature 427:649
26. Mao H, Arias-Gonzalez JR, Smith SB, Tinoco I Jr, Bustamante C (2005) Temperature control methods in a laser tweezers system. Biophys J 89:1308
27. Marko JF, Siggia ED (1995) Stretching DNA. Macromolecules 28:8759
28. Perkins TT, Dalal RV, Mitsis PG, Block SM (2003) Sequence-dependent pausing of single lambda exonuclease molecules. Science 301:1914
29. Rouzina I, Bloomfield VA (1999) Heat capacity effects on the melting of DNA. 1. General aspects. Biophys J 77:3242
30. Rouzina I, Bloomfield VA (1999) Heat capacity effects on the melting of DNA. 2. Analysis of nearest-neighbor base pair effects. Biophys J 77:3252
31. Sarkar A, Leger JF, Chatenay D, Marko JF (2001) Structural transitions in DNA driven by external force and torque. Phys Rev E Stat Nonlin Soft Matter Phys 63:051903-1
32. Shokri L, Marintcheva B, Eldib M, Hanke A, Rouzina I, Williams MC (2008) Kinetics and thermodynamics of salt-dependent T7 gene 2.5 protein binding to single- and double-stranded DNA. Nucleic Acids Res 36:5668
33. Smith SB, Finzi L, Bustamante C (1992) Direct mechanical measurements of the elasticity of single DNA molecules by using magnetic beads. Science 258:1122
34. Smith SB, Cui Y, Bustamante C (1996) Overstretching B-DNA: the elastic response of individual double-stranded and single-stranded DNA molecules. Science 271:795
35. Smith SB, Cui Y, Bustamante C (2003) Optical-trap force transducer that operates by direct measurement of light momentum. Methods Enzymol 361:134
36. Suzuki N, Miyata H, Ishiwata S, Kinosita K Jr (1996) Preparation of bead-tailed actin filaments: estimation of the torque produced by the sliding force in an in vitro motility assay. Biophys J, vol 70
37. Svoboda K, Schmidt CF, Schnapp BJ, Block SM (1993) Direct observation of kinesin stepping by optical trapping interferometry. Nature 365:721
38. Svoboda K, Block SM (1994) Force and velocity measured for single kinesin molecules. Cell 77:773
39. Thomen P, Bockelmann U, Heslot F (2002) Rotational drag on DNA: a single molecule experiment. Phys Rev Lett 88:248102-1
40. van Mameren J, Gross P, Farge G, Hooijman P, Modesti M, Falkenberg M, Wuite GJ, Peterman EJ (2009) Unraveling the structure of DNA during overstretching by using multicolor, single-molecule fluorescence imaging. Proc Natl Acad Sci USA 106:18231
41. Veigel C, Bartoo ML, White DC, Sparrow JC, Molloy JE (1998) The stiffness of rabbit skeletal actomyosin cross-bridges determined with an optical tweezers transducer. Biophys J 75:1424
42. Visscher K, Schnitzer MJ, Block SM (1999) Single kinesin molecules studied with a molecular force clamp. Nature 400:184
43. Wang MD, Yin H, Landick R, Gelles J, Block SM (1997) Stretching DNA with optical tweezers. Biophys J 72:1335
44. Wenner JR, Williams MC, Rouzina I, Bloomfield VA (2002) Salt dependence of the elasticity and overstretching transition of single DNA molecules. Biophys J 82:3160
45. Whitelam S, Pronk S, Geissler PL (2008) There and (slowly) back again: entropy-driven hysteresis in a model of DNA overstretching. Biophys J 94:2452
46. Williams MC, Wenner JR, Rouzina I, Bloomfield VA (2001) Effect of pH on the overstretching transition of double-stranded DNA: evidence of force-induced DNA melting. Biophys J 80:874
47. Williams MC, Wenner JR, Rouzina I, Bloomfield VA (2001) Entropy and heat capacity of DNA melting from temperature dependence of single molecule stretching. Biophys J 80:1932
48. Yin H, Wang MD, Svoboda K, Landick R, Block SM, Gelles J (1995) Transcription against an applied force. Science 270:1653

Investigating Adhesion Proteins by Single Cell Force Spectroscopy

Laura Andolfi and Marco Lazzarino

Abstract The ability to sense and measure adhesion forces by using force spectroscopy techniques has opened new perspectives in the field of mechanobiology. Single-cell force spectroscopy enables to directly measure interactive forces of single living cell with extracellular environment (i.e., cell, proteins and tissue) with extremely high resolution (single-protein level). Cell adhesion processes rely on the interaction of adhesion proteins with their environment. Cells sense and recognize the specific forces that are generated by the interaction with the environment, and transduce them into biochemical signals by which the cells evolve, move and grow. Single-cell force spectroscopy is the ideal tool to measure these forces and investigate the cellular response from its origin.

1 Introduction

The imaging and mechanical characterization of biological systems at a molecular level has become possible with the invention of atomic force microscope (AFM). The AFM initially was applied mainly in hard sciences, such as physics and engineering; however, soon it became clear that it would have a revolutionary impact also in the field of biochemistry and cell biology, since it allowed the imaging and manipulation of biological samples, in physiological condition, at the nanoscale [1]. This technique is able to sense and apply a wide range of forces

L. Andolfi (✉) · M. Lazzarino
CBM S.c.r.l. Basovizza, Area Science Park, 34149 Trieste, Italy
e-mail: laura.andolfi@cbm.fvg.it

M. Lazzarino
e-mail: lazzarino@iom.cnr.it

L. Andolfi · M. Lazzarino
CNR-IOM Basovizza, Area Science Park, 34149 Trieste, Italy

F. Benfenati et al. (eds.), *Novel Approaches for Single Molecule Activation and Detection*, Advances in Atom and Single Molecule Machines,
DOI: 10.1007/978-3-662-43367-6_8, © Springer-Verlag Berlin Heidelberg 2014

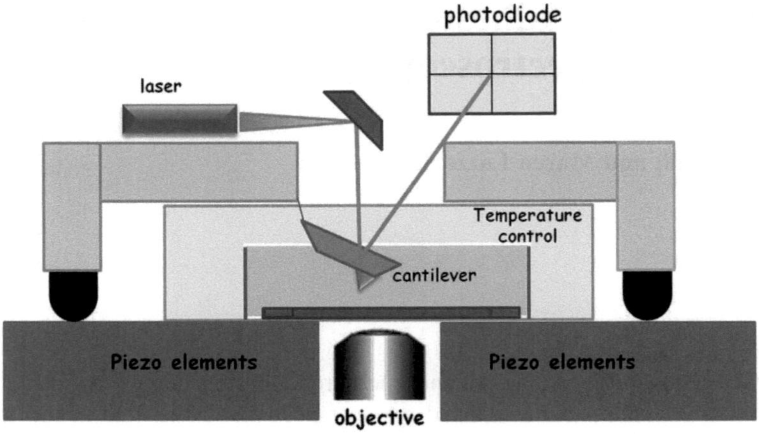

Fig. 1 Sketch of the experimental setup of AFM including conical probe at the end of a cantilever. A laser is reflected to back of the cantilever toward a photodiode that detects changes in cantilever deflection. Nanometer precision movements are regulated by piezo-elements that can be positioned below sample stage. Measurements can be carried out in liquid cell at controlled temperature suitable for cell culture. Frequently for biological studies AFM is coupled with an optical microscopy working in transmission mode

(1 pN–100 nN) that corresponds to the forces that dominate the biological phenomena from the molecular to the single cell scale. Moreover, the nanometer scale precision in positioning the tip relative to the surface and ability to operate in aqueous physiological conditions made the technique capable to follow biological process in situ. The core of this instrumentation is represented by a micromachined cantilever that may deflect upon interacting with the sample surface. Cantilever deflection is detected by a laser beam reflected from the free end of the cantilever into a photodiode (Fig. 1) with a sub-nm precision. Usually a sharp tip is located at the free end of the cantilever and is used to investigate the sample property with in-plane nm resolution. The tip is usually made of silicon or silicon nitride and has a radius of curvature of about 10 nm. However, other materials can be used for higher spatial resolution (carbon nanotubes) such as increased stiffness (diamond whiskers) and electrical conductivity (tungsten carbide). For biological applications, chemical functionalization can be applied to the tip to exploit specific interactions; for force mapping or force spectroscopy, cantilevers may be used flat without sharp tips or with micrometer-sized beads. In image mode, the initial application of AFM, the tip/cantilever is scanned across the sample surface, while the tip senses the local forces that are used for feedback control. In biological applications, these forces may have biological, chemical or physical origins. A piezoelectric scanner allows high-resolution 3D positioning (1 Å) of the tip. Measurements can be performed into a liquid cell where temperature can be controlled and can be combined with the other imaging techniques used in biology [2]. AFM was used for imaging fixed cells [3] living cells [4], protein structures [5]

down to individual proteins [6, 7]. An AFM can be also operated in different fashion, in which the tip is kept fixed at a given position and the deflection versus distance is recorded. Thanks to the elastic properties of the cantilever, a force-distance curve is recorded that provides useful information on the mechanical and dynamic properties of the sample. This approach, known as force-spectroscopy, allows to probe biological, chemical and physical interaction forces of individual molecules with pN resolution. The ability to work in force-spectroscopy mode opened up the possibility to analyze the kinetics and the interaction strength of single proteins even in living cells [1, 8]. Single-molecule force spectroscopy (SMFS) measures the adhesive interactions between biological molecules: the force required to break a single intermolecular bond was measured directly by separating a protein linked to the AFM tip from that fixed on the substrate. The bond rupture force measured at different loading rates produces information about kinetic barriers, binding constants, binding mechanisms and free energy [9]. Whereas in molecular recognition mapping (MRM) a probe molecule bound to the tip is made interact with its receptor, generally embedded in membrane living cells, exposing an active recognition site to the outer environment [10, 11] mapping the interaction forces reproduces the receptor distribution on the cell surface, while the intrinsic force dependence of the mechanism is used to separate the contribution of different kind of receptors. Another application of AFM in the field of the biomechanics is so-called single-cell force spectroscopy (SCFS), in which a single living cell is first attached to a tipless AFM cantilever and then is brought into contact with a substrate, with another cell grown of a solid substrate or with a tissue, providing information about the cell adhesion properties as a whole and on the molecular mechanism of cell adhesion [12–14].

Cell adhesion is a complex biological process that plays a central role in regulating numerous fundamental physiological and pathological cellular activities such as proliferation, migration, differentiation, metastasis, immunological response and communication [15]. The process is mediated by focal adhesion points located on and close to the cell membrane. They are assembled by a core made of a large family of multidomain transmembrane proteins and a network of cytoplasmatic proteins that ensure the mechanical link to the cell cytoskeleton and are responsible of the regulation of a number of signaling pathways that mediate the cell adhesion [16–18]. These membrane areas include integrins, cadherins, immunoglobulin superfamily and selectins. They have an extracellular domain able to bind either to components of extracellular matrix or other cells and an intercellular domain that interacts with cytoskeleton directly via cytoskeleton-associated proteins (i.e., vinculin, talin) or indirectly (i.e., Rho-family GTPase, protein kinase, paxillin), which are involved in the transduction of the signaling process and regulation of cytoskeleton reorganization [17] (see Fig. 2). Usually, cell adhesion is investigated by using various adhesion assays: washing assay [19], spinning disk [20, 21] and centrifugation assay [19] which have disclosed important features of the key components regulating adhesion mechanism. The role of adhesion proteins is also investigated by several biochemical assays as immunoblotting that offers information about the different expression of adhesion proteins [22, 23] or by high-

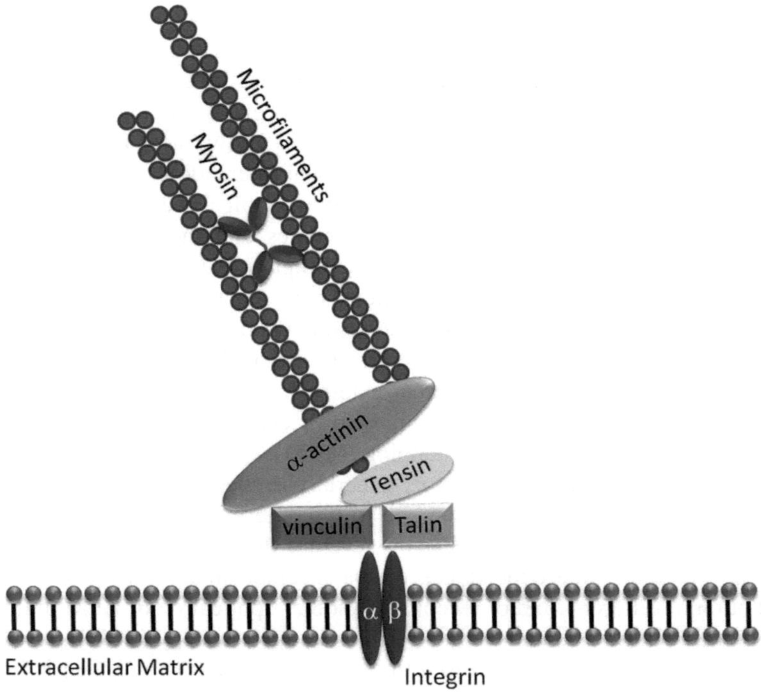

Fig. 2 Simplified scheme of a focal adhesion site that depicts interaction of transmembrane adhesion proteins (integrin) with some of the cytoplasmatic proteins included in the transduction machinery that mediate the adhesion process

resolution fluorescence microscopy that can reveal their distribution and clustering at the cell membrane [24–26]. However, these methodologies have some important flaws: mechanical assays provide averaged information about cellular behavior while immunoblotting and immunofluorescence are useful to investigate expression and distribution of adhesion proteins but do not give a direct proof of binding; moreover, neither of them can be used to investigate the dynamics of the adhesion process.

The combination of AFM imaging with optical and fluorescence microscopy represents a very attractive tool for high-resolution study of cellular adhesion process [8, 27, 28]. An example of such a correlated fluorescence-AFM study is the high-resolution imaging of proteins localized within focal adhesion sites interacting with microfilaments of cytoskeleton. The high resolution of the AFM topographs overcomes that of the light microscope images and structural information about the 3D organization of microfilaments in focal adhesion areas can be provided without using complex protocol and in physiological environment [27]. More recently, a new cantilever holder which enables a side view of the cantilever tip-sample interaction was designed and commercialized by JPK Instrument (Fig. 3a). The

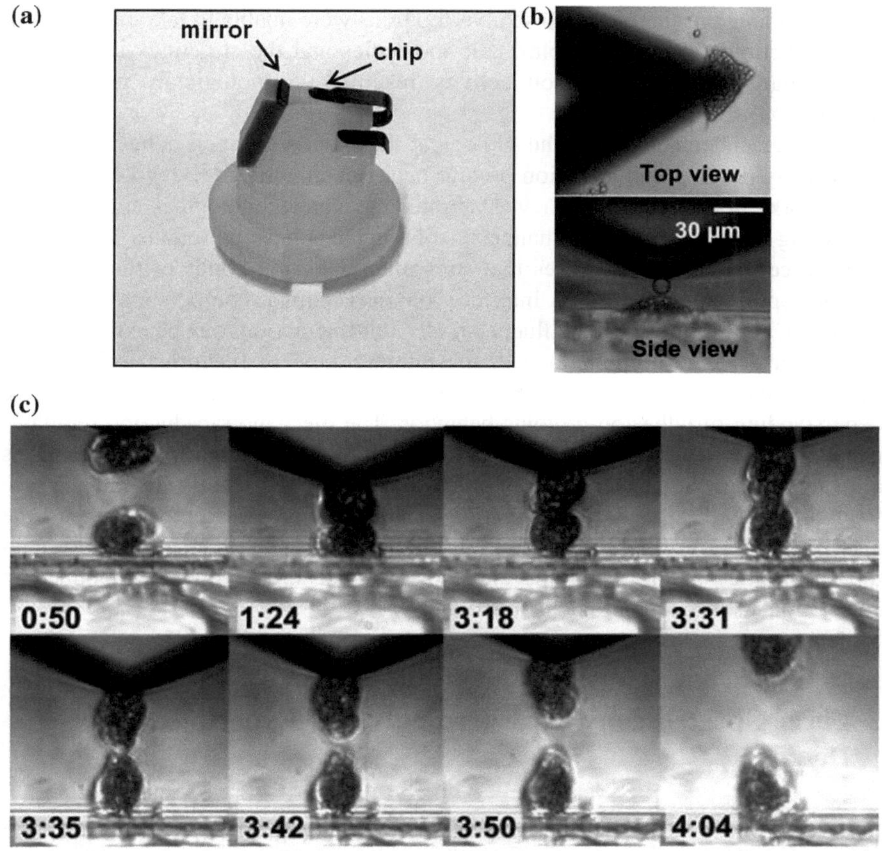

Fig. 3 Side view of a SCFS measurement. **a** The special design of the cantilever with a 45° mirror makes possible to follow and visualize cell detachment with an optical or fluorescence microscopy. **b** *Top* and *side* view image show a colloidal nanoparticle in contact with a fixed neural crest cell of a Xenopus leavis. **c** An example of optically monitoring cell–cell detachment is shown. All images are data courtesy of Dr C. Franz KIT (Germany)

special design integrated with the transmission light techniques allows to fully monitor the cell membrane adherence points during the detachment process as shown in Fig. 3b–c.

Besides molecular resolution imaging of adhesion-proteins SCFS provides quantitative information of the cell adhesion behavior. Initially, the interaction strength of adhesion proteins at single-molecule level was widely investigated by SMFS. The dynamic of interaction process of cell adhesion proteins, including selectins [9], cadherins [29, 30] and integrins [7] disclosed new insight into the association and dissociation behavior of these proteins. However, these analyses were conducted on purified proteins or on isolated domains, removed from their biological context: therefore, these investigations were unable to take into account the interactions with the cytoplasmatic molecules and the structures involved in the regulation of cell adhesion activity resulting in biologically nonrelevant information.

A better understanding of the biological mechanisms of cell adhesion can be attained by probing cell-adhesion protein behavior directly in living cells. In this framework, SCFS represents a very promising nanotechnological approach for dissecting localized signaling changes involving the adhesion proteins. This means that the cell signaling processes that strengthen adhesion bonds or the action of forces applied to cell-surface interface on intracellular components to trigger biochemical processes or the influence molecular interactions can be explored with nanometer scale resolution. Indeed, the interpretation of functional role of signaling adhesive components in the mechanosensing process requires a quantitative understanding of adhesion-proteins behavior. The mechanotransduction regulates gene expression and cell fate [31] and that it is now increasingly evident its relevant role in the cancer cell migration and diffusion [32]. In this chapter, we focus our attention on the investigation of adhesion proteins embedded into the membrane of a living cell at single molecule level by using SCFS, and we discuss the capabilities of the technique and the relevance of the findings obtained in correlation with their biological function.

2 Force-Spectroscopy Measurements and Experimental Details

In the case of SMFS and MRM, a conical/pyramidal tip positioned at the end of the cantilever is chemically functionalized with an adhesion binding molecule and is brought into contact with the adhesion receptor (SMFS) or living cell (MRM) immobilised on a supporting substrate. Thanks to the sharp tip of the AFM probe only one or few adhesion proteins interact with the ligand in each experiment. After a certain contact time, the cantilever is withdrawn at a constant speed or, correcting for the cantilever deflection, at a constant loading rate. The force acting on the ligand-receptor connection increases with time, thus reducing the connection lifetime until the interaction bonds break. The detachment force is measured by detecting the deflection of the cantilever by the photodiode which is converted in force using the cantilever elastic constant. A typical single molecule force-distance spectrum obtained by pulling a ligand-bound tip from its receptor counterpart on the cell membrane is displayed in Fig. 4 The result of such MRM experiments is a bidimensional map of unbinding events performed over the cell surface that gives a clear visualization of the receptor spatial distribution [33]. While SMFS measurements and analysis allow the estimation of binding affinity, rate constants and structural data of the binding pocket of ligand-receptor at single-molecule level of adhesion proteins [29, 34].

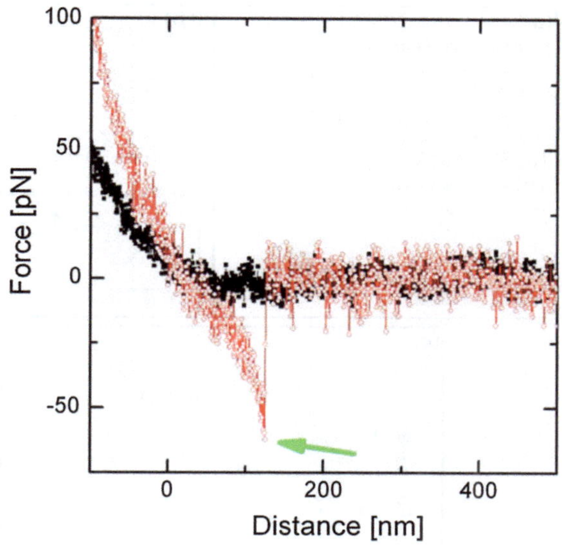

Fig. 4 An example of force-distance *curve* obtained when pulling a receptor ligand (leuprorelin acetate) bound to the tip from its receptor counterpart embedded into the cell membrane of prostate cancer cells until the binding between ligand and receptor is broken. When receptor-ligand binding interactions are not detected (*black square*) and when interactions are detected (*red circle*). The unbinding event (or *rupture point*) is indicated by the *arrow* (Reprinted with permission from Lama et al. [33] copyright 2013 PlosOne)

In SCFS, a tipless cantilever is used for attaching a living cell. Even in this case, the stable binding of the cell is assured by functionalization of the cantilever. The choice of the protein for functionalization is very important to attain a firm cell binding without influencing cell state during the measurements. Usually proteins that are components of the extracellular matrix (i.e., concavalin-A, laminin, fibronectin, collagen, etc.) which have high affinity for receptors present on the cell membrane are used for functionalization.

Before attaching cell to the cantilever, the sensitivity and spring constant of the cantilever have to be calibrated. The deflection of an AFM cantilever is measured by the position of the reflected laser beam on the photodetector. Accordingly, the units of measurement are volts and to convert the units into newtons it is necessary to determine the cantilever sensitivity (i.e., the relationship between the output of the photodiode in volts and the deflection of the cantilever in nm) and the spring constant (which converts deflection in nm to force in nN). First, the sensitivity is determined from a force-distance (F-D) curve recorded by pressing the cantilever on a stiff surface. When cantilever and surface are in contact, the deflection of the cantilever is proportional to the vertical movement of the AFM piezo element. Most AFM softwares offer an option to measure the thermal noise of the cantilever and apply the equipartition theorem to calculate the cantilever spring constant [35].

Afterward, cells are introduced in the AFM fluid cell chamber and the cantilever is gently pushed with a force load of less than 1 nN for several seconds onto a cell selected by optical microscope in an area of the substrate coated with protein having very low affinity for integrin binding, bovine serum albumin (BSA) is commonly used [Fig. 5a (i, ii)]. The cell bound to the cantilever is separated from the support and allowed to establish firm adhesion for several minutes [Fig. 5a(iii)]. The

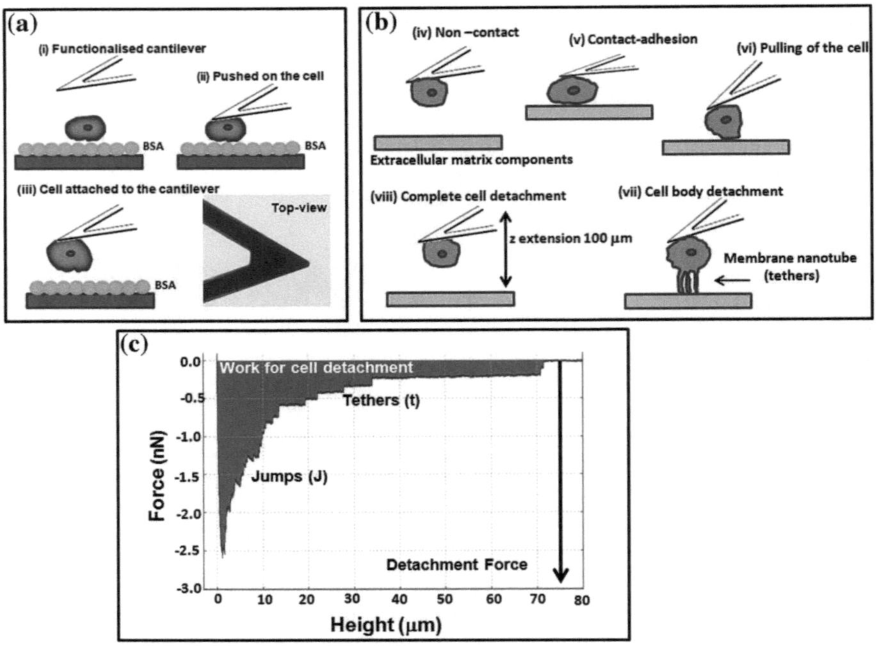

Fig. 5 Scheme of cell capture by tipless cantilever (**a**). The functionalized cantilever is positioned over a cell in suspension at close proximity to the surface (i). Then the cantilever is gently pushed for a few seconds onto the cell (ii). After this, the cantilever-bound cell is separated by the support (iii) besides a phase-contrast microscopy image of cell firmly immobilized on the cantilever. Sketch illustrating a single SCFS measurement (**b**). The cell attached to the cantilever is approached to the substrate to establish cell-surface interaction (iv, v). After a predefined contact time, the cell is retracted and the cantilever bends because of the adhesive strength between the cell and the substrate. Once the force of the cantilever exceeds the strength of the interactions between cell and substrate, the cell starts to detach (vi) forming membrane nanotube (vii) up to a complete detachment (viii). **c** The resulting F-D retraction curves from the SCFS measurement provides quantitative measure of different cell adhesion stages

sequence of these events can be followed by an optical microscope that allows also to control the cell state during SCFS measurements. A phase-contrast image of a cell bound to a tipless cantilever is shown in Fig. 5a. The sketch in Fig. 5b shows the cycle of a SCFS measurement. The cell attached to the cantilever is approached either with a specific substrate or with a target cell until a predefined repulsive contact force is established Fig. 5b (iv). This contact force is held constant for a given time to allow establishment of cell adhesion [Fig. 5b (v)]. Contact time may range from milliseconds to several minutes. Different adhesion regime can be investigated by setting different contact times so that the signaling cascade pathway that leads to cell adhesion could be obtained. Upon cantilever retraction, the cantilever force exceeds the strength of the interactions between cell and substrate and the cell starts to detach (vi) forming membrane nanotube (vii). During this procedure, the force as a function of the distance of the cantilever is recorded until contact with the cells is broken [Fig. 5b (viii)]. The result of this cycle is a F-D curve where the retraction curve represents a full characterization of the cell adhesion as shown in Fig. 5c. Care must be taken in order to avoid that the cell detaches from the cantilever. The complete cell separation from substrate requires a longer z range extension. The CellHesionTM module (JPK Instruments, Berlin, Germany) is a technical solution for this problem. The module contains a sample stage that is fitted with piezo-elements that have a 100 μm range for moving the sample in the z direction.

It is worth to mention that SCFS measurements can be performed also in presence of blocking agents (i.e., antibody or small RGB peptide) able to block the activity of the adhesion proteins on cell membrane [36–38]. Moreover, since measurements are performed on living cells, it can be possible to use blocking agents that can diffuse into the cell membrane and interact with specific intracellular intermediate of the transduction machinery [36]. This allows to molecularly associate interaction forces with proteins that mediate the signal between adhesion proteins and cytoskeleton. Even in this case, the coupling of SCFS measurements with high-resolution fluorescence microscopy can provide considerable advantages. The use of GFP-labeled adhesion proteins conjugated with fluorescent dye would reveal more detailed information on the adhesion characteristics of the cell in different conditions.

3 Analysis of Force-Distance Curve

The analysis of the F-D retraction curve points out three main features (see Fig. 5c): the peak minimum value (F detachment) that is the measurement of the maximum force exerted to detach the cell from substrate (i.e., adhesion force); a train of saw-like peaks following the detachment force involving receptors that remain anchored to cell cytoskeleton and unbind as force increases (named jumps); finally long plateau where receptors anchoring is lost and membrane nanotubes are pulled out of the cell (tethers). The area below the retraction curve represents the mechanical work done by the cantilever during the whole detachment processes or, in other words, the total adhesion energy. More detailed information about adhesion mechanism can be obtained by discriminating jumps from tethers. Both jumps and tethers correspond to small discrete rupture events, which result from unbinding of single receptor-ligand pairs that remain anchored to the cytoskeleton (jumps) or to the membrane (tethers) [8, 39]. Those that remain bound to the membrane lead to the formation of membrane nanotubes (tethers) that can be of potential physiological relevance, since this kind of membrane structures are generally related to cellular attachment, migration and communication [40].

The analysis of jumps and tethers features can reveal energetic and kinetic properties of adhesion proteins embedded into a living cell (Fig. 6) [8]. When the cantilever starts pulling the cell out of contact, if the anchor strength to the

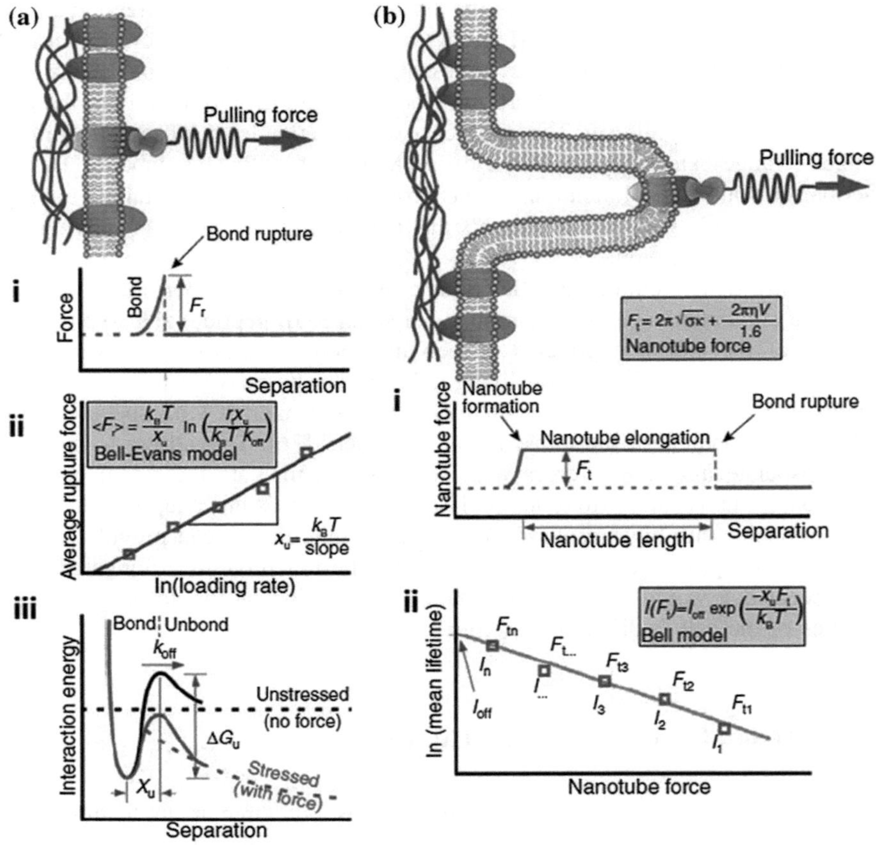

Fig. 6 Energetic and kinetic properties of force bonds at cell surfaces for different adhesion proteins. **a** Cytoskeleton-bound adhesion proteins. **a** (i) The receptor-ligand bond is mechanically stressed until it ruptures at a force (Fr). **a** (ii) According to the Bell-Evans model, the average rupture force <Fr> of the bond linearly increases with the loading rate (rf). **a** (iii) Interpretation of how an externally applied force reduces the unbinding barrier and increases the unbinding rate of the probed bonds. **b** Membrane-bound adhesion protein presents at the tip of a membrane nanotube. The force required to extend a nanotube (Ft) depends on some factors: extension speed (V), isotropic tension (σ), bending rigidity (κ) and viscosity (η) of the cell membrane. **b** (i) The force to extend the nanotube remains constant at constant extension speed. The extension speed and length of the nanotube can be used to calculate the bond's lifetime. **b** (ii) The measurements of bond's lifetime for different extension speeds analyzed by the Bell model can provide information about the dynamics of these bonds (Reprinted with permission from [8] copyright 2009 Nature Publishing Group)

cytoskeleton is greater than that of adhesion, the transmembrane adhesion proteins will likely unbind first [Fig. 6a (i)]. In such a case, the force vs. distance curve shows a typical elastic character and the Bell-Evans model [Fig. 6a (ii)] [41] may be applied for data analysis. According to this model, the force at which a single bond unbinds increases as the rate at which the force on the bonds is applied (loading rate, rf) increases. The loading rate is controllable, as it depends on the speed at which the AFM cantilever is retracted. For most receptor-ligand bonds, their rupture force increases linearly with the logarithm of the loading rate [Fig. 6a (ii)]. By measuring the most probable rupture force over a range of loading rates, the unbinding rate (koff), the distance to the transition state (xu) and the free energy (ΔGu) of the bond can be estimated [Fig. 6a (iii)]. In the opposite case, when the link of the focal adhesion points with the cytoskeleton is weaker, the transmembrane adhesion proteins are pulled away from the cytoskeleton, and remain localized at the tip of a membrane nanotube (Fig. 6b). The measured forces (Ft) have a typical viscous character depend on plasma membrane properties and increase with extension speed (V). As nanotube length has very little effect on extension force, cell membranes establish constant force clamps that can be used to measure the lifetime (ln) of receptor-ligand bonds under force [Fig. 6b (i)]. The measure of the mean lifetime of a bond at different forces (i.e., extension speeds) allows the lifetime at equilibrium (loff) and distance to the transition state (xu) to be determined [Fig. 6b (ii)] [42]. By analyzing nanotube extension forces, plasma membrane properties such as the extent of anchoring to the cytoskeleton and viscosity can also be characterized. It is worth to stress here that in both cases the observed force jumps refer to the transmembrane adhesion protein-ligand bond rupture, while the forces between the focal adhesion structures and the cytoskeleton is not investigated by SCFS.

4 The Advantages of SCFS in the Study of Single Adhesion Proteins

Besides biochemical and structural description of adhesion proteins, quantitative data about interaction forces involved in the binding with their surrounding environment are fundamental to understand the mechanisms that guide and regulate cell adhesion in physiological conditions and diseases. Numerous of SCFS capabilities can strongly improve the understanding of adhesion mechanisms. In the following list they are briefly summarized.

- Quantification of the interacting forces of adhesion proteins with their natural ligands in living cells [43];
- short-term investigation of adhesion steps and understanding of the activity of factors affecting adhesion formation [44];
- the ability to resolve interactions of individual proteins and examine their individual contribution instead of describing an average behavior [45];
- coupling with genetic manipulation [36, 44, 46];
- the possibility of modulating interaction parameters as contact force, contact time, speed retraction of the cantilever and also biochemical factors (i.e., pH, ion concentration of functional relevance, external stimuli, etc.) allows

studying the role of individual adhesion proteins and intermediate proteins regulating cell adhesion within a single SCFS experiment [36, 47].

The technique also opens up the possibility to study the role of adhesion proteins in various conditions that have great interest in the biomedical field, as for example:

- evaluation of adhesion-blocking drugs [48];
- investigation of initial steps in the interaction of cells with artificial surfaces of medical interest, such as different culture supports or receptor mediated adhesion of particles for phagocytosis.

Through its numerous advantages, the technique has a main drawback in biological field. A single SCFS experiment is unfortunately confined to a reduced number of cells, and statistics is still a time-consuming procedure.

5 Quantifying Cell Adhesion at Single Molecule Scale

The advantages of SCFS have allowed of getting new insights about the complex protein machinery that regulates the cell adhesion down to single-protein resolution.

One of the first examples that demonstrate the capability of SCFS to measure adhesion force with molecular resolution is represented by the quantification of the adhesion strength of single glycoprotein contact site A (csA) expressed in aggregating cells of *Dictyostelium discoideum*. This protein is differently expressed depending on cellular aggregation stage (growth-phase and developing cells). The analysis of data obtained by performing SCFS measurements as function of contact force and time in two states of cellular development are shown in Fig. 7. During growth phase, a low number of steps in F-D curves were observed in agreement with the lower expression of the csA protein. Steps increase considerably for developing cells even at when contact force is reduced to decrease the contact area. By lowering the contact time, the number of steps is further reduced. By plotting the steps force values as histograms, a peak at 23 $\pm$ 8 pN can be observed in all graphs. This value is attributed to the force of single csA proteins, while the other peaks of the distribution seem to be related to multiples of the basic quantum of 23 pN. Genetic manipulation confirmed that this force value was associated with csA expression [45].

The role and the dynamics of the $\alpha 2\beta 1$ integrin in binding and spreading onto collagen type I matrix has been clarified by SCFS [49]. By comparing the adhesion strength of wild type-chinese hamster ovary cells (CHO) with that of (CHO)-A2 $\alpha 2\beta 1$-expressing cells as function of short contact time (5 s), it was found that the mean detachment force (189 $\pm$ 12 pN) of CHO-A2 cell was almost 4 times higher than that of CHO-WT cells (49 $\pm$ 7 pN). Moreover, measurements in absence of Mg^{2+} definitively confirmed the principal role of $\alpha 2\beta 1$ integrin in the interaction

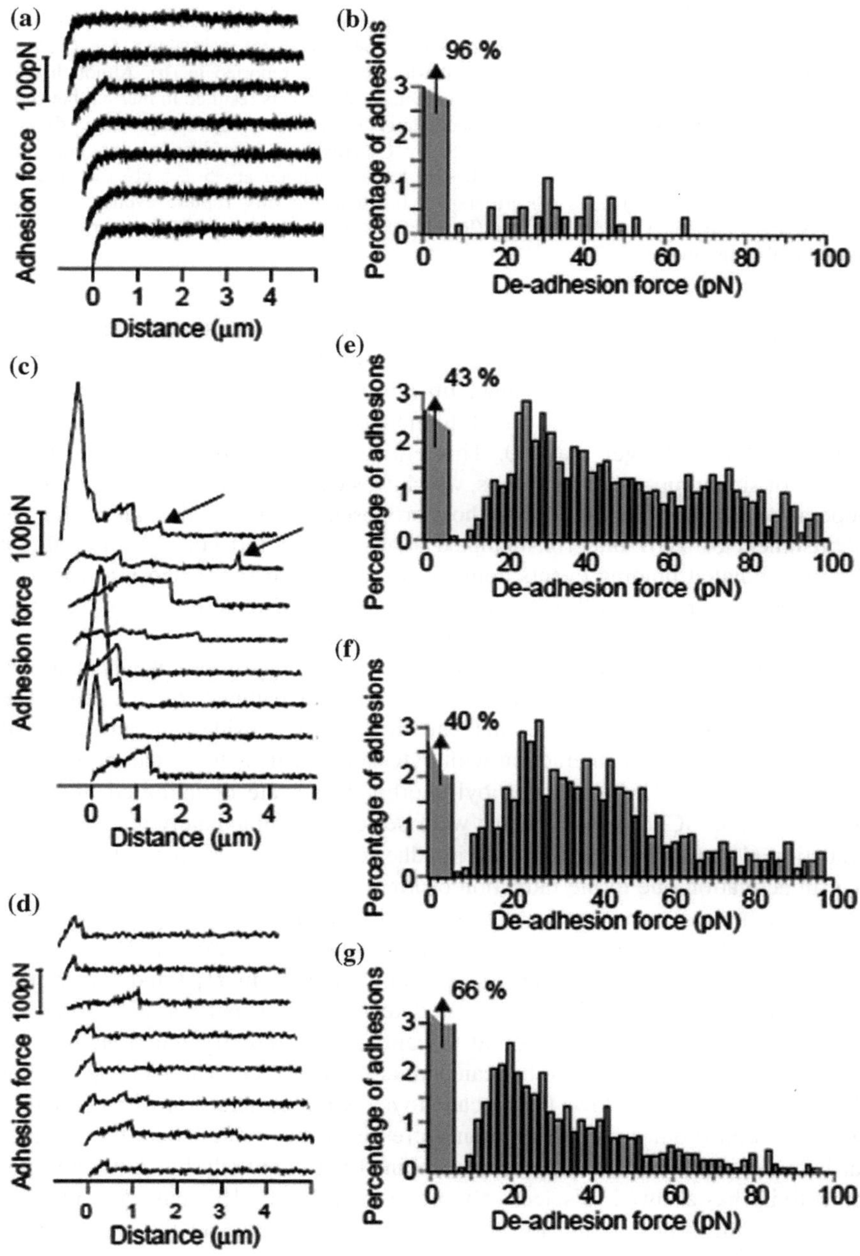

(a)
100pN
Adhesion force
0 1 2 3 4
Distance (µm)
(b)
96 %
Percentage of adhesions
0 1 2 3
0 20 40 60 80 100
De-adhesion force (pN)
(c)
100pN
Adhesion force
0 1 2 3 4
Distance (µm)
(d)
100pN
Adhesion force
0 1 2 3 4
Distance (µm)
(e)
43 %
Percentage of adhesions
0 1 2 3
0 20 40 60 80 100
De-adhesion force (pN)
(f)
40 %
Percentage of adhesions
0 1 2 3
0 20 40 60 80 100
De-adhesion force (pN)
(g)
66 %
Percentage of adhesions
0 1 2 3
0 20 40 60 80 100
De-adhesion force (pN)

◀**Fig. 7** Force spectra for stable adhesion of undeveloped and developed cells. **a** F-D curves for adhesions of growth-phase cells where cell–cell contacts were maintained for 0.2 s at 90 ± 10 pN. The resulting histogram of de-adhesion forces is reported in (**b**). Although high contact force (90 pN) is applied, only a small percentage of contacts resulted in measurable cell–cell adhesion. F-D curves for adhesions of discoideum cells at the developed stage (**c** and **d**) with cell–cell contacts for 2 s (**c**) or 0.2 s (**d**) at 35 ± 5 pN. Arrows indicate force steps for complete rupture. The resulting histograms of de-adhesion forces obtained at 35 ± 5 pN for different contact time 2 s (**e**) 1 s (**f**) 0.2 s (**g**) are shown. In all histograms, rupture events occurring at <7 pN are represented by the first bar (Reprinted with permission of [45] copyright 2000 Nature Publishing Group)

with collagen I. In fact, this metal binds to the domain of α2 activating the protein. Additionally in this work, adhesion measurements were performed as function of contact time and a minimum contact time was found for the activation of the integrin-based adhesion. Indeed, a tenfold increase of adhesion force was observed with contact time longer than 60. This effect was accompanied by a parallel increase of single rupture event jumps, which was explained in term of the onset of cooperative receptor binding. The authors proposed a two-step mechanism for the establishment of α2β1 integrin mediated adhesion: at short contact times single integrin–mediated binding events dominates the cell-substrate adhesion; for contact times longer than 120 s strong adhesive interactions involving receptor cooperativity and actomyosin contractility are observed.

The possibility to explore the adhesion over short range time reveals also new important details about the mechanism of activation pathway of integrins. The protein kinase C is an enzyme that can mediate the inside–out activation of integrins α2β1 and their interaction with cytoskeleton. Its activity can be induced by the presence of 12-O-tetradecanoyl-phorbol-13-acetate or inhibited by bisindolylmaleimide. SCFS measurements were performed immediately after activator incubation (10 min) show an increase in adhesion strength which results from α2 integrin activation due to the induction of kinase enzyme activity, thus excluding the genetic activation or protein expression in this condition and providing clear evidences for the role of the kinase activity in integrin activation [44].

The integrin cross talk between collagen-binding integrin α1β1 and fibronectin-binding integrin α5β1 in Hela cells was detected by a small variation in SCFS protocol. The Hela cells were bound to cantilever differently functionalized by using integrin binding proteins that can activate different cascade pathways. After 10 min incubation that allows the cascade to activate, the adhesion with collagen I and fibronectin (both proteins commonly present in the mammalian extracellular matrix) was measured. This approach combined with antibody blocking the integrin activity has allowed to selectively probe the role of α1β1 and α5β1 integrin receptors in the adhesion and to provide clear evidences about the cross talk of these adhesion proteins that is found to be unidirectional from integrin α1β1 to integrin α5β1. The cross talk seems to function through the regulation of integrin α5β1 endocytosis [47].

Multi-protein adherens junction that links cell–cell contact to the actin cytoskeleton and various signaling molecules play an important role also in adhesion

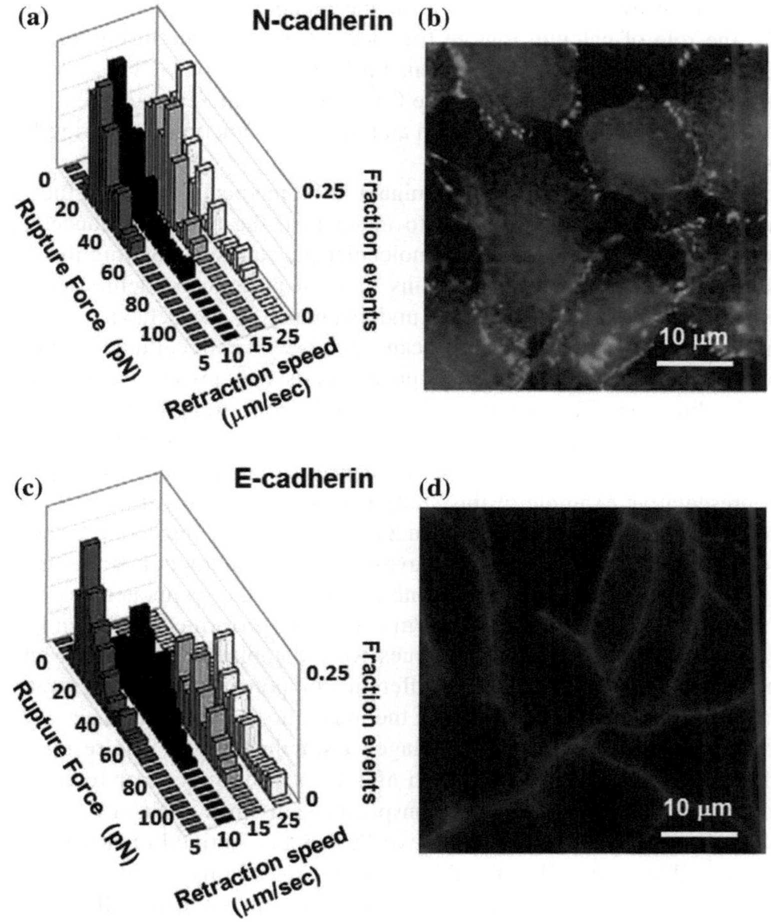

Fig. 8 Rupture forces of a single cadherin–cadherin bond as a function of retraction speed. **a** Distribution of rupture forces to break a single E-cadherin–E-cadherin bond between apposing cells obtained by using different retraction speeds. **c** Distribution of rupture forces to break a single N-cadherin–N-cadherin bond at different retraction speeds. Immunofluorescence images facilitate the visualization of N-cadherin **b** E-cadherin **d** distribution in the CHO cells. Data courtesy of Prof. K. Konstantopoulos, The Johns Hopkins University (USA)

and intercellular communication. SCFS measurements demonstrate that type I N- and E-cadherins establish intercellular bonds that despite having a similar biological role, exhibit significantly different kinetic and micromechanical properties. This is highlighted by performing SCFS as function of detachment speed, which shows that the rupture forces of the two proteins are considerably different (30 pN for N-cadherins against 73 pN for E-cadherins) and also their dissociation rates ($0.98 \pm 0.46\ s^{-1}$ for N-cadherins against $1.09 \pm 0.35\ s^{-1}$for E-cadherins) strongly vary (see Fig. 8). The combination with immunofluorescence was very useful to demonstrate the distribution of the receptors on the membrane. Additionally, the role of calcium ions in the mechanism of cadherin binding on cells was also confirmed in such experiment [36]. Analogously, SCFS provides convincing evidences that activation of the Ca^{2+} receptors increases expression of the epithelial adhesion proteins E-cadherin and increases functional tethering between β-cells in pancreas [23].

The use of SCFS can be also determinant in biomedical field. The study of cell adhesion mechanism is fundamental to understand the ability of cancer cells to migrate and diffuse into tissues. The molecular mechanisms defining the different modes of cancer cell migration remains in most parts to be delineated [50]. A fundamental step in this study is the understanding of the behavior of adhesion proteins that mediate the interaction of cancer cells with extracellular matrix of cells within different tissue. In this way, cancer cells can invade tissue leading to metastasis diffusion. The quantification of adhesion strength can provide new insights into such mechanism, particularly because it is observed that the mechanical properties of the matrix could strongly influence the fate of the cells [32].

A representative example of this study is reported by Sariisik et al. [51]. They dealt with the problem of prostate cancer cells diffusion into bone tissue where these cells can form metastasis. The investigation of interaction strength of bone marrow-specific prostate cancer cell line (PC3) with the components of the bone tissue (collagen I and mesenchymal stem cells) was performed as shown in Fig. 9. In these measurements, SCFS techniques were combined with optical and fluorescence microscopy to select the different components of the substrate. This substrate organization allowed to probe the interaction of prostate cancer cells with both mesenchymal stem cells and collagen I simultaneously. The result findings provided quantitative data for the high affinity of prostate cancer for these bone components as compared to other nonspecific cell lines (LNCaP). These measurements coupled with semi-quantitative PCR data identified integrins $\alpha 1\beta 1$ and $\alpha 2\beta 1$ as possible responsible for this high-binding affinity. Analogously, higher binding affinity of breast cancer cells for mineralized extracellular matrices secreted by primary human osteoblasts was quantified and associated to $\beta 1$ integrins which are critical for bone metastasis [52].

A good knowledge of cell interaction mechanism with the external environment is also a key requirement for numerous applications in biomedical field (i.e., tissue engineering, electrical and mechanical stimulation of cells, etc.). In such case, it is very important to identify the interaction mechanisms that facilitate cell proliferation, differentiation and migration on such materials. There are some very exemplificative examples of the SCFS application and how the technique enable the study, the involvement and the role of adhesion proteins in the interaction with various nanostructured materials. Such findings provided information about the adhesion behavior of cells with nanopattern substrates differently functionalized [53], the importance of nanopattern elasticity in guiding neural precursor adhesion and differentiation [54] and also implants formation [55].

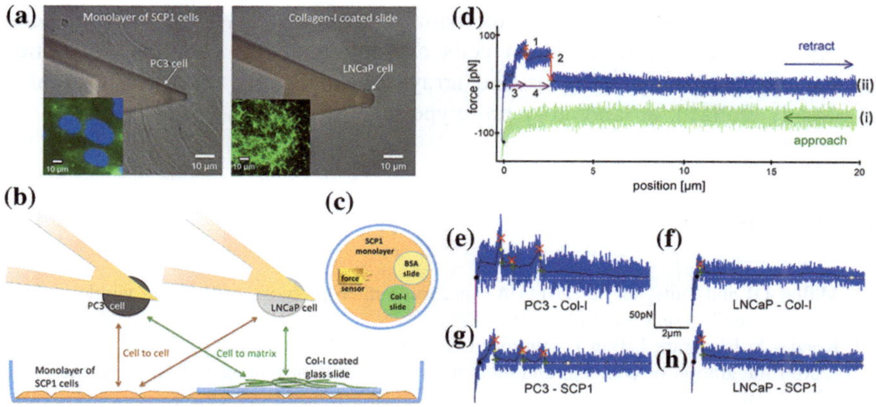

Fig. 9 **a** Phase contrast images of a prostate cancer of two cell lines (PC_3 and LNCaP) attached to the cantilever indicated by the arrows above an SCP1 (mesenchymal stem cell) monolayer (*left*) and a Collagen-I-coated slide (*bone matrix*) (*right*). On the *lower left* corners immunofluorescence images are inserted illustrating cell and collagen distribution. Collagen I, labeled with AlexaFluor488 fluorescence dye appears in green and cell nuclei, stained with DAPI in blue. **b** Schematic illustration that depicts the SCFS experiment along with a schematic *top view* of the glass dish with a BSA-coated glass cover slip (as substrate for fishing prostate cancer cell) and a Collagen-I coated glass cover slip and the monolayer of mesenchymal stem cells (**c**). Representative force-distance curves are shown in (**d**) in *green* the approach of the prostate cancer cell to the substrate (i) and in *blue* the retraction (ii). The *black line* is the *smoothed curve* and the *red crosses* indicate detected de-adhesion steps. A *force curve* obtained from a PC3-cell interacting with Collagen I is used to illustrate the adhesion force evaluation (**d**): *Red arrow* (1) step height of the first de-adhesion event in the retraction *curve*; step height of the second de-adhesion event after a force plateau of 0.9 µm in length (2); step position of the first de-adhesion event (3); step position of the second de-adhesion event (4). Typical retraction curves from each of the four different experiments are shown: (**e**) PC3 on Col-I, (**f**) PC3 on SCP1 monolayer, (**g**) LNCaP on Col-I and (**h**) LNCaP on SCP1 monolayer (Reprinted with permission from modified [51] copyright 2013 PlosOne)

6 Future Perspectives

SCFS is a powerful technique recently introduced in the realm of biology, biophysics and medical science. Initially designed for characterize cell adhesion interaction, it was used to shed light on several different open issues, ranging from molecular biology, to biophysics, tissue engineering and biomedicine. Recently, application slightly different from the original ideas was proposed. It is worth to mention here the experiment described by Stewart et al. [56] in which the hydrostatic pressure exerted inside a cell during the mitosis circle is measured using a SCFS set up in constant height clamp configuration. In future, SCFS set up will be applied to investigate other biological issues, wherever forces are relevant. The most important limitation of these techniques is the relatively low throughput, due to the sequential process adopted in measurement collection. However, due to the increased interest in the technique and the possibility to combine it with other techniques, we expect that semi-automatized instruments will be soon delivered by AFM manufacturer, in which cells are automatically identified by optical microscopy and image analysis and arrays of cantilevers are used in parallel, as already realized, at laboratory prototype level [57].

References

1. Müller DJ and Dufrêne YF (2011) Atomic force microscopy: a nanoscopic window on the cell surface. Trends Cell Biol 21:461–469
2. Kondra S, Laishram J, Ban J, Migliorini E, Di Foggia V, Lazzarino M, Torre V, Ruaro ME (2009) Integration of confocal and atomnic force microscopy images. J Neuroscience Meth 177:94–107
3. Ban J, Migliorini E, Di Foggia V, Lazzarino M, Ruaro ME, Torre V (2011) Fragmentation as a mechanism for growth cone pruning and degeneration. Stem Cells Dev 20:1031–1041
4. Lasalvia M, Perna G, Mezzenga E, Migliorini E, Lazzarino M, L'Abbate N, Capozzi V (2011) Atomic force microscopy investigation of morphological changes in living keratinocytes treated with HgCl2 at not cytotoxic doses. J Microscopy 243:40–46
5. Polano M, Bek A, Benetti F, Lazzarino M, Legname G (2009) Structural insights into alternate aggregated prion protein forms source. J Mol Biol 393:1033–1042
6. Müller DJ, Janovjak H, Lehto T, Kuerschner L, Anderson K (2002) Observing structure, function and assembly of single proteins by AFM. Prog Biophys Mol Biol 79:1–43
7. Nordin D, Donlon L, Frankel D (2012) Characterising single fibronectin-integrin complexes. Soft Matter 8:6151–6160
8. Müller DJ, Helenius J, Alsteens D, Dufrêne YF (2009) Force probing surfaces of living cells to molecular resolution. Nat Chem Biol 5:383–390
9. Raman PS, Alves CS, Wirtz D, Konstantopoulos K (2011) Single-molecule binding of CD44 to fibrin versus P-selectin predicts their distinct shear-dependent interactions in cancer. J Cell Sci 124:1903–1910
10. Raab A, Han W, Badt D, Smith-Gill S-J, Lindsay SM, Schindler H, Hinterdorfer P (1999) Antibody recognition imaging by force microscopy. Nat Biotechnol 17:901–905
11. Hinterdorfer P and Dufrêne YF (2006) Detection and localization of single molecular recognition events using atomic force microscopy. Nat Methods 3:347–355
12. Puech P-H, Poole K, Knebel D, Müller DJ (2006) A new technical approach to quantify cell-cell adhesion forces by AFM. Ultramicroscopy 106:637–644
13. Helenius J, Heisenberg C-P, Gaub HE, Müller DJ (2008) Single-cell force spectroscopy. J Cell Sci 121:1785–1791
14. Friedrichs J, Legate KR, Schubert R, Bharadwaj M, Werner C, Müller DJ, Benoit M (2013) A practical guide to quantify cell adhesion using single-cell force spectroscopy. Methods 60:169–178
15. Manzanares M-V, Webb DJ, Horwitz AR (2005) Cell migration at a glance. J Cell Sci 118:4917–4919F
16. Hynes RO (1992) Integrins: Versatility, modulation and signaling in cell adhesion. Cell 69:11–25
17. Huttenlocher A and Horwitz AR (2011) Integrings in cell migration. Cold Spring Harb Perspect Biol 3:a0055074
18. Campbell ID and Humphries MJ (2011) Integrin structure, activation and interactions. Cold Spring Harb Perspect Biol 3:a004994
19. García AJ and gallant ND (2003) "Stick and Grip": measurement system and quantitative analyses of integrin-mediated cell adhesion strength. Cell Biochem Biophys 39:61–74

20. Garcia AJ, Ducheyne P, Boettinger D (1997) Quantification of cell adhesion using a spinning disc device and application to surface-reactive materials. Biomaterials 18:1091–1098
21. Gallant ND, Michael KE, Garcia AJ (2005) Cell Adhesion Strengthening: Contributions of Adhesive area, integrin binding and focal adhesion assembly. Mol Biol Cell 16:4329–4340
22. Owen GRH, Meredith DO, Gwynn I, Richards RG (2005) Focal adhesion quantification- A new assay of material biocompatibility? Review European Cells and Materials 9:85–96
23. Hills EC, Mustafa YYG, Bennett J, Siamantouras E, Liu K-K, Squires PE (2012) Calcium-sensing receptor activation increases cell-cell adhesion and β-cell function. Cell Physiol Biochem 30:575–586
24. Mack PJ, Kaazempur-Mofrad MR, Karcher H, Lee RT, Kamm RD (2004) Force-induced focal adhesion translocation: effects of force amplitude and frequency. Am J Physiol Cell Physiol 287:C954–C962
25. Estevez M, Fernandez-Ulibarri I, Martinez E, Egeac G and Samitiera J (2010) Changes in the internal organizzation of the cell by microstructured substrates Soft Matter 6:582–590
26. Bakker GJ, Eich C, Torreno-Pina JA, Diez-Ahedoa R, Perez-Samper G, Van Zanten TS, Figdor CG, Cambi A, and Garcia-Parajo MF (2012) Lateral mobility of individual integrin nanoclusters orchestrates the onset for leukocyte adhesion. PNAS 109:4869–4874
27. Franz CM and Müller DJ (2005) Analysing focal adhesion structure by atomic force microscopy. J Cell Sci 118:5315–5323
28. Dufrêne YF and Pelling AE (2013) Force nanoscopy of cell mechanics and cell adhesion. Nanoscale 5:4094–4104
29. Baumgartner W, Hinterdorfer P, Ness W, Raab, Vestweber D, Scindler H and Drenckhahn (2000) Cadherin interaction probed by atomic force microscopy. PNAS 97:4005–4010
30. Wojcikiewicz EP, Abdulreda MH, Zhang X, Moy VT (2006) Force spectroscopy of LFA-1 and its ligand ICAM-1 ICAM-2. Biomacromolecules 7:3188–3195
31. Provenzano PP and Keely PJ (2011) Mechanical signaling through the cytoskeleton regulates cell proliferation by coordinated focal adhesion and Rho GTPase signaling. J Cell Sci 124:1195–1205
32. Wirtz D, Konstantopoulos K, Searson PC (2011) The physics of cancer: the role of physical interactions and mechanical forces in metastasis. Nat Rev 11:512–522
33. Lama G, Papi M, Angelucci C, Maulucci G, Sica G, De Spirito M (2013) Leuprorelin acetate long-lasting effects on GnRH receptors of prasttae cancer cells: an atomic force microscopy study of agonist/receptor interaction. PlosOne 8:e52530
34. Puntheeranurak T, Neundlinger I, Kinne RKH, Hinterdorfer P (2011) Single molecule recognition force spectroscopy of trasmembrane transporters on living cells. Nat Protocols 6:1443–1452
35. Hutter J and Bechhoefer J (1993) Calibration of atomic force microscopy tips. Rev Sci Instrum 64:1868–1873
36. Panorchan P, Thompson MS, Davis KJ, Tseng Y, Konstantopoulos K, Wirtz D (2005) Single-molecule analysis of cadherin-mediated cell-cell adhesion. J Cell Sci 119:66–74
37. Hoffmann SC, Cohnen A, Ludwig T, Watzl C (2011) 2B4 engagement mediates rapid LFA-1 and actin-dependent NK cell adhesion to tumor cell as measured by single cell force spectroscopy. J Immunol 186:2757–2764
38. Lim TS, Hao Goh JK, Mortellaro A, Lim CT, Hämmerling GJ, Ricciardi-Castagnoli P (2012) CD80 and CD86 differentially regulate mechanical interactions of T-cells with antigen-presenting dendritic cells and B-cells. PlosOne 7:e45185
39. Evans EA and Calderwood DA (2007) Forces and bond dynamics in cell adhesion. Sci 316:1148–1153
40. Sherer NM (2013) Long-distance relationships: do membrane nanotubes regulate cell-cell communication and disease progression? Mol Biol Cell 24:1095–1098
41. Evans E and Ritchie K (1997) Dynamic strength of molecular adhesion bonds. Biophys J 72:1541–1555
42. Davis DM and Sowinski S (2008) Membrane nanotubes dynamic long-distance connections between animal cells. Nat Rev Mol Cell Biol 9:431–436

43. Wojcikiewicz EP, Zhang X, Moy VT (2003) Force and compliance measurements on living cells using atomic force microscopy (AFM). Biol Proced Online 6:1–9
44. Tulla M, Helenius J, Jokinen J, Taubenberger A, Müller DJ, Heino J (2008) TPA primes α2β1 integrins for cell adhesion. FEBS letters 582:3520–3524
45. Benoit M, Gabriel D, Gerisch G and Gaub HE (2000) Discrete interactions in cell adhesion measured by single-molecule force spectroscopy. Nat Cell Biol 2:313–317
46. Hoffmann S, Hosseini BH, Hecker M, Louban I, Bulbuc N, Garbi N, Wabnitz GH, Samsag Y, Spatz JP, Hämmerling GJ (2011) Single cell force spectroscopy of T cells recognizing a myelin-derived peptide on antigen presenting cells. Immunol Lett 136:13–20
47. Friedrichs J, Helenius J, Müller DJ (2010) Stimulated single-cell force spectroscopy to quantify cell adhesion receptor crosstalk. Proteomics 10:1455–1462
48. Hills CE, Jin T, Siamantouras E, Liu I K-K, Jefferson KP, Squires PE (2013) "Special K" and a loss of cell-to-cell adhesion in proximal tubule-derived epithelial cells: modulation of the adherens junctions complex by ketamine. PlosOne 8:e71819
49. Taubenberger A, Cisneros DA, Friedrichs J, Puech P-H, Müller DJ, Franz CM (2007) Revealing early steps of 21 integrin-mediated adhesion to collagen type I by using single-cell force spectroscopy. Mol Biol Cell 18:1634–1644
50. Yilmaz M and Christofori G (2010) Mechanism of motility in metastasizing cells. Mol Cancer Res 8:629–642
51. Sariisik E, Docheva D, Padula D, Popov C, Opfer J, Schieker M, Clausen-Schaumann H, Benoit M (2013) Probing the interaction forces of prostate cancer cells with collagen I and bone marrow derived stem cells on the single cell level. PlosOne 8:e57706
52. Taubenberger AV, Quent VM, Thibaudeau L, Clements JA, Hutmacher DW (2013) Delineating breast cancer cell interactions with engineered bone microenvironments. J Bone Miner Res 28:1399–1411
53. Canale C, Petrelli A, Salerno M, Diaspro A, Dante S (2013) A new quantitative experimental approach to investigate single cell adhesion on multifunctional substrates. Biosens Bioelectron 48:172–179
54. Migliorini E, Ban J, Grenci G, Andolfi L, Pozzato A, Tormen M, Torre V, Lazzarino M (2013) Nanomechanics controls neuronal precursors adhesion and differentiation. Biotechnol Bioeng 110:2301–2310
55. Bertoncini P, Le Chevalier S, Lavenus S, Layrolle P and Louarn G (2012) Early adhesion of human mesenchymal stem cells on TiO2 surfaces studied by single-cell force spectroscopy measurements. J Mol Recogn 25:262–269
56. Stewart MP, Helenius J, Toyoda Y, Ramanathan SP, Müller DJ, Hyman AA (2011) Hydrostatic pressure and the actomyosin cortex drive mitotic cell rounding. Nat 469:226–230
57. Favre M, Polesel-Maris J, Overstolz T, Niedermann P, Dasen S, Gruener S, Ischer R, Vettinger P, Liley M, Heinzelmann H, Meister A (2011) Parallel AFM imaging and force spectroscopy using two-dimensional probe arrays for application in cell biology. J Mol Recogn 24:4446–4452

Photoswitchable Ion Channels and Receptors

Antoni Bautista-Barrufet, Mercè Izquierdo-Serra and Pau Gorostiza

Abstract The development of photochromic and photoswitchable ligands for ion channels and receptors has made important contributions to optopharmacology and optogenetic pharmacology. These compounds provide new tools to study ion channel proteins and to understand their function and pathological implications. Here, we describe the design, operation, and applications of the available photoswitches, with special emphasis on ligand- and voltage-gated channels.

1 Introduction

One of the major aims of neuroscience is to develop tools to measure and manipulate the activity of neurons in the brain with cellular and action-potential resolution. The synergy of functional mapping and control of neural circuits in the brain will lead to understanding their operation and regulation, and their dysfunction in neurological diseases [1]. Ion channels and ionotropic receptors are membrane proteins that directly regulate the excitability of neural cells in response to changes

Antoni Bautista-Barrufet and Mercè Izquierdo-Serra contributed equally to this work

A. Bautista-Barrufet · M. Izquierdo-Serra · P. Gorostiza (✉)
Institut de Bioenginyeria de Catalunya (IBEC), Barcelona, Spain
e-mail: pau@icrea.cat

A. Bautista-Barrufet
Institut Català d'Investigació Química (ICIQ), Tarragona, Spain

P. Gorostiza
Centro de Investigación Biomédica en Red en Bioingeniería Biomateriales y Nanomedicina (CIBER-BBN), Zaragoza, Spain

P. Gorostiza
Institució Catalana de Recerca i Estudis Avançats (ICREA), Barcelona, Spain

F. Benfenati et al. (eds.), *Novel Approaches for Single Molecule Activation and Detection*, Advances in Atom and Single Molecule Machines,
DOI: 10.1007/978-3-662-43367-6_9,

in membrane potential and to binding of a variety of ligands. While neuronal channels and receptors are normally not sensitive to light, it has proven very useful to artificially photoregulate them in order to remotely control neuron excitability with temporal and spatial precision [2]. This aim can also be achieved by optogenetics, i.e., expressing non-neuronal natural photoswitchable channels [3].

In this chapter, we review synthetic photoswitchable ion channels and receptors, and the strategies employed to obtain them, with special emphasis on recent reports. Several reviews have been published that discuss photostimulation methods [4, 5], design and mechanism of photoswitches [6], applications to photocontrol cell signaling and neuronal activity [2, 5, 7], or specifically focused on ligand-gated ion channels [8] and potassium ion channels [9].

Light sensitivity can be achieved by using protein ligands that can be switched between an active and an inactive form with light. Ligands include pore blockers, orthosteric agonists and antagonists, and allosteric modulators. They are linked to a light-regulated moiety formed by a photochromic group like azobenzene, which changes in polarity, geometry, and end-to-end distance upon isomerization. This light-regulated change can be designed to alter the ligand properties (affinity, agonist/antagonist character) or the accessibility of the ligand to its binding site in the protein. Synthetic photoswitches can be classified as soluble photochromic ligands (PCLs, Table 1) if they are free in solution or as photoswitchable tethered ligands (PTLs, Table 2) if they are covalently attached to the protein. The development of these light-regulated ligands has given rise, respectively, to the fields of optopharmacology (including compounds that can act on endogenous receptors) and optogenetic pharmacology (when the PTL is covalently attached to mutated residues in the protein, usually cysteine) [5]. Here, we have classified photoswitches according to the target channel family, and in each case, we specify the methodology used and the reported applications of light regulation.

2 Pentameric Ligand-Gated Receptors

2.1 Introduction

The pentameric ligand-gated receptor family (also known as "Cys-loop" receptors because they have a conserved extracellular region with a disulfide bridge) is a superfamily of receptors located at the postsynaptic cell membrane, showing both excitatory and inhibitory effects on the mammalian central nervous system (CNS) [10]. This superfamily includes nicotinic acetylcholine receptors (nAChr), serotonin receptors (5-HT_3), γ-aminobutyric acid receptors ($GABA_A$), and glycine receptors (GlyR) [11].

The receptors of this family are organized as homo or heteropentamers, with up to four different subunits α, β, γ, δ, and different stoichiometries. In heteromers, the neurotransmitter binding sites are usually located at the interface between α and β subunits. The structure of some prokaryotic homologs of these receptors is known

Table 1 Properties of the PCLs and its targeted receptors and ion channels

	Photoswitch	Target	Ligand	Activity	Azobenzene $\lambda_{trans \rightarrow cis}/\lambda_{cis \rightarrow trans}$ (nm)	Active conformation	Reference
PCL	BisQ	nAChR	Quaternary ammonium	Agonist	380/500	*trans*	[17]
	MPC088	$GABA_A$ receptor	propofol	Potentiates, allosteric ligand	500/380	*trans*	[28]
	AP-2	$GABA_A$ receptor	propofol	Potentiates, allosteric ligand	404/fast thermal relaxation	*trans*	[27]
	Gluazo	Kainate receptor	Glutamate	agonist	380/500	*trans*	[38]
	ATA	AMPA receptor	BnTetAMPA	agonist	480/fast thermal relaxation	*trans*	[41]
	AAQ	Shaker $K_v1.2$, $K_v1.3$, $K_v1.4$, $K_v2.1$, $K_v3.3$, $K_v4.2$	Quaternary ammonium	pore blocker	380/500	*trans*	[63, 64]
	BzAQ	Shaker	Quaternary ammonium	pore blocker	380/500	*trans*	[64]
	PrAQ	Shaker	Quaternary ammonium	pore blocker	380/500	*cis*	[64]
	DENAQ	$K_v3.1$	Quaternary ammonium	pore blocker	480/fast thermal relaxation	*trans*	[66]
	PhENAQ	Shaker K_v	Quaternary ammonium	pore blocker	480/fast thermal relaxation	*cis*	[66]
	QAQ	Shaker Kv Na_v Ca_v	Quaternary ammonium	pore blocker	380/500	*trans*	[65]
	2,2'-dimethoxy-QAQ	Shaker K_v Na_v	Quaternary ammonium	pore blocker	420/fast thermal relaxation	*trans*	[67]
	2,2'-dimethyl-QAQ and 2,2',6,6'-tetramethyl-QAQ				380/500		
	ABCTC	TRPV1	BCTC	voltage antagonist	370/420	*cis*	[29]
	AC4	TRPV1	capsazepin	Capsaicin antagonist voltage antagonist	360/440	*cis* *trans*	[29]

Table 2 Properties of the PTLs and its targeted receptors and ion channels

	Photoswitch	Target	Ligand	Activity	Ligand reactive moiety	Azobenzene $\lambda_{trans \rightarrow cis}/\lambda_{cis \rightarrow trans}$ (nm)	Active conformation	Reference
	QBr	Wild-type AChR	Quaternary Ammonium	agonist	Benzyl bromide group	380/500	*trans*	[17]
	MAACh	LinAChRs α3β4-E61C α4β2-E61C	AC-5	agonist	Maleimide	380/500	*cis*	[22]
	MAHoCh	LinAChRs α3β4-E61C α4β2-E61C	HoChPE and MG-624	antagonist	Maleimide	380/500	*cis*	[22]
	MPC100	γ-79C $GABA_A$Rs	propofol	Potenciates, allosteric ligand	Maleimide	380/500	*trans*	[28]
PTL	MAG	LiGluR GluK2-L439C HyLighter iGluR6-p0-C-L439C	Glutamate	agonist	Maleimide	380/500 820 nm for two-photon stimulation	*cis*	[35, 50] [42]
	L-$MAG0_{460}$	LiGluR GluK2-L439C	Glutamate	agonist	Maleimide	445/fast thermal relaxation	*cis*	[37]
	MAG_{2p}	LiGluR GluK2-L439C	Glutamate	agonist	Maleimide	425/fast thermal relaxation 900 for two -photon stimulation	*cis*	[50]
	$MAGA_{2p}$	LiGluR GluK2-L439C	Glutamate	agonist	Maleimide	425/fast thermal relaxation 740 and 880 for two-photon stimulation	*cis*	[50]

(continued)

Table 2 Properties of the PTLs and its targeted receptors and ion channels

	Photoswitch	Target	Ligand	Activity	Ligand reactive moiety	Azobenzene $\lambda_{trans \to cis}/\lambda_{cis \to trans}$ (nm)	Active conformation	Reference
	D-MAG-1	LimGluR2-block mGluR2-L302C LimGluR6-block mGluR6-K306C	Glutamate	antagonist	Maleimide	380/500	*cis*	[43]
	D-MAG-0	LimGluR2 mGluR2-L300C LimGluR3 mGluR3-Q306C	Glutamate	agonist	Maleimide	380/500	*cis*	[43]
	MEA-TMA	P2X2-I328C	-	With ATP: pore blocker Without ATP: channel gating	Quaternary ammonium and maleimide	365/525	*trans*	[59]
PTL	BMA	P2X2-P329C P2X3-P320C ASICS1-G430C and I428C	-	Channel gating	Maleimide and maleimide	360/440	*trans*	[58]
	MAQ	SPARK Shaker-E422C	Quaternary ammonium	Pore blocker	Maleimide	380/500	*trans*	[60]
		D-SPARK Shaker-V443Q and E422C						[61]
		Kv1.3- H401Y and P374C Kv3.1-E308C Kv3.4-D420C Kv7.2-E257C SK2-Q339C						[62]
		TREK1-K231C TREK-PSC TREK1ΔC-S121C					*trans* *cis*	[69]

and was reviewed recently [12]. Three conserved domains can be distinguished in this protein family. First is a large extracellular N-terminal domain including residues responsible for ligand-type specificity, which differs between the family subgroups. The second part is the transmembrane domain with four hydrophobic entities (M1–M4) formed by β-sheet segments. M2 lines the ion channel pore and is responsible for ion permeation and selectivity. And third, a hydrophilic domain between M3 and M4, exposed to the cytoplasm [13, 14].

Allosteric ligands bind to a different protein spot than the neurotransmitter (orthosteric ligand) and regulate the activity of the receptor. In pentameric receptor-channels, the three structural domains described above are thought to host allosteric binding sites. Allosteric ligands of nAChrR include ions (Ca^{2+} and Zn^{2+}), ivermectin, and steroids. Benzodiazepines and the anesthetic propofol are allosteric modulators of $GABA_A$ receptors [14, 15].

2.2 Nicotinic Acetylcholine Receptors (nAChR)

nAChRs and serotonin receptors (5-HT_3) are ligand-gated, cation-permeable channels with excitatory activity, responsible for the fast synaptic transmission in muscles or nerves [10]. nAChr was the first ligand-gated protein channel to be isolated and greatly helped to understand neurotransmission mechanisms [16]. This protein was also studied with several chemical biology tools, including PCLs, which originated the first light-gated receptor [17]. This pioneering contribution introduced the PCL and PTL concepts using an azobenzene unit linked to a quaternary ammonium, a receptor agonist. When the quaternary ammonium is present in both sides of the azobenzene, the compound (Bis-Q) acts as a PCL. This symmetry confers cholinergic activity to the *trans* isomer and reduced agonism in *cis*, presumably due to steric hindrance caused by the bent configuration of the azobenzene group. Those results were obtained by measuring the transmembrane potential of electric eel electroplax organs. In order to demonstrate the PTL approach, they used a compound in which azobenzene was flanked by one quaternary ammonium salt and a benzyl bromide group (QBr). Benzyl bromide reacts with a well-characterized native cysteine residue close to the binding site [18] and covalently attaches QBr. In this way, QBr produces the same photoregulated activity than a PCL, but with persistent activation–deactivation cycles [17]. Reversible photocontrol of nAChRs was demonstrated using rat myoballs, electric eel electroplaques, and frog muscles and was extensively studied in the following years [19, 20]. However, in order to make the target cysteine available for conjugation to the PTL, the use of a strong reducing agent was required [21], which limited cellular applications to mechanistic channel gating studies.

The photoswitchable nAChR was recently revisited [22] taking advantage of the structural, genetic, and computational methods not available at the time of the original design. PTL-anchoring cysteines were selected based on structural homology models and simulations and introduced by site-directed mutagenesis in nAChRs ($\alpha_3\beta_4$ and $\alpha_4\beta_2$ heteromers). Two PTLs including the agonist acetylcholine and the antagonist homocholine were tested in order to, respectively, photoactivate and photoinhibit the receptors [22]. Position E61C at the β subunit of $\alpha_4\beta_2$ receptors led to activation of the receptor using the agonist maleimide-azobenzene-acylcholine (MAACh), which can be photoswitched reversibly between violet (380 nm) and green illumination (500 nm). Attaching the antagonist maleimide-azobenzene-homocholine (MAHoCh) at the same residue produced photoswitchable inhibition between violet and green lights. Thus, two different nAChR PTLs were rationally designed and synthesized with analogous photoregulation but opposite physiological responses, by varying the linker length and replacing an ester functional group by an ether. This methodology opened a general pathway to photosensitizing pentameric ligand-gated ion channels.

2.3 Ionotropic γ-Aminobutyric Acid Receptor ($GABA_A$)

The receptors of γ-aminobutyric acid ($GABA_A$ and $GABA_C$) and glycine (GlyRs) are chloride-selective channel proteins and thus have an inhibitory effect on the neuronal activity in the CNS. They are important targets for anxiety treatment and anesthesia [10], and controlling their activity with light is attractive because it would allow inhibiting neuronal tissue using designed spatiotemporal patterns, for example, to functionally isolate specific neuronal circuits. This aim can be achieved with natural light-gated ion pumps like halorhodopsin or archaerhodopsin [23, 24], but requires continuous illumination, which is not convenient for certain applications [25]. In principle, photoswitchable glycine or $GABA_A$ receptors with slow relaxation lifetimes could be stimulated with a short light pulse, and the open channel would inhibit neurons in the dark. Tethered glycine or GABA receptor ligands are required to design PTLs of these receptors, but adding a tether reduces the affinity in the case of $GABA_A$ agonists [26]. However, a photoswitchable competitive inhibitor has been used to regulate $GABA_A$ receptors with light [5].

As an alternative to the orthosteric site, an allosteric binding site like that of propofol [14, 15] can be the target of photoswitch design. Two recent articles reported a PCL [27, 28] and a PTL [28] that potentiate GABA currents in a light-dependent fashion. A series of azo-derived propofol ligands were developed and applied to $\alpha_1\beta_{2/3}\gamma_2$ heteropentamers. These compounds are active in *trans* and inactive in *cis*, but differ in photostability, therefore in their potential applications. The photochromic AP-2 ligand, synthesized using the classical diazo coupling, presents a red-shifted *cis* isomerization (404 nm) and a fast *trans* restoration due to para-substituted electrodonating groups that decrease the *cis* thermal stability. The compound was tested in tadpoles where it controlled the anesthetic action of propofol using one wavelength [27]. On the other hand, the PTL MPC088 [28] shows *cis–trans* reversibility between 365 and 440 nm, respectively, and allows modulating GABA currents in retinal neurons and cerebellar Purkinje neurons. Spatiotemporal control of neuronal spike rate was demonstrated using this allosteric PTL.

3 Tetrameric Ligand-Gated Receptors

3.1 Introduction

Glutamate is the main excitatory neurotransmitter in the mammalian CNS. It binds and activates a highly conserved orthosteric ligand-binding domain (LBD) that is present in both ionotropic and metabotropic glutamate receptors (mGluRs) [30]. Ionotropic glutamate receptors (iGluRs) are assembled as tetramers where each subunit is composed of four conserved domains: an extracellular amino-terminal domain, the LBD, the transmembrane domain that includes the ionic pore, and an intracellular C-terminal domain. Depending on the pharmacology and structural homology, glutamate receptors are divided in four classes: (1) N-methyl-D-aspartate (NMDA) receptors (NMDARs) have high affinity for NMDA except for the glycine-binding GluN1 subunit, (2) α-Amino-3-hydroxy-5-methyl-4-isoxazolepropionic acid (AMPA) receptors (AMPARs) are activated by AMPA and include GluA1–GluA4 subunits; (3) kainate receptors (KARs) that can be activated with AMPA and have high affinity for kainate include GluK1–GluK5 subunits; and finally (4), the less known family of δ receptors [31].

The LBD of iGluRs and mGluRs are evolutionary related, and they might have originated from the same precursor, the periplasmic binding proteins of gram-negative bacteria. The common structure of this module is characterized by two large polypeptide lobes connected via a flexible linker, and the ligand-binding site that is formed by the cleft between the lobes. Upon ligand binding, there is a conformational change and the two lobes twist and close entrapping the ligand, hence the name Venus flytrap [32, 33].

In the case of the iGluRs, the two lobes are composed by sequentially separated segments. Closing of the two lobes upon ligand binding induces conformational changes on the transmembrane region, which lead to opening of the pore and receptor activation. The degree of LBD closure is proportional to pore opening, and it depends on the ligand bound to the receptor: agonists allow total closure of the lobes, partial agonists induce a more open conformation of the LBD and antagonists stabilize the open state of LBD [32, 33]. Similar conformational changes to the LBD occur upon ligand binding to mGluRs. These receptors have seven transmembrane helices with the LBD located at the extracellular N-terminus of the protein. They assemble in dimers whose LBDs rest in the open state (in the absence of ligand or stabilized by an antagonist) or in the closed state (stabilized in the presence of an agonist). The change in the relative orientation of the two LBDs in the closed state is determinant for the activation of these receptors [34].

Glutamate receptors play crucial roles in brain function. They support glutamatergic neurotransmission and are involved in many other fundamental processes such as regulation of synaptic strength, plasticity (memory and learning), and brain development [30]. Glutamate receptors have been the target of several approaches to control their activity with light, with the aim of controlling and understanding these critical processes.

3.2 Kainate Receptors

The application of PTL approach to kainate receptors led to the development of the light-gated glutamate receptor (LiGluR) [35]. An agonist (glutamate) was covalently tethered to the protein via an azobenzene. The photoswitch maleimide-azobenzene-glutamate (MAG) was conjugated to a cysteine introduced in the LBD of the kainate receptor subunit GluK2. The agonist was able to bind and activate the receptor (open the channel) under violet illumination (380 nm), and green (500 nm) light caused the ligand to unbind and close the channel. Photoswitching occurs in a step function manner due to the slow thermal relaxation of MAG: The channel remains open in the dark for minutes until it is illuminated again with green light. The mechanism of activation and deactivation involves changes in the local effective concentration of the glutamate moiety near its binding site, associated to the isomerization of MAG. The conjugation of MAG to the receptor is mediated by the affinity of the ligand and can be modulated with illumination [36].

Recently, a new PTL has been developed for LiGluR (L-MAG-0_{460}) with modified functional groups that allow *cis–trans* isomerization by visible light (400–520 nm) and millisecond thermal relaxation in the absence of illumination. This novel MAG derivative enables single wavelength operation of LiGluR [37].

The PCL strategy was also applied to KARs and allowed to control the activity of native receptors using GluAzo [38]. The design of this KAR-selective ligand is based on the KAR agonist LY339434. *Trans*-GluAzo has a higher affinity to GluK1 and GluK2 than the *cis* isomer, such that receptor-mediated currents can be photoswitched in a certain GluAzo concentration range. This approach has been used to reversibly trigger neuronal firing with light in non-transfected hippocampal neurons [38].

Computational models of ligands docked in their receptors have been extensively used to understand the action mechanism of photoswitches and to design new actuators [22, 39]. However, the crystal structure of a receptor bound to its photoswitchable ligand has been resolved only recently, revealing their detailed molecular interactions [40]. The structure of the planar *trans*-GluAzo bound to GluK2-LBD shows a similar interaction as free glutamate, while the bent *cis*-GluAzo might interfere with LBD closure, and subsequent ion channel opening. The molecular contacts show in striking detail how the azobenzene tether of *trans*-GluAzo is enclosed within an "exit tunnel" of the GluK2-LBD. This structure is similar to that of the LBD-domoate complex and displays a conformation of the two lobes that is not fully closed, which explains the partial agonist character of GluAzo.

3.3 AMPA Receptors

A PCL of AMPARs has also been developed [41]. The LBD of AMPARs and KARs are homologous, but in AMPARs, the LBD closes more tightly, which poses difficulties to design a PCL based on GluAzo. The X-ray structure of the

AMPAR LBD bound to a modified ATPA agonist, BnTetAMPA, reveals a cleft for ligand exit that is different from that used by GluAzo. A series of PCLs named azobenzene tetrazolyl ATPA (ATA) was designed to accommodate to this cleft, thus allowing the AMPA receptor clamshell to close for receptor activation. They display red-shifted isomerization and fast thermal relaxation. *Trans*-ATA isomers are strong agonists of GluA2 receptors, and agonism is reduced with 480 nm light (*cis* configuration). These compounds were used in pyramidal neurons to trigger trains of action potentials with light [41].

3.4 Bacterial iGluRs

An alternative approach to generate new light-activated channels is to take advantage of the modularity of glutamate receptor domains. This strategy was used to generate a K^+selective light-activated channel termed HyLighter [42] in which the LBD of GluK2 (including the PTL attachment site) was combined with the pore region of sGluR0 (a K^+selective prokaryotic receptor homolog to iGluRs). As it happens with LiGluR [39], *cis*-MAG-0 binds to the LBD and activates HyLighter. In this case, however, photocurrents are mainly due to K^+ and cause the hyperpolarization of the cell membrane. Since the *cis* isomer is thermally stable, continuous illumination is not needed to silence neurons as in the case of light-gated ion pumps. After a brief violet light pulse, HyLighter keeps the membrane hyperpolarized until 500 nm illumination inactivates the receptor.

3.5 Metabotropic Glutamate Receptors

Since glutamate-binding domains are highly conserved among all glutamate receptors, in principle, few structural changes are required in order to extend the PTL approach from the LBD of ionotropic receptors to that of mGluRs. In this way, the PTL strategy was extended to the optical control of a G protein-coupled receptor [43]. Two distinct enantiomeric photoswitches of MAG-0 and MAG-1 were synthesized, in which the stereochemistry of the glutamate carbon was changed from 4–L to 4–D. In this way, the D-tethers D-MAG-0 and D-MAG-1 (D-maleimide-azobenzene-glutamate) were obtained that better accommodate in the LBD of mGluRs. These photoswitches were tested on mGluR2 by means of a cysteine introduced in its LBD. The longer photoswitch, D-MAG-1, antagonized glutamate under 380 nm light in the mGluR2-L302C mutant, which they called LimGluR2-block. In contrast, the short PTL, D-MAG-0, successfully activated the receptor when it was conjugated to mGluR2-L300C mutant (LimGluR) under 380 nm light. These agonist and antagonist photoswitches can also be used to photocontrol the activity of the mGluR3-Q306C (LimGluR3) and mGluR6-K306C (LimGluR6-block) [43].

3.6 Applications

LiGluR has been applied to photocontrol a wide variety of systems such as cultured neurons, astrocytes, chromaffin cells, mouse retina, and zebrafish locomotive responses. LiGluR contains the non-selective cationic pore of GluK2, and its sodium permeability allows depolarizing neurons and precisely controlling their activity with light. In cultured hippocampal neurons expressing LiGluR, violet light triggers trains of action potentials that are maintained in the dark until the channel is closed with green light. Millisecond photostimulation of LiGluR evokes single action potentials that precisely correlate with the photoswitch pattern up to a frequency of 50 Hz [44]. LiGluR has also been used in vivo to investigate the function of neuronal subpopulations of the zebrafish spinal cord [45]. The activation by means of LiGluR of each subset of neurons induced different locomotive responses enabling to understand their underlying circuitry.

The relatively high calcium permeability of GluK2 makes LiGluR an effective tool to optically induce calcium currents and trigger calcium-dependent processes. Light induced intracellular calcium elevation on LiGluR-expressing astrocytes and a delayed activation of the neighboring cells and was used to study astrocyte-to-astrocyte communication [46]. LiGluR also allows to increase intracellular calcium concentration with light and directly triggered exocytosis in chromaffin cells independently of voltage-gated calcium channel activation [47]. In this way, the calcium influx through LiGluR can be regulated with the wavelength of illumination and has been used to trigger and modulate neurotransmission in cultured hippocampal neurons [48].

The therapeutic applications of LiGluR have also been explored. In retinal degeneration, which is characterized by the death of photoreceptors, the surviving non-photoreceptor cells can be utilized to restore visual responses. These cells have been photosensitized by using electrode arrays in humans or gene therapy with opsins in mice. In an alternative gene therapy approach, Caporale and co-workers inserted LiGluR to retinal ganglion cells in a mice model of blindness and showed that after MAG administration, the animals responded to violet light [49]. New red-shifted PTL L-$MAG0_{460}$ might then be used to restore the response to visible light [37].

Recently, two-photon stimulation of MAG and two derivatives (MAG_{2p} and $MAGA_{2p}$) was reported to enable the activation of astrocytic and neuronal cells with cellular and subcellular resolution [50]. Versatile synthetic strategies were described to enhance the photochemical responses of azobenzene-based photoswitches in order to exploit the high spatiotemporal precision and tissue penetration of two-photon stimulation.

PCLs of glutamate receptors have also been applied to study neuronal systems. The kainate photoswitchable agonist GluAzo allowed photomodulating oscillations in cerebellar Purkinje neurons, a continuously firing neuron that is thought to be an integrator for the motor control system of the brain. When the photoswitchable agonist is added to the system in dark, it induces astability (oscillation between rapid firing and quiescence) in a similar way to other kainate agonists. However, application of 500/380 nm light pulses enabled tuning of astable oscillations by switching on and off the neuronal bursts [51].

On the other hand, the light-gated, K^+permeable channel HyLighter has been used to silence neuronal activity. It was tested in transgenic zebrafish expressing HyLighter in spinal cord motor neurons and enabled to reversibly inhibit escape responses when HyLighter was activated under violet light [42].

4 Trimeric Ligand-Gated Receptors

Among purinergic receptors, which are activated by ATP or adenosine, P2X receptors are cation-permeable non-selective (Na^+, K^+, Ca^{2+}) ion channels, with a trimeric structure that was solved recently [52]. They are gated by extracellular adenosine-5-triphosphate (ATP) that promotes a conformational rearrangement and pore opening, rendering an excitatory, depolarizing effect on cells. These receptors are involved in various physiological processes having a role in neurotransmission and neurosecretion and are associated with sensory, motor, and cognitive tasks.

These receptor-channels are structurally organized in three conserved domains: (1) a pore domain with two transmembrane segments that shares similarities with inward rectifying K^+ channels, (2) an unusually large extracellular fragment containing the ATP binding site and a Cys-loop region similar to that of pentameric receptors, and (3) a short intracellular domain [53]. Several allosteric ligands have been described for these receptors [52, 54].

In a series of pioneering optogenetic experiments to manipulate animal behavior, Gero Miesenböck used P2X channels to photocontrol membrane potential using ATP uncaging [55, 56], which had some advantages over his rhodopsin-based charge system [57].

More recently and guided by the available atomic structure, both PTL and azo-crosslinking strategies have been used to control trimeric ligand-gated receptors with light. In P2X receptors, a symmetric maleimide crosslinker (BMA, (4,4́-Bis(maleimide)azobenzene)) was introduced between cysteines, changing the conformation of the channel protein in milliseconds upon 440 nm illumination, and reversing the effect at 360 nm [58]. However, this modification abolishes ATP dependence. The same crosslinker was applied in another trimeric channel, the acid-sensing ion channel (ASICs), which is activated by the acidity of the extracellular medium via structural transitions similar to those of P2X. In ASIC1, BMA was covalently conjugated between introduced cysteines and resulted in *trans* activated, partial photocurrents [58].

P2X receptors have also been photocontrolled using a compound with PTL architecture: MEA-TMA has a charged, unspecific "ligand" that is able to modulate channel gating with light of 365 and 525 nm. Two photocontrol mechanisms were found: a photomodulation of ATP response and direct photoactivation in the absence of ATP [59]. Photoswitchable P2X allowed photocontrolling action potentials in hippocampal neurons and opened the way to investigate other P2X receptors involved in pain and inflammatory pathologies.

5 Potassium Channels

5.1 Introduction

Potassium channels constitute a big and diverse family of ion channels proteins. Among other criteria, they can be classified in three groups by their general architecture: (1) the family of inward rectifying K^+channels (K_{ir}) are tetramers formed by subunits with two transmembrane domains that are activated by hyperpolarization, (2) the two-pore (2P) potassium channels are dimers formed by subunits with four transmembrane domains that assemble forming a two-pore ion channel, and (3) the voltage-gated potassium channel family. This group is the biggest and is evolutionary related to voltage-gated Na^+ (Na_v) and Ca^{2+} (Ca_v) channels. All these voltage-gated channels have a conserved motif composed of six transmembrane segments (S1–S4 form the voltage sensor domain and S5–S6 form the pore domain). In voltage-gated potassium channels, each motif corresponds to one subunit, and they are assembled as tetramers, whereas in Na_v and Ca_v channels, these motifs are repeated four times in a single protein. K_v channels, the small conductance Ca^{2+}activated potassium channels (SK), and cyclic nucleotide-gated (CNG) channels are members of this family. Voltage-gated potassium channels are responsible for many cellular functions like maintaining the resting membrane potential or repolarizing and hyperpolarizing the cell membrane during the propagation of action potentials in neurons.

5.2 K_v Family

The potassium channel blocker tetraethyl ammonium (TEA) was used as the ligand to develop a PTL of potassium channels [60]. A cysteine introduced at the extracellular loop of the Drosophila Shaker K^+channel allowed the covalent attachment of the PTL by means of a maleimide reactive group [maleimide-azobenzene-quaternary ammonium salt (MAQ)]. The channel pore is blocked in the *trans* configuration, under 500 nm illumination, and this effect is reversed with 380 nm light. This engineered channel was named Synthetic Photoisomerizable Azobenzene-Regulated K^+ (SPARK) and is both voltage-gated and light-controlled. It has been used for optically silencing neuronal activity. Introduction of point mutations in the pore of SPARK alters cation selectivity and yielded a depolarizing version of this channel

(D-SPARK), which was developed based on the same PTL and allows firing action potentials with light [61].

MAQ was also applied to photosensitize other potassium channels by introducing cysteine mutations at the equivalent position of SPARK [62]. The PTL principle was thus extended to four mammalian potassium channels with different cell-type distribution, gating properties, kinetics, and modulatory control. The diverse physiological functions of K^+channels suggest that overexpression of light-sensitive versions could be used not only to suppress action potentials with light, but also to fine-tune several aspects of cellular electrophysiology. For example, light-regulated $K_v7.2$ channels might be useful to control resting potential, $K_v1.3$ to control repolarization of action potentials, $K_v3.1$ to control accommodation, and SK2 to control the size of excitatory postsynaptic potentials (EPSPs).

PCLs have also been employed to photocontrol native potassium channels, avoiding the cysteine insertion into the protein. Maintaining the azo-unit linked to the pore blocker "head" (TEA), several versions of PCLs were synthesized that differ in the "tail" group: The first PCL had an acrylamide moiety [acrylamide-azobenzene-quaternary ammonium (AAQ)] that blocked the pore of Shaker and various native K_v channels in its *trans* configuration [63]. This PCL acts from the internal TEA binding site, whereas MAQ blocks the ion pore from an external binding site [64]. Since it is an intracellular blocker, the functionality of AAQ depends on its ability to reach the cytoplasm. To promote partitioning of the PCL into the membrane, two new molecules were synthesized [64]: a hydrophobic benzylamide group (BzAQ) and a propyl group at the end of the molecule (PrAQ) that, respectively, act as *trans*- and *cis*-blockers. The QAQ molecule has symmetric quaternary ammonium groups and is able to control the excitability of cells by blocking at 500 nm a wide variety of voltage-gated ionic channels (Shaker, K_v, Na_v, and Ca_v) [65]. As a double-charged molecule, it does not partition in the cell membrane but in analogy to the analgesic lidocaine, QAQ permeates through nociceptive pore channels such as TRPV1 or P2X. The accumulation of QAQ in neurons, expressing nociceptive channels like dorsal root ganglion neurons from the spinal cord, offers opportunities to use this photoswitch as a potential drug to control pain. Implanting fiber-optic systems on spinal cords would allow to spatiotemporally regulate the activity of these PCLs [65].

The photoisomerization properties of azobenzene can be adjusted by chemical substitutions like the introduction of electro-donating substituents, and in this way, several red-shifted and fast-relaxing PCLs have been described: two AAQ derivatives named DENAQ and PhENAQ [66] and two derivatives of QAQ [67].

A therapeutic application of K^+-channel PCLs is the restoration of visual responses in degenerated retina, where photoreceptor cells have died. AAQ photosensitizes surviving retinal neurons and allows to control their firing activity [68]. AAQ treatment of degenerated retina has the advantage over LiGluR [49] that photosensitization is restored without gene therapy.

5.2.1 2P-Potassium Channels

The 2P-potassium channel TREK1 was also rendered photoswitchable using the blocker MAQ and a cysteine introduced at the extracellular side by homology with SPARK [69]. Depending on the cysteine position, the PTL blocked the pore in *trans* (TREK1-K231C-MAQ) and a *cis* (TREK1-S212C-MAQ). This channel allowed defining a novel scheme to photocontrol native proteins using a "photoswitchable conditional subunit" (PCS), which contains the PTL-anchoring site and a mutation that blocks plasma membrane expression. In cells not expressing native subunits for the protein, the PCS is not functional, but when native subunits are expressed, they coassemble with the PCS, traffic to the plasma membrane, and place the native protein under optical control. Overcoming the lack of specific TREK1 blockers to pharmacologically study their function in neurons, expression of the photoswitchable PCS subunit TREKΔC-S121C allowed to specifically characterize the function of native TREK channels independently of other expressed potassium channels and related them with GABAergic signaling [69].

6 Transient Receptor Potential Channels

Transient receptor potential (TRP) channels are a superfamily of cationic channels characterized by their multiple activation mechanisms and roles in sensory physiology. TRP channels have a six-transmembrane structure like voltage-gated cation channels, but they can be considered as integrators of multiple signals because the response of one input can be modified by another. The capsaicin receptor TRPV1 is one of the best studied channels of this diverse superfamily. TRPV1 responds to several stimuli from heat to proinflammatory agents, pH, other chemicals [endocannabinoid, anandamide, camphor, piperine (from pepper), and allicin (from garlic)] or exocytosis [70].

Photocontrol of TRPV1 channels using agonist uncaging was used in early studies to manipulate neurons and behavior with light [55]. More recently, two PCLs were developed to reversibly control TRPV1, consisting of azobenzene derivatives of capsazepines and BCTC (AC4 and ABCTC) [29]. These compounds displayed potent voltage antagonism at complementary conformations, in *trans* for AC4 (440 nm) and in *cis* for ABCTC (370 nm). In the presence of the agonist capsaicin, AC4 was able to strongly inhibit TRPV1 responses in the *cis* state and this effect was reversible in *trans* [29].

7 Future Challenges

Although the first synthetic photoswitches of ion channel activity were reported more than 40 years ago [17], this field has boomed in the last decade owing to the availability of pharmacology, high-resolution protein structures, recombinant

DNA technology, heterologous expression systems, and molecular dynamics simulations, which have allowed the rational development of photoswitches for almost every class of ion channel present on neurons. Photocontrol of voltage-gated calcium channels and receptors of NMDA, glycine, and serotonin has not yet been described, but will be facilitated by the reported examples of structurally related proteins.

PCLs act on endogenous proteins and can be useful therapeutically (opto-pharmacology) and to study the physiology of intact tissues and organisms. These applications would benefit from the availability of PCLs for the abovementioned channels and receptors, and from improved photochemical and pharmacological properties (selectivity, potency). In this way it would be possible to study the roles played by each channel in its complex, undisturbed physiological context.

On the other hand, although PTLs provide robust photoswitching that is not subject to diffusion, the current PTL conjugation methodology requires mutating the channel to introduce cysteines and thus does not allow gaining direct control over the activity of endogenous channels. Channels that have been engineered to conjugate to a PTL must be overexpressed in tissue or a transgenic or knock-in animal must be developed. Therefore, the interest of PTLs would increase if a conjugation methodology could be developed for endogenous proteins.

An outstanding advantage of synthetic photoswitches is that they are based on chromophores that act on the protein surface, in contrast to the tight-binding pockets found in natural light-sensitive proteins. Synthetic photoswitches thus offer excellent opportunities for rationally tuning their photochemical characteristics by chemical substitutions that do not affect the functional properties of the protein. The absorption spectra, photostationary state, and relaxation lifetime have been adjusted in several photoswitches [37, 66, 67], but multiphoton stimulation has only been recently investigated despite the advances of neurotransmitter uncaging and optogenetics using pulsed illumination [50]. Two-photon switching with near-infrared light is attractive to stimulate ion channels deep into tissue [71], to achieve high three-dimensional resolution [72] and patterned illumination [73, 74]. Finally, while all synthetic photoswitches reported for ion channels so far are based on azobenzene, different photoisomerizable groups have been successfully applied to other proteins [75] and have advantages of their own that could be exploited with channels.

References

1. Alivisatos AP, Chun M, Church GM, Deisseroth K, Donoghue JP, Greenspan RJ, McEuen PL, Roukes ML, Sejnowski TJ, Weiss PS, Yuste R (2013) Neuroscience. The brain activity map. Science 339:1284–1285
2. Gorostiza P, Isacoff EY (2008) Optical switches for remote and noninvasive control of cell signaling. Science 322:395–399
3. Tye KM, Deisseroth K (2012) Optogenetic investigation of neural circuits underlying brain disease in animal models. Nat Rev Neurosci 13:251–266

4. Gorostiza P, Isacoff E (2007) Optical switches and triggers for the manipulation of ion channels and pores. Mol BioSyst 3:686–704
5. Kramer RH, Mourot A, Adesnik H (2013) Optogenetic pharmacology for control of native neuronal signaling proteins. Nat Neurosci 16:816–823
6. Gorostiza P, Isacoff EY (2008) Nanoengineering ion channels for optical control. Physiology (Bethesda) 23:238–247
7. Szobota S, Isacoff EY (2010) Optical control of neuronal activity. Annu Rev Biophys 39:329–348
8. Szobota S, McKenzie C, Janovjak H (2013) Optical control of ligand-gated ion channels. Methods Mol Biol 998:417–435
9. Mourot A, Fehrentz T, Kramer RH (2013) Photochromic potassium channel blockers: design and electrophysiological characterization. Methods Mol Biol 995:89–105
10. Connolly CN, Wafford KA (2004) The Cys-loop superfamily of ligand-gated ion channels: the impact of receptor structure on function. Biochem Soc Trans 32:529–534
11. Ortells MO, Lunt GG (1995) Evolutionary history of the ligand-gated ion-channel superfamily of receptors. Trends Neurosci 18:121–127
12. Hilf RJ, Dutzler R (2009) A prokaryotic perspective on pentameric ligand-gated ion channel structure. Curr Opin Struct Biol 19:418–424
13. Brejc K, van Dijk WJ, Klaassen RV, Schuurmans M, van der Oost J, Smit AB, Sixma TK (2001) Crystal structure of an ACh-binding protein reveals the ligand-binding domain of nicotinic receptors. Nature 411:269–276
14. Changeux JP, Edelstein SJ (1998) Allosteric receptors after 30 years. Neuron 21:959–980
15. Yip GM, Chen ZW, Edge CJ, Smith EH, Dickinson R, Hohenester E, Townsend RR, Fuchs K, Sieghart W, Evers AS, Franks NP (2013) A propofol binding site on mammalian $GABA_A$ receptors identified by photolabeling. Nat Chem Biol 9:715–720
16. Hille B (2001) Ion channels of excitable membranes. Sinauer Associates, Inc., Sunderland
17. Bartels E, Wassermann NH, Erlanger BF (1971) Photochromic activators of the acetylcholine receptor. Proc Natl Acad Sci USA 68:1820–1823
18. Silman I, Karlin A (1969) Acetylcholine receptor: covalent attachment of depolarizing groups at the active site. Science 164:1420–1421
19. Chabala LD, Lester HA (1986) Activation of acetylcholine receptor channels by covalently bound agonists in cultured rat myoballs. J Physiol 379:83–108
20. Lester HA, Chang HW (1977) Response of acetylcholine receptors to rapid photochemically produced increases in agonist concentration. Nature 266:373–374
21. Barrantes FJ (1980) Modulation of acethylcholine receptor states by thiol modification. Biochemistry 19:2957–2965
22. Tochitsky I, Banghart MR, Mourot A, Yao JZ, Gaub B, Kramer RH, Trauner D (2012) Optochemical control of genetically engineered neuronal nicotinic acetylcholine receptors. Nat Chem 4:105–111
23. Chow BY, Han X, Dobry AS, Qian X, Chuong AS, Li M, Henninger MA, Belfort GM, Lin Y, Monahan PE, Boyden ES (2010) High-performance genetically targetable optical neural silencing by light-driven proton pumps. Nature 463:98–102
24. Zhang F, Wang LP, Brauner M, Liewald JF, Kay K, Watzke N, Wood PG, Bamberg E, Nagel G, Gottschalk A, Deisseroth K (2007) Multimodal fast optical interrogation of neural circuitry. Nature 446:633–639
25. Kasparov S, Herlitze S (2013) Optogenetics at a crossroads? Exp Physiol 98:971–972
26. Raster P, Späth A, Bultakova S, Gorostiza P, König B, Bregestovski P (2013) New GABA amides activating $GABA_A$ receptors. Beilstein J Org Chem 9:406–410
27. Stein M, Middendorp SJ, Carta V, Pejo E, Raines DE, Forman SA, Sigel E, Trauner D (2012) Azo-propofols: photochromic potentiators of $GABA_A$ receptors. Angew Chem Int Ed Engl 51:10500–10504
28. Yue L, Pawlowski M, Dellal SS, Xie A, Feng F, Otis TS, Bruzik KS, Qian H, Pepperberg DR (2012) Robust photoregulation of $GABA_A$ receptors by allosteric modulation with a propofol analogue. Nat Commun 3:1095

29. Stein M, Breit A, Fehrentz T, Gudermann T, Trauner D (2013) Optical control of TRPV1 channels. Angew Chem Int Ed Engl 52:9845–9848
30. Gereau RW, Swanson GT (2008) The glutamate receptors. Humana Press, New York
31. Traynelis SF, Wollmuth LP, McBain CJ, Menniti FS, Vance KM, Ogden KK, Hansen KB, Yuan H, Myers SJ, Dingledine R (2010) Glutamate receptor ion channels: structure, regulation, and function. Pharmacol Rev 62:405–496
32. Felder CB, Graul RC, Lee AY, Merkle HP, Sadee W (1999) The Venus flytrap of periplasmic binding proteins: an ancient protein module present in multiple drug receptors. AAPS PharmSci 1:E2
33. Sobolevsky AI, Rosconi MP, Gouaux E (2009) X-ray structure, symmetry and mechanism of an AMPA-subtype glutamate receptor. Nature 462:745–756
34. Doumazane E, Scholler P, Fabre L, Zwier JM, Trinquet E, Pin JP, Rondard P (2013) Illuminating the activation mechanisms and allosteric properties of metabotropic glutamate receptors. Proc Natl Acad Sci USA 110:1416–1425
35. Volgraf M, Gorostiza P, Numano R, Kramer RH, Isacoff EY, Trauner D (2006) Allosteric control of an ionotropic glutamate receptor with an optical switch. Nat Chem Biol 2:47–52
36. Gorostiza P, Volgraf M, Numano R, Szobota S, Trauner D, Isacoff EY (2007) Mechanisms of photoswitch conjugation and light activation of an ionotropic glutamate receptor. Proc Natl Acad Sci USA 104:10865–10870
37. Kienzler MA, Reiner A, Trautman E, Yoo S, Trauner D, Isacoff EY (2013) A red-shifted, fast-relaxing azobenzene photoswitch for visible light control of an ionotropic glutamate receptor. J Am Chem Soc 135:17683–17686
38. Volgraf M, Gorostiza P, Szobota S, Helix MR, Isacoff EY, Trauner D (2007) Reversibly caged glutamate: a photochromic agonist of ionotropic glutamate receptors. J Am Chem Soc 129:260–261
39. Numano R, Szobota S, Lau AY, Gorostiza P, Volgraf M, Roux B, Trauner D, Isacoff EY (2009) Nanosculpting reversed wavelength sensitivity into a photoswitchable iGluR. Proc Natl Acad Sci USA 106:6814–6819
40. Reiter A, Skerra A, Trauner D, Schiefner A (2013) A photoswitchable neurotransmitter analogue bound to its receptor. Biochemistry 52:8972–8974
41. Stawski P, Sumser M, Trauner D (2012) A photochromic agonist of AMPA receptors. Angew Chem Int Ed Engl 51:5748–5751
42. Janovjak H, Szobota S, Wyart C, Trauner D, Isacoff EY (2010) A light-gated, potassium-selective glutamate receptor for the optical inhibition of neuronal firing. Nat Neurosci 13:1027–1032
43. Levitz J, Pantoja C, Gaub B, Janovjak H, Reiner A, Hoagland A, Schoppik D, Kane B, Stawski P, Schier AF, Trauner D, Isacoff EY (2013) Optical control of metabotropic glutamate receptors. Nat Neurosci 16:507–516
44. Szobota S, Gorostiza P, Del Bene F, Wyart C, Fortin DL, Kolstad KD, Tulyathan O, Volgraf M, Numano R, Aaron HL, Scott EK, Kramer RH, Flannery J, Baier H, Trauner D, Isacoff EY (2007) Remote control of neuronal activity with a light-gated glutamate receptor. Neuron 54:535–545
45. Wyart C, del Bene F, Warp E, Scott EK, Trauner D, Baier H, Isacoff EY (2009) Optogenetic dissection of a behavioural module in the vertebrate spinal cord. Nature 461:407–410
46. Li D, Herault K, Isacoff EY, Oheim M, Ropert N (2012) Optogenetic activation of LiGluR-expressing astrocytes evokes anion channel-mediated glutamate release. J Physiol 590:855–873
47. Izquierdo-Serra M, Trauner D, Llobet A, Gorostiza P (2013) Optical control of calcium-regulated exocytosis. Biochim Et Biophys Acta-Gen Subj 1830:2853–2860
48. Izquierdo-Serra M, Trauner D, Llobet A, Gorostiza P (2013) Optical modulation of neurotransmission using calcium photocurrents through the ion channel LiGluR. Front Mol Neurosci 6:1–8

49. Caporale N, Kolstad KD, Lee T, Tochitsky I, Dalkara D, Trauner D, Kramer R, Dan Y, Isacoff EY, Flannery JG (2011) LiGluR restores visual responses in rodent models of inherited blindness. Mol Ther 19:1212–1219
50. Izquierdo-Serra M, Gascón-Moya M, Hirtz JJ, Pittolo S, Poskanzer KE, Ferrer E, Alibés E, Busqué F, Yuste R, Hernando J, Gorostiza P (2014) Two-photon neuronal and astrocytic stimulation with azobenzene-based photoswitches. J Am Chem Soc (Accepted Manuscript)
51. Abrams ZR, Warrier A, Wang Y, Trauner D, Zhang X (2012) Tunable oscillations in the Purkinje neuron. Phys Rev E: Stat Nonlin Soft Matter Phys 85:041905
52. Kawate T, Michel JC, Birdsong WT, Gouaux E (2009) Crystal structure of the ATP-gated P2X(4) ion channel in the closed state. Nature 460:592–598
53. Conley EC (1996) The ion channel facts book I: extracellular ligand-gated ion channels. Academic Press Limited, London
54. Valera S, Hussy N, Evans RJ, Adami N, North RA, Surprenant A, Buell G (1994) A new class of ligand-gated ion channel defined by P2X receptor for extracellular ATP. Nature 371:516–519
55. Lima SQ, Miesenböck G (2005) Remote control of behavior through genetically targeted photostimulation of neurons. Cell 121:141–152
56. Zemelman BV, Nesnas N, Lee GA, Miesenbock G (2003) Photochemical gating of heterologous ion channels: remote control over genetically designated populations of neurons. Proc Natl Acad Sci USA 100:1352–1357
57. Zemelman BV, Lee GA, Ng M, Miesenböck G (2002) Selective photostimulation of genetically ChARGed neurons. Neuron 33:15–22
58. Browne LE, Nunes JP, Sim JA, Chudasama V, Bragg L, Caddick S, Alan North R (2014) Optical control of trimeric P2X receptors and acid-sensing ion channels. Proc Natl Acad Sci USA 111:521–526
59. Lemoine D, Habermacher C, Martz A, Méry PF, Bouquier N, Diverchy F, Taly A, Rassendren F, Specht A, Grutter T (2013) Optical control of an ion channel gate. Proc Natl Acad Sci USA 110:20813–20818
60. Banghart M, Borges K, Isacoff E, Trauner D, Kramer RH (2004) Light-activated ion channels for remote control of neuronal firing. Nat Neurosci 7:1381–1386
61. Chambers JJ, Banghart MR, Trauner D, Kramer RH (2006) Light-induced depolarization of neurons using a modified Shaker K(+) channel and a molecular photoswitch. J Neurophysiol 96:2792–2796
62. Fortin DL, Dunn TW, Fedorchak A, Allen D, Montpetit R, Banghart MR, Trauner D, Adelman JP, Kramer RH (2011) Optogenetic photochemical control of designer K^+ channels in mammalian neurons. J Neurophysiol 106:488–496
63. Fortin DL, Banghart MR, Dunn TW, Borges K, Wagenaar DA, Gaudry Q, Karakossian MH, Otis TS, Kristan WB, Trauner D, Kramer RH (2008) Photochemical control of endogenous ion channels and cellular excitability. Nat Methods 5:331–338
64. Banghart MR, Mourot A, Fortin DL, Yao JZ, Kramer RH, Trauner D (2009) Photochromic blockers of voltage-gated potassium channels. Angew Chem Int Ed Engl 48:9097–9101
65. Mourot A, Fehrentz T, le Feuvre Y, Smith CM, Herold C, Dalkara D, Nagy F, Trauner D, Kramer RH (2012) Rapid optical control of nociception with an ion-channel photoswitch. Nat Methods 9:396–402
66. Mourot A, Kienzler MA, Banghart MR, Fehrentz T, Huber FM, Stein M, Kramer RH, Trauner D (2011) Tuning photochromic ion channel blockers. ACS Chem Neurosci 2:536–543
67. Fehrentz T, Kuttruff CA, Huber FM, Kienzler MA, Mayer P, Trauner D (2012) Exploring the pharmacology and action spectra of photochromic open-channel blockers. Chem Bio Chem 13:1746–1749
68. Polosukhina A, Litt J, Tochitsky I, Nemargut J, Sychev Y, de Kouchkovsky I, Huang T, Borges K, Trauner D, van Gelder RN, Kramer RH (2012) Photochemical restoration of visual responses in blind mice. Neuron 75:271–282

69. Sandoz G, Levitz J, Kramer RH, Isacoff EY (2012) Optical control of endogenous proteins with a photoswitchable conditional subunit reveals a role for TREK1 in $GABA_B$ signaling. Neuron 74:1005–1014
70. Venkatachalam K, Montell C (2007) TRP channels. Annu Rev Biochem 76:387–417
71. Papagiakoumou E, Begue A, Leshem B, Schwartz O, Stell BM, Bradley J, Oron D, Emiliani V (2013) Functional patterned multiphoton excitation deep inside scattering tissue. Nat Photon 7:274–278
72. Denk W, Strickler JH, Webb WW (1990) Two-photon laser scanning fluorescence microscopy. Science 248:73–76
73. Oron D, Papagiakoumou E, Anselmi F, Emiliani V (2012) Two-photon optogenetics. Prog Brain Res 196:119–143
74. Watson BO, Nikolenko V, Yuste R (2009) Two-photon imaging with diffractive optical elements. Front Neural Circ 3:6
75. Szymanski W, Beierle JM, Kistemaker HAV, Velema WA, Feringa BL (2013) Reversible photocontrol of biological systems by the incorporation of molecular photoswitches. Chem Rev 113:6114–6178

The Use of Light-Sensitive Organic Semiconductors to Manipulate Neuronal Activity

Duco Endeman, Paul Feyen, Diego Ghezzi, Maria Rosa Antognazza, Nicola Martino, Elisabetta Colombo, Guglielmo Lanzani and Fabio Benfenati

Abstract Organic semiconducting polymers display several beneficial properties to interface with biological substrates. These materials have been employed successfully for cellular interfaces such as neural probes, biosensors, and actuators for drug release. Recent experiments demonstrate that they can also be used to optically modulate the membrane potential of cultured neurons and astrocytes. Moreover, application of an organic light-sensitive semiconductor has been shown to restore light sensitivity in degenerated retinas, suggesting the use of conjugated polymers as active materials in retinal prosthesis.

1 Introduction

Modulation of neuronal activity is a principle strategy in neuroscience research and in treatment of neurological and psychiatric disorders. Presently, a diverse range of techniques is available to implement this strategy. In neuroscientific experiments, they include the use of stimulation electrodes, application of neurotransmitters, and transcranial magnetic stimulation. Administration of certain

D. Endeman (✉) · P. Feyen · D. Ghezzi · E. Colombo · F. Benfenati
Department of Neuroscience and Brain Technologies, Istituto Italiano di Tecnologia, Genoa, Italy
e-mail: duco.endeman@iit.it

M. R. Antognazza · N. Martino · G. Lanzani
Center for Nano Science and Technology @PoliMi, Istituto Italiano di Tecnologia, Milan, Italy

G. Lanzani
Department of Physics, Politecnico di Milano, Milan, Italy

N. Martino · F. Benfenati
Department of Experimental Medicine, University of Genova, Genoa, Italy

F. Benfenati et al. (eds.), *Novel Approaches for Single Molecule Activation and Detection*, Advances in Atom and Single Molecule Machines,
DOI: 10.1007/978-3-662-43367-6_10,

pharmaceuticals (e.g., L-dopa or selective serotonin reuptake inhibitors), cochlear implants, and deep brain stimulation are prevalent examples of clinical applications, which modulate neuronal activity.

Optical techniques, aimed at manipulating neuronal activity, comprise a distinct class, characterized by the use of targeted light stimuli to initiate neuronal responses. This class offers several important advantages over other techniques, such as chemical or electrical stimulation, serving the same purpose. First, today's optical technology provides the possibility to apply stimuli with a high level of spatial and temporal resolution. Second, both light intensity and wavelength can be precisely regulated, and thus, diverse stimulation can be applied. Third, optical techniques are generally noninvasive. Finally, in some cases, stimulation can be targeted to specific cell types.

Recently, a novel approach has been added to the class of optical techniques. This approach involves the use of light-sensitive, semiconducting polymers to excite neurons, which are in close vicinity to the material. Biocompatibility properties of specific polymers have been widely reported in literature; moreover, thanks to their inherent softness, they are specifically suitable for several in vivo applications, limiting the risk of rejection and/or tissue inflammation. Current research aims at using organic polymers in prosthetics to restore vision in patients suffering from retinal degenerative diseases. After a short overview of currently available optical techniques, this chapter will describe the use of light-sensitive organic semiconductors to manipulate neuronal activity in general and their application as retinal prosthetics in particular.

2 Optical Techniques Modulating Neuronal Activity

In 1957, Pierce and Giese were the first to demonstrate that neuronal activity could be controlled optically. Application of UV light to frog and crab nerves reduced the amplitude of action potentials, an effect that could subsequently be reversed with white light [1]. Subsequent advancements in laser technology made it possible to photostimulate neurons with single-cell resolution using stimuli with a wavelength of 488 [2] and 532 nm [3]. The use of two-photon irradiation with a pulsed infrared laser [4] yields better tissue penetration due to the long wavelengths employed and confines the stimulus locus to a single focal plane. Presumably, the methods described above act mainly through the generation of reactive oxygen species (ROS) induced by the excitation of endogenous chromophores. The mechanism by which ROS generation results in cell polarization most likely involves inhibition of voltage-gated K^+ channels [5] (Fig. 1a), although alternative mechanisms may also be involved [3, 4, 6]. The main disadvantage of this class of photostimulation is that the generated ROS rapidly lead to toxicity and cell death, which limits the number of times the cells can be stimulated. In addition, the uncertainty concerning the signaling cascades here involved further complicates its use in neuroscience research.

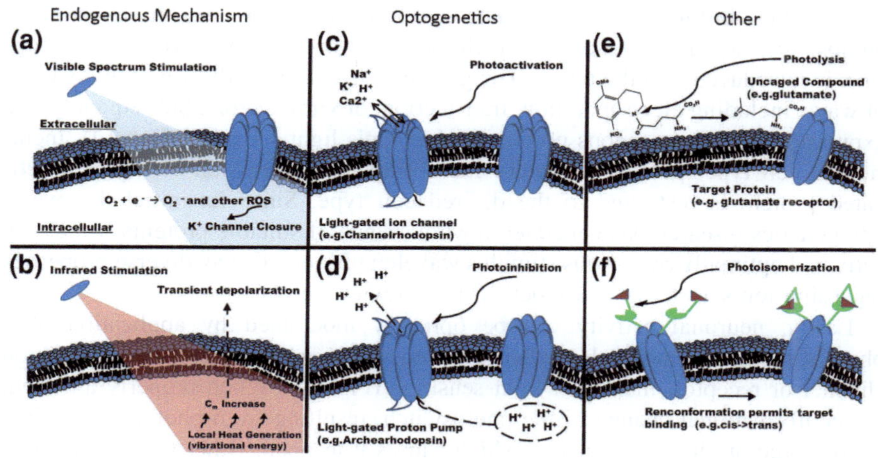

Fig. 1 Overview of various optical techniques and their working mechanisms for modulation of neuronal activity. **a** High-energy visible light stimulation generates reactive oxygen species (*ROS*), which in turn close K^+ channels. **b** Infrared stimulation leads to local temperature increase, which modulates membrane capacitance (C_m). **c** Light stimulation induces a conformational change of a light-gated channel (e.g., channelrhodopsin) to its conductive state, which leads to a depolarization. **d** Light stimulation induces a hyperpolarization of the cell membrane by activating a proton pump (e.g., bacteriorhodopsin). **e** Light stimulation uncages a compound by photolysis of a cleaved group. Its active site thus exposed, the compound can act on its target protein, e.g. a glutamate receptor. **f** A photoswitch can be used to present a ligand to a ligand-gated receptor. The conformational change induced by light stimulation results in binding of the ligand to the receptor

Infrared neural stimulation (INS) uses infrared light pulses to modulate neuronal activity [7] (Fig. 1b). INS relies on the high absorption of infrared light by water, which is abundantly present in biological tissue. A short pulse of infrared light leads to a local transient rise in temperature [8], which reversibly alters the electrical capacitance of the plasma membrane, resulting in cell depolarization [9]. INS facilitates long-lasting stimulation of a single nerve, without affecting neighboring structures, but does not provide single-cell spatial resolution or cell type specificity.

Light stimulation can also be used to biochemically activate a signaling molecule, such as a neurotransmitter or intracellular calcium, which was inert due to the presence of photocleavable shielding group (Fig. 1c). The initially inactive compound becomes functional when the shielding group is cleaved through photolysis by illumination with a specific wavelength. Subsequently, with its active region exposed, the compound can act on its cellular target. Caged glutamate, GABA, calcium, and other compounds have successfully been applied to modulate neuronal activity (for a review see [10]).

The discovery of light-sensitive ion channels from algae [11, 12] was the starting point for the development of a novel photostimulation method, termed optogenetics. In this technique, a gene coding for light-activated ion channels or pumps, such as the excitatory channelrhodopsins and the inhibitory halorhodopsins, is introduced into the tissue. The gene construct can be delivered in a variety of ways, including electroporation, transfection, or viral vectors. Subsequently, the expression of the exogenous protein renders cells light sensitive (Fig. 1d). Inclusion of cell type-specific promoters can ensure that expression of the light-activated protein is restricted to the desired cell type. Since its initial discovery, optogenetics research has generated a whole range of suitable proteins, which are activated optimally by various stimulus wavelengths and display diverse properties regarding ion selectivity and kinetics (for a review see [13]).

Lastly, neuronal activity can be optically modulated by application of a photoisomerizable molecule, called a photoswitch, which binds to a native ion channel or receptor, making it light sensitive (Fig. 1e). When properly attached, the conformational change of the photoswitch, resulting from photon absorption, exerts force on the ion channels, which causes it to open. Other types of photoswitches can regulate the activity of ligand-gated channels or receptors. In this case, photoisomerization of the photoswitch leads to delivery/removal of the ligand to its binding site on the channel or receptor. Photoswitches can be applied both in a genetically targeted and nongenetic fashion, and their target proteins can be endogenous and/or exogenous. Photoswitches have successfully been used to regulate voltage-gated K^+ channels, acetylcholine receptors, and kainate-type glutamate receptors [14–17].

3 Organic Semiconductive Polymers

Conducting polymers (CPs) consist of chains of carbon-based molecules. The presence of alternating single and double bonds in the polymer backbone, called a conjugated system, renders these materials electrically conductive. Although the level of conductivity of CPs, at room temperature, is not sufficient to be of practical use, it can be substantially increased by means of doping. During the doping process, a neutral polymer is supplied with a counter anion or cation through oxidization (*n*-doping) or reduction (*p*-doping), respectively. The π-electrons, delocalized along the backbone, induce strong coupling with light and make CPs intrinsically light sensitive. Moreover, chemical engineering of the molecules constituting CPs allows tuning of their optical properties, such as absorbance spectrum and stimulation bandgap. In addition, tremendous progress has been made in developing CPs, in the last few years, with improved efficiency in converting photons to an electric field.

Photoexcitation in CPs generates a bound electron–hole pair, called an exciton [18]. To give rise to a photocurrent, excitons need to be converted to free charge carriers. The dissociation efficiency can be greatly enhanced by blending the CP, which is typically *p*-doped and thus acts as an electron donor, with an electron acceptor [19]. The combination of an electron donor with an electron

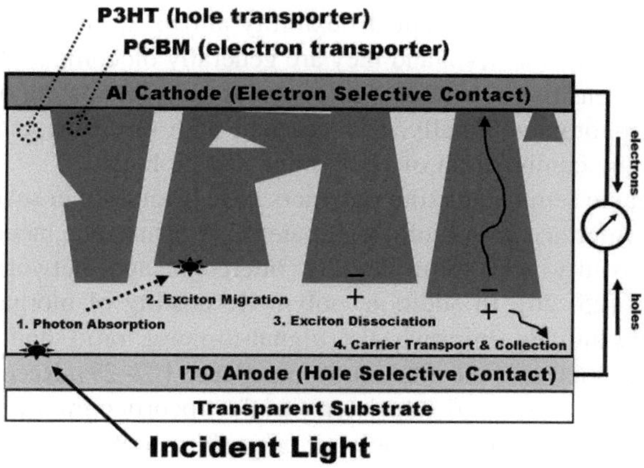

Fig. 2 Standard configuration and operational principle of a bulk heterojunction organic photovoltaic cell. The electron donor, poly(3-hexylthiophene) (*P3HT*), and acceptor, phenyl-C_{61}-butyric acid methyl ester (*PCBM*), materials in the blend are positioned between the indium tin oxide (*ITO*) anode and aluminum oxide cathode of the device. Photoexcitation of the material occurs through a four-step mechanism: (*1*) After passing through the transparent anode, an incident photon is absorbed by P3HT causing exciton generation. (*2*) The exciton migrates to the interface of the bulk heterojunction, (*3*) Here, it dissociates due to interfacial charge transfer into a pair of free carriers consisting of a hole in the polymer (P3HT) and an electron in the fullerene (PCBM). (*4*) Carriers migrate to the electrode, passing charge to external circuit

acceptor is called bulk heterojunction [20] and is the typical way to produce polymer-based organic photovoltaic devices (Fig. 2).The most common electron acceptors are fullerenes, such as the buckminsterfullerene (C60) or one of its soluble derivatives. Excitons can only travel a short distance in the polymer before decaying to the ground state [21]. Efficient charge extraction, therefore, requires that the polymer and the fullerene form a closely intermingled network [19] and that electrons and holes are able to flow to charge-selective contacts, where they are extracted in the external circuit.

Interfacing an organic photovoltaic device with the biological tissue poses several constraints in terms of degradation, in particular for what concerns the electrodes. A possible solution relies on replacing the electron-selective contact by an electrolyte solution. Several reports demonstrate that it is possible to generate a photocurrent in this configuration [22–24]. The main difference between this, so-called hybrid, and conventional, solid-state, configuration is related to interface phenomena between the polymer and the electrolyte solution. In the hybrid device, the conductance combines both electronic and ionic features, whereas it remains purely electronic in the solid-state device.

Several important features make CPs attractive for biological applications. First, synthesis and processing of these materials is relatively easy and inexpensive. Second, they can be made in the form of semitransparent, thin films with proper thickness, which is an important factor in microscopic and optical applications. Third, their mechanical flexibility allows an intimate interface with tissue both in vitro and in vivo, and they are generally biocompatible. Fourth, they are intrinsically sensitive to visible light, opening thus interesting opportunities in the field of biophotonics. Finally, CPs conduct both electrons and ions, making them ideal for the combination of electronics and biology.

To date, organic semiconducting polymers have been used in several nonoptical applications to interface with biological material. For instance, neurons have been cultured successfully and form densely interconnected networks on organic semiconductors [25, 26]. In addition, polymeric coating of inorganic microelectrodes can substantially improve the signal-to-noise ratio and decreases the occurrence of artifacts and the amount of impedance [27–29]. Recently, the use of polymer-based electrodes and transistors in electrocorticography has been demonstrated, showing that they are able to outperform conventional electrodes of similar geometry [30, 31]. In addition, organic transistors can modulate the electrical activity of cells, both for extracellular stimulation and recording [26, 32]. Furthermore, organic semiconductors are currently employed as biosensors to detect and analyze biologically relevant substances, ranging from ions and small molecules and metabolites to complex biomolecules such as proteins or DNA (for a review see [33, 34]). As a final example, they are also used as actuators for the controlled release of substances, such as neurotransmitters [35, 36].

4 Modulation of Membrane Potential Using Light-Sensitive Polymers

In 2011, Ghezzi and coworkers were the first to demonstrate the proof of principle for light-induced modulation of neuronal activity by means of an organic, light-sensitive polymer [37]. In this case, the photovoltaic device consisted of a prototypical blend of poly(3-hexylthiophene) (P3HT), functioning as the electron donor and phenyl-C_{61}-butyric acid methyl ester (PCBM) acting as the electron acceptor. It was observed that this device, completely submerged in electrolytic solution, maintained its photoconversion capability for over 28 days. Furthermore, after a transient decay at stimulation onset, the generated photocurrent remained constant for more than 28 h of continuous illumination.

Primary neurons were grown on the organic polymer blend to test its biocompatibility. In this configuration, cells could be kept in culture for many weeks. Growing patterns, neuronal adhesion, cell viability, and mortality were comparable between cultures on the photovoltaic device and on control substrates. Moreover, the examination of the physiological properties of cultured neurons by means of patch-clamp and microelectrode arrays did not reveal any significant differences regarding resting membrane potential, the frequency and amplitude of excitatory postsynaptic currents, and spontaneous firing rates, between the two culture conditions. These results demonstrate that the organic polymer, on which neurons were cultured, is highly biocompatible and neuron friendly.

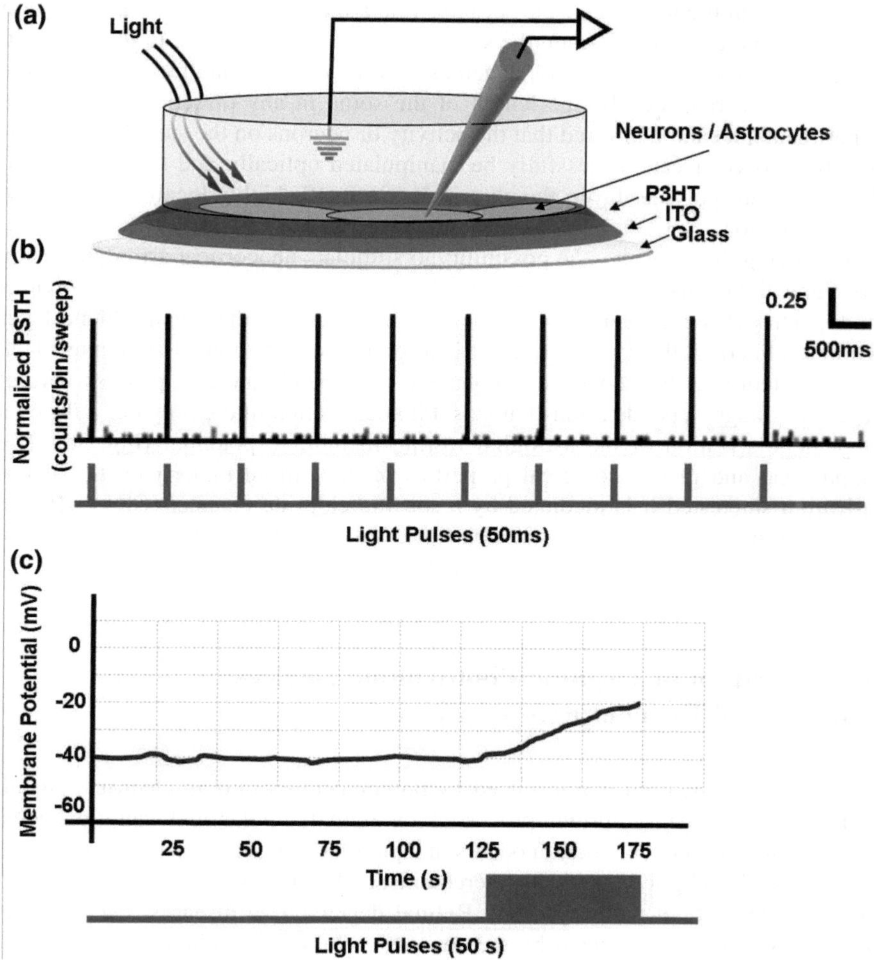

Fig. 3 The membrane potential of cells grown on a P3HT: PCBM device can be modulated using light. **a** Experimental configuration for patch-clamp recordings of cells grown on the organic photovoltaic (*OPV*) device. **b** Action potentials of neurons can be accurately and reliably elicited with short (20 ms) light pulses. **c** Astrocytes stimulated with long light pulses (50 s) respond with increasing depolarization of the membrane potential during the light pulse. Adapted from [37, 38]

To test whether the photovoltaic device was capable in manipulating neuronal activity, whole-cell patch-clamp recordings were made on cultured neurons that were illuminated by spatially restricted light stimuli. Application of a 50-ms pulse of light was successful in generating an action potential in neurons. Using a 1-Hz train of 20-ms light pulses, neuronal spike trains could be elicited, with only a small percentage of failures (Fig. 3b). In these experiments, a close relation was found between the location of the stimulus relative to the cell soma and the action potential generation. Only when the stimulus was directly targeted to the soma did spike generation occur reliably, whereas it was virtually ineffective when the stimulus was translated by the length of the soma in any direction. From these experiments, it can be inferred that the activity of neurons on the surface of a light-sensitive polymer can successfully be manipulated optically and that, in spite of the continuous polymer layer, the photovoltaic effect is highly localized to the site of illumination.

Following these results, the possibility to stimulate neocortical astrocytes using the same light-sensitive polymeric blend was reported [38]. As in the previous study, astrocytes could successfully be kept in culture on the organic blend layer without affecting their passive electrical properties. Although astrocytes lack action potentials, they do have a polarized cell membrane and possess several types of voltage-dependent ion channels. Photoexcitation of the polymer induced a large depolarization of the astroglial resting membrane potential (Fig. 3c). The biophysical and pharmacological properties of the current underlying this depolarization suggested it is mediated by a specific chloride channel (ClC-2). These results strengthen and broaden the concept of using organic polymers to produce biocompatible, photosensitive devices for cell stimulation.

5 Application of Organic Photovoltaic Devices in Retinal Degenerative Diseases

In healthy retinas, light is transduced by photoreceptors into an electrical signal, which is subsequently further processed by retinal bipolar and ganglion cells. Processing of visual information occurs in a parallel fashion, meaning that specific features, such as light decrements, increments, and stimulus direction, are encoded in segregated retinal pathways [39]. Retinal degenerative diseases, e.g., retinitis pigmentosa, are characterized by the successive loss of the photoreceptive outer segments of both types of photoreceptors—rods and cones. Beginning with the loss of night vision or other visual impairments, the progressive disease can ultimately lead to complete blindness.

Given the ability of organic polymers to modulate neuronal activity following light stimulation, this class of materials may find a suitable application in the restoration of vision in retinal degenerative diseases. To test the feasibility of this approach, Ghezzi and colleagues examined whether the retina from an animal model of photoreceptor degeneration could be resensitized by interfacing it with a single-component organic film of P3HT [40]. To this end, multi-unity activity of ganglion cells was recorded in explanted degenerated retinas in a subretinal configuration, i.e., the degenerate photoreceptor layer contacted the polymer. It was found that light-elicited spiking activity of ganglion cells in degenerate retinas on polymer was comparable with that in control retinas on glass. By contrast, at

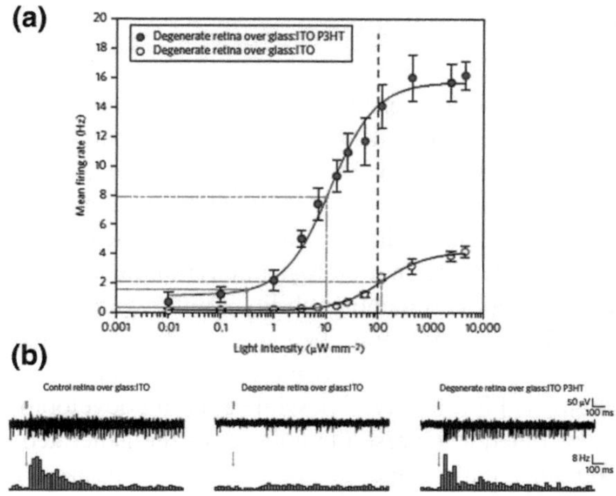

Fig. 4 An organic photovoltaic device can restore light responsiveness in degenerated retinas. **a** Relation between the mean firing rate of ganglion cells (*GCs*) versus stimulus intensity measured in degenerated retinas on top of P3HT-coated glass with an ITO anode (*red dots*) or glass plus ITO anode alone (*open dots*). **b** Extracellular recordings of light-induced GC activity together with resulting poststimulus time histograms from a healthy control retina on glass coverslip with an ITO anode, a degenerated retina on the same substrate, and a degenerated retina on a glass coverslip coated with P3HT and ITO. Reproduced with permission from [40]

these intensities, light stimuli failed to elicit spiking activity in degenerated retinas on glass (Fig. 4a). These findings suggest that photovoltaic polymeric devices are able to restore light sensitivity in degenerated retinas. Presently, research is being conducted to further explore this possibility by subretinally implanting these devices in an animal model of retinal degenerative diseases, such as the Royal College of Surgeons rats bearing a mutation in the *Mertk* gene.

Several other approaches to treat degenerative diseases of the outer retina have previously been proposed. These include regeneration of lost photoreceptors by transplantation [41, 42], gene therapy [43], optogenetic approaches [44, 45], injection of a photoswitch [46], and the use of metal-/silicon-based prosthetics [47, 48]. The optogenetic approaches yield promising results in animal models, but the expression of inserted genes remains limited in humans. Moreover, gene transplantation is costly and can involve several clinical risks. Metal-/silicon-based retinal prosthetics are currently in the phase of clinical trial, with one model recently gaining FDA approval. Both the epiretinal Argus device and the subretinal Alpha device have been able to partially restore vision in blind patients [47, 48]. However, these devices do still need to overcome several important issues, such as the need for external power supply, poor spatial resolution, and low biocompatibility.

6 Future Research Direction

The significant progress made over the last 3 years in coupling semiconductors to biological tissue is readily evident from the studies described above. However, this newly emerging field must still overcome several challenges as well as employ novel developments to further establish its position as a member of the neuro-modulatory toolkit. The main challenges facing this application are the long-term durability of polymeric devices in a biological environment and a more extensive investigation of the materials' biocompatibility. With regard to their use in retinal prosthetics, conductive polymers may benefit from the developments of devices capable of multi-chromatic detection to more closely match the natural trichromatic vision of humans [49]. Lastly, a possible exciting development may emerge if the device can be tuned to reverse the polarity of the generated charge, thereby allowing both photoexcitation and photoinhibition of tissue. Here, we briefly address these future research directions that will require a close collaboration between divergent fields of study such as electronics, physics, materials science, chemistry, biology, neuroscience, and medicine.

The retina is an exquisitely tuned tissue, evolved to efficiently amplify incoming photons, preprocess stimulus information, and transmit the information to the brain for further processing. Interestingly, the natural amplification carried out by the photoreceptive layer of the retina was shown to be closely matched by the device presented by Ghezzi et al. [40]. The light sensitivity that the group described corresponds to a stimulation threshold of 0.3 mW/mm^2, a level 30-fold lower than the ocular safety limit for continuous exposure to visible light, and at the low end of daytime levels of retinal irradiance experienced outdoors. However, the device may also be found to be susceptible to degradation following long-term exposure to aqueous medium. As such, the chemical engineering of encapsulation layers to protect the exposed polymer will be a high-priority challenge for the development of prosthesis based on organic polymers.

Maintaining the focus on the prosthesis application, future devices may benefit from the incorporation of multiple conductive polymers sensitive to different wavelengths of visible light, such as the primary colors. In such a device, the polymers can be chemically tuned to have bandgaps matching the natural bandgap of the different photoreceptive cone cells found in the retina [50]. A careful patterning of the different polymers, matching the distribution of cone subtypes, opens the possibility of stimulating intact retinal layers in degenerative pathology to recover not only light sensitivity, but potentially even trichromatic sensitivity. However, color vision is a complex phenomenon relying not only on the retina, but also on the brain and its high-level processing of visual signals to carry out three sets of discrimination: light from dark, yellow from blue, and red from green (for extensive overview of color vision see [51]). Achieving a restoration of trichromatic sensitivity may thus require more than the proper engineering of the implant itself, such as the involvement of biological processes such as neural plasticity in the brain.

A final area of interest for future research comes from the finding described by Gautam et al. [24]. In their paper, the group describes that the polarity of the photovoltage generated by their bulk heterojunction photovoltaic device is a function of the polymer layer thickness and that the threshold for reversal lays around 300 nm thickness. So far, publications have only described the effect of photostimulation of cells grown on devices with smaller thickness and have described photoexcitation (depolarization of membrane potential). The finding by Gautam et al. [24] might translate to the possibility of photoinhibition using organic semiconductors as culture substrates.

7 Conclusions

Techniques modulating neuronal activity are instrumental to unravel neuronal working mechanisms and treat diseases. Light-induced modulation of neuronal activity is the hallmark feature of a class of optical techniques, which offers several important advantages over others. During the past decade, this class has seen a great expansion in its repertoire, among which the application of optogenetics and photoswitches.

The use of photovoltaic polymers can be a valuable addition to optical techniques aimed at modulating neuronal activity. The number of reports using this approach currently available remains very limited, and further research is required to fully determine its applicability in neuroscience research. Nevertheless, the experiments described in this chapter illustrate that it represents a straightforward way of photostimulating neurons with a high temporal and spatial resolution. In addition, their organic nature combined with the ability to conduct both electrons and ions makes these materials suitable to interact with biological tissue. On the other hand, this technique does not discriminate between neuronal types and presently only provides excitatory regulation.

An obvious clinical relevance of photovoltaic polymers lies in the field of visual prosthetics. Organic polymers can act as artificial photoreceptors to restore vision in degenerated retinas and might be an attractive alternative for existing metal-/silicon-based retinal implants. The main benefits of organic polymers in this application are their intrinsic sensitivity to visible light, biocompatibility, spatial selectivity, functioning in photovoltaic mode, reduced impact of reduction/oxidation processes, and low costs. Future research must establish whether this implementation will be successful for rescuing vision in patients affected by photoreceptor degeneration.

References

1. Pierce S, Giese AC (1957) Photoreversal of ultraviolet injury to frog and crab nerves. J Cell Comp Physiol 49:303–317
2. Fork RL (1971) Laser stimulation of nerve cells in Aplysia. Science 171:907–908
3. Reece PJ, Dholakia K, Thomas RC, Cottrell GA (2008) Green laser light (532 nm) activates a chloride current in the C1 neuron of Helix aspersa. Neurosci Lett 433:265–269
4. Hirase H, Nikolenko V, Goldberg JH, Yuste R (2002) Multiphoton stimulation of neurons. J Neurobiol 51:237–247
5. Duprat F, Guillemare E, Romey G, Fink M, Lesage F, Lazdunski M, Honore E (1995) Susceptibility of cloned K^+ channels to reactive oxygen species. Proc Natl Acad Sci USA 92:11796–11800
6. Kourie JI (1998) Interaction of reactive oxygen species with ion transport mechanisms. Am J Physiol 275:C1–C24
7. Wells J, Kao C, Jansen ED, Konrad P, Mahadevan-Jansen A (2005) Application of infrared light for in vivo neural stimulation. J Biomed Opt 10:064003
8. Wells J, Kao C, Konrad P, Milner T, Kim J, Mahadevan-Jansen A, Jansen ED (2007) Biophysical mechanisms of transient optical stimulation of peripheral nerve. Biophys J 93:2567–2580
9. Shapiro MG, Homma K, Villarreal S, Richter C-P, Bezanilla F (2012) Infrared light excites cells by changing their electrical capacitance. Nat Commun 3:736
10. Ellis-Davies GCR (2007) Caged compounds: photorelease technology for control of cellular chemistry and physiology. Nat Methods 4:619–628
11. Nagel G, Szellas T, Huhn W, Kateriya S, Adeishvili N, Berthold P, Ollig D, Hegemann P, Bamberg E (2003) Channelrhodopsin-2, a directly light-gated cation-selective membrane channel. Proc Natl Acad Sci USA 100:13940–13945
12. Nagel G, Ollig D, Fuhrmann M, Kateriya S, Musti AM, Bamberg E, Hegemann P (2002) Channelrhodopsin-1: a light-gated proton channel in green algae. Science 296:2395–2398
13. Fenno L, Yizhar O, Deisseroth K (2011) The development and application of optogenetics. Annu Rev Neurosci 34:389–412
14. Bartels E, Wassermann NH, Erlanger BF (1971) Photochromic activators of the acetylcholine receptor. Proc Natl Acad Sci 68:1820–1823
15. Banghart M, Borges K, Isacoff E, Trauner D, Kramer RH (2004) Light-activated ion channels for remote control of neuronal firing. Nat Neurosci 7:1381–1386
16. Chambers JJ, Banghart MR, Trauner D, Kramer RH (2006) Light-induced depolarization of neurons using a modified Shaker K(+) channel and a molecular photoswitch. J Neurophysiol 96:2792–2796
17. Volgraf M, Gorostiza P, Numano R, Kramer RH, Isacoff EY, Trauner D (2006) Allosteric control of an ionotropic glutamate receptor with an optical switch. Nat Chem Biol 2:47–52
18. Lanzani G (2012) The photophysics behind photovoltaics and photonics. Wiley-VCH, Germany
19. Sariciftci NS, Smilowitz L, Heeger AJ, Wudl F (1992) Photoinduced electron transfer from a conducting polymer to buckminsterfullerene. Science 258:1474–1476
20. Yu G, Gao J, Hummelen JC, Wudl F, Heeger AJ (1995) Polymer photovoltaic cells: enhanced efficiencies via a network of internal donor-acceptor heterojunctions. Science (80-) 270:1789–1791
21. Nelson J (2004) The physics of solar cells. Imperial College Press, London
22. Antognazza MR, Ghezzi D, Musitelli D, Garbugli M, Lanzani G (2009) A hybrid solid-liquid polymer photodiode for the bioenvironment. Appl Phys Lett 94:243501
23. Lanzarini E, Antognazza MR, Biso M, Ansaldo A, Laudato L, Bruno P, Metrangolo P, Resnati G, Ricci D, Lanzani G (2012) Polymer-based photocatalytic hydrogen generation. J Phys Chem C 116:10944–10949

24. Gautam V, Bag M, Narayan KS (2010) Dynamics of bulk polymer heterostructure/electrolyte devices. J Phys Chem Lett 1:3277–3282
25. Bystrenova E, Jelitai M, Tonazzini I, N. Lazar A, Huth M, Stoliar P, Dionigi C, Cacace MG, Nickel B, Madarasz E, Biscarini F (2008) Neural networks grown on organic semiconductors. Adv Funct Mater 18:1751–1756
26. Cramer T, Chelli B, Murgia M, Barbalinardo M, Bystrenova E, de Leeuw DM, Biscarini F (2013) Organic ultra-thin film transistors with a liquid gate for extracellular stimulation and recording of electric activity of stem cell-derived neuronal networks. Phys Chem Chem Phys 15:3897–3905
27. Abidian MR, Ludwig KA, Marzullo TC, Martin DC, Kipke DR (2009) Interfacing conducting polymer nanotubes with the central nervous system: chronic neural recording using Poly(3,4-ethylenedioxythiophene) nanotubes. Adv Mater 21:3764–3770
28. Cui X, Lee VA, Raphael Y, Wiler JA, Hetke JF, Anderson DJ, Martin DC (2001) Surface modification of neural recording electrodes with conducting polymer/biomolecule blends. J Biomed Mater Res 56:261–272
29. Cui X, Wiler J, Dzaman M, Altschuler RA, Martin DC (2003) In vivo studies of polypyrrole/peptide coated neural probes. Biomaterials 24:777–787
30. Khodagholy D, Doublet T, Gurfinkel M, Quilichini P, Ismailova E, Leleux P, Herve T, Sanaur S, Bernard C, Malliaras GG (2011) Highly conformable conducting polymer electrodes for in vivo recordings. Adv Mater 23:H268–H272
31. Khodagholy D, Rivnay J, Sessolo M, Gurfinkel M, Leleux P, Jimison LH, Stavrinidou E, Herve T, Sanaur S, Owens RM, Malliaras GG (2013) High transconductance organic electrochemical transistors. Nat Commun 4:2133
32. Benfenati V, Toffanin S, Bonetti S, Turatti G, Pistone A, Chiappalone M, Sagnella A, Stefani A, Generali G, Ruani G, Saguatti D, Zamboni R, Muccini M (2013) A transparent organic transistor structure for bidirectional stimulation and recording of primary neurons. Nat Mater 12:672–680
33. Gerard M, Chaubey A, Malhotra BD (2002) Application of conducting polymers to biosensors. Biosens Bioelectron 17:345–359
34. Arshak K, Velusamy V, Korostynska O, Oliwa-Stasiak K, Adley C (2009) Conducting Polymers and their applications to biosensors: emphasizing on foodborne pathogen detection. IEEE Sens J 9:1942–1951
35. Simon DT, Kurup S, Larsson KC, Hori R, Tybrandt K, Goiny M, Jager EWH, Berggren M, Canlon B, Richter-Dahlfors A (2009) Organic electronics for precise delivery of neurotransmitters to modulate mammalian sensory function. Nat Mater 8:742–746
36. Richardson RT, Wise AK, Thompson BC, Flynn BO, Atkinson PJ, Fretwell NJ, Fallon JB, Wallace GG, Shepherd RK, Clark GM, O'Leary SJ (2009) Polypyrrole-coated electrodes for the delivery of charge and neurotrophins to cochlear neurons. Biomaterials 30:2614–2624
37. Ghezzi D, Antognazza MR, Dal Maschio M, Lanzarini E, Benfenati F, Lanzani G (2011) A hybrid bioorganic interface for neuronal photoactivation. Nat Commun 2:166
38. Benfenati V, Martino N, Antognazza MR, Pistone A, Toffanin S, Ferroni S, Lanzani G, Muccini M (2014) Photostimulation of whole-cell conductance in primary rat neocortical astrocytes mediated by organic semiconducting thin films. Adv Healthc Mater 3:392–399
39. Masland RH (2001) The fundamental plan of the retina. Nat Neurosci 4:877–886
40. Ghezzi D, Antognazza MR, Maccarone R, Bellani S, Lanzarini E, Martino N, Mete M, Pertile G, Bisti S, Lanzani G, Benfenati F (2013) A polymer optoelectronic interface restores light sensitivity in blind rat retinas. Nat Photonics 7:400–406
41. MacLaren RE, Pearson RA, MacNeil A, Douglas RH, Salt TE, Akimoto M, Swaroop A, Sowden JC, Ali RR (2006) Retinal repair by transplantation of photoreceptor precursors. Nature 444:203–207
42. Pearson RA, Barber AC, Rizzi M, Hippert C, Xue T, West EL, Duran Y, Smith AJ, Chuang JZ, Azam SA, Luhmann UFO, Benucci A, Sung CH, Bainbridge JW, Carandini M, Yau K-W, Sowden JC, Ali RR (2012) Restoration of vision after transplantation of photoreceptors. Nature 485:99–103

43. Koch S, Sothilingam V, Garcia Garrido M, Tanimoto N, Becirovic E, Koch F, Seide C, Beck SC, Seeliger MW, Biel M, Mühlfriedel R, Michalakis S (2012) Gene therapy restores vision and delays degeneration in the CNGB1(-/-) mouse model of retinitis pigmentosa. Hum Mol Genet 21:4486–4496
44. Bi A, Cui J, Ma Y-P, Olshevskaya E, Pu M, Dizhoor AM, Pan Z-H (2006) Ectopic expression of a microbial-type rhodopsin restores visual responses in mice with photoreceptor degeneration. Neuron 50:23–33
45. Busskamp V, Duebel J, Balya D, Fradot M, Viney TJ, Siegert S, Groner AC, Cabuy E, Forster V, Seeliger M, Biel M, Humphries P, Paques M, Mohand-Said S, Trono D, Deisseroth K, Sahel JA, Picaud S, Roska B (2010) Genetic reactivation of cone photoreceptors restores visual responses in retinitis pigmentosa. Science 329:413–417
46. Polosukhina A, Litt J, Tochitsky I, Nemargut J, Sychev Y, De Kouchkovsky I, Huang T, Borges K, Trauner D, Van Gelder RN, Kramer RH (2012) Photochemical restoration of visual responses in blind mice. Neuron 75:271–282
47. Ahuja AK, Dorn JD, Caspi A, McMahon MJ, Dagnelie G, Dacruz L, Stanga P, Humayun MS, Greenberg RJ (2011) Blind subjects implanted with the Argus II retinal prosthesis are able to improve performance in a spatial-motor task. Br J Ophthalmol 95:539–543
48. Zrenner E, Bartz-Schmidt KU, Benav H, Besch D, Bruckmann A, Gabel V-P, Gekeler F, Greppmaier U, Harscher A, Kibbel S, Koch J, Kusnyerik A, Peters T, Stingl K, Sachs H, Stett A, Szurman P, Wilhelm B, Wilke R (2011) Subretinal electronic chips allow blind patients to read letters and combine them to words. Proc Biol Sci 278:1489–1497
49. Martino N, Ghezzi D, Benfenati F, Lanzani G, Antognazza MR (2013) Organic semiconductors for artificial vision. J Mater Chem B 1:3768
50. Antognazza MR, Scherf U, Monti P, Lanzani G (2007) Organic-based tristimuli colorimeter. Appl Phys Lett 90:163509
51. Gegenfurtner KR, Sharpe LT (2001) Color vision: from genes to perception. Cambridge University Press, Cambridge

Probing the Lateral Diffusion of Individual Neurotransmitter Receptors

Enrica Maria Petrini and Andrea Barberis

Abstract Synaptic transmission is a highly dynamic and regulated process in which electrical information is transferred between two neurons by means of a chemical transmitter that diffuses from the presynaptic element to reach, bind, and activate the neurotransmitter receptors located at postsynaptic side. Traditionally, postsynaptic receptors have been considered fixed in front of the releasing site, but over the last decade, compelling evidence has shown that they diffuse in the plane of the neuronal membrane, thus adding a further level of complexity to synaptic neurotransmission. The development of new technologies that allow a close inspection of the diffusive properties of receptors at synapses have revealed that the receptors dynamics is not only part of a "constitutive recycling" but also is responsible for the fast tuning of the receptor number at synapses both in basal conditions and in response to external stimuli, being therefore a major determinant of synaptic plasticity. In this section, we will review the techniques used to study the lateral mobility of individual receptors and the recent advances in the comprehension of the role of receptor diffusion in neuronal synaptic computation.

1 Introduction

Although sometimes referred to as a "background noise," Brownian diffusion is crucial for the actuation of most biological processes. Indeed, many biochemical reactions taking place in the cytosol and in the nucleus of living cells occur thanks to diffusion that ensures the interaction between molecules. At the synapse, for instance, the sequential events of synaptic transmission are dominated by "controlled" diffusive processes. In the presynaptic terminal, after the invasion of the

E. M. Petrini · A. Barberis (✉)
Department of Neuroscience and Brain Technologies, Italian Institute of Technology (IIT), Via Morego 30, 16163 Genoa, Italy
e-mail: andrea.barberis@iit.it

F. Benfenati et al. (eds.), *Novel Approaches for Single Molecule Activation and Detection*, Advances in Atom and Single Molecule Machines,
DOI: 10.1007/978-3-662-43367-6_11,

action potential, indeed, the entry and diffusion of calcium ions trigger the release of synaptic vesicles, hence the diffusion of neurotransmitter in the synaptic cleft [5, 17, 25]. The opening of receptor channels induced by neurotransmitter binding, in turn, elicits the diffusion of ions in the postsynaptic element, leading to changes in the postsynaptic membrane potential and/or to the activation of intracellular signaling cascades. More recently, it has been shown that diffusion at the synapse is not limited to ions and small molecules but also involves proteins such as neurotransmitter receptors that are the main players in the transduction between chemical to electrical signals at the postsynaptic element [71]. Neurotransmitter receptors are enriched at postsynaptic sites by means of anchoring proteins that bind receptors to the cytoskeleton. Due to the stability of postsynaptic receptor clusters, receptors were assumed to be immobile and rigidly connected with the neuronal backbone. However, it is now clear that receptors can diffuse at the surface of the neuronal membrane and that scaffolding proteins provide "diffusion traps" that hinder Brownian free diffusion, thus ensuring the formation and maintenance of receptor clusters at the postsynaptic side ([6], 56). The discovery of neurotransmitter receptors diffusion imposed a revision in the existing view of synaptic processes. The traditional vision "receptors are static until external forces change their status" was replaced by "receptors naturally diffuse until external forces limit their mobility." After the first direct demonstrations of individual surface receptor lateral diffusion and of receptor reversible "stop-and-goes" at synaptic sites ([44], 10), clear hints that receptors exchanges between synaptic and extrasynaptic compartments contributed to the regulation of receptor number at synapses derived from the analysis of the mechanisms underlying the fast remodeling of the postsynaptic receptor organization during synaptic plasticity [14, 41]. In particular, changes of synaptic receptor number observed in response to plasticity induction were demonstrated to depend on the changes of exocytosis and/or endocytosis rates, suggesting the presence of a "dynamic equilibrium" between surface receptors and a pool of intracellular receptors [11, 12, 61]. The observation that both exocytosis and endocytosis occur outside the synaptic area [8, 9, 52] implied that receptors have to laterally diffuse in the plane of the membrane to be inserted to or removed from synapses, thus introducing the concept of "lateral diffusion" in the rapid adjustment of receptor number at synapses [14]. It is now clear that the dynamic interaction of receptors with scaffold proteins underlies several forms of postsynaptic long-term potentiation or depression (LTP, LTD), thus influencing the strength of synaptic contacts and consequently the functioning of brain microcircuits [70, 71]. In addition to its role in long-term plasticity, receptor lateral diffusion has been also implicated in the fast tuning of synaptic signals by modulating the availability of functional receptors at synapses during synaptic transmission in the millisecond range [27, 71]. This highlights the role of receptor diffusion in tuning synaptic transmission and discloses the general paradigm that the dynamic interaction between molecules and the membrane organization at nanoscale range are central determinants of the computational properties of the neuron.

2 Measuring the Lateral Diffusion of Neurotransmitter Receptors

2.1 Population Diffusion Measurements

Over the last two decades, the detection and the analysis of receptor diffusion at the neuronal surface has been a major technical and intellectual challenge. Converging evidence obtained with different approaches unanimously suggested that neurotransmitter receptors undergo an intense diffusive activity that is much higher and complex than that expected by simple "constitutive receptor recycling." The first attempts to study the dynamics of receptors at synapses were made by video imaging of receptors tagged with fluorescent proteins. For instance, by performing two-photon imaging of AMPA receptors tagged with GFP in organotypic cultures, Shi and colleagues [61] found that high-frequency stimulation induced the enhancement of GFP fluorescence at glutamatergic spines, thus indicating an activity-dependent AMPA receptors fast redistribution between the synaptic and extrasynaptic areas at the neuronal surface. The fluorescence recovery after photo-bleaching (FRAP) is another approach that allowed studying the diffusion properties of receptors tagged with fluorescent proteins at population level [3]. Indeed, after photo-bleaching a small area containing fluorescent-tagged receptors with a high-intensity laser illumination, neighboring unbleached fluorescent-tagged receptors will diffuse in the bleached area and replenish it with fluorescence ([33, 53], 28). The study of the fluorescence recovery dynamics in the bleached area allows inferring the diffusive properties of the receptors. As a general rule, the fluorescence recovery rate depends on the receptor diffusion coefficient, whereas the fraction of fluorescence recovery at steady state sheds light on the mobile population of the receptors. However, although conceptually straightforward, it has to be pointed out that the "fluorescence replenishment" after the photo-bleaching is a complex phenomenon, not only due to the lateral diffusion of unbleached receptors but also dependent, for instance, on the rate at which the photo-bleached receptors leave the bleached area and/or on possible direct receptor exocytosis in the bleached area [28]. A similar technique to FRAP is the fluorescence loss in photo-bleaching (FLIP), which can be analogously used to monitor the mobility of populations of fluorescence-tagged receptors and exchanges between compartments. With this approach, consequent to a continuous bleaching in a defined area, protein mobility will be deciphered as the fluorescence loss in the surrounding area. The fading of fluorescence outside the bleached area will be due to the invasion by lateral diffusion of bleached proteins [51]. The application of FRAP and FLIP approaches to receptor-tagged pH-sensitive fluorescent proteins, such as pHluorin or SuperEcliptic pHluorin (SEP), represents a technical advantage to restrain the observation to surface receptors, being the fluorescence of these proteins quenched in acidic intracellular compartments [2]. Moreover, the restriction of photo-bleaching to synaptic areas by means of diffraction-limited laser spots allows studying the dynamics of synaptic receptors [27, 53]. More recently, the advent of photo-activated and photo-convertible fluorescent proteins able to change their emission spectral properties upon specific illumination allowed a further advancement in the monitoring of receptor dynamics in living cells [42]. Indeed, by tagging receptors with these proteins, it is possible to "photo-activate" and "photo-convert" them in a specific region of interest and to observe the displacements of "photo-activated/converted receptors" in neighboring areas [42]. These approaches not only allow inferring receptor diffusion properties but also are instrumental for studying the spatial patterns of receptor accumulation in specific subcellular domains [21, 47]. Another "bulk measurement" of molecule diffusion can be achieved by using the fluorescence correlation spectroscopy (FCS), a technique that reveals the mobility of the molecule of interest by its time of residence in small detection volumes [58].

In addition to the aforementioned imaging techniques, pharmacological tools have also proved useful to probe receptor "population diffusion." For instance, Tovar et al. [68] revealed the diffusion of NMDA receptors at glutamatergic synapses by studying the recovery of NMDA currents following irreversible block of the NMDA receptors by "open-channel" blocker MK-801. This pharmacological agent, indeed, selectively exerts its antagonism on synaptic receptors and not on extrasynaptic ones, since receptor "opening" would selectively occur at synapses. The unexpected current recovery in the presence of MK-801 clearly indicated the replacement of synaptic "blocked NMDA receptors" with naïve extrasynaptic receptors, thus implying the existence of receptor exchanges between synaptic and extrasynaptic compartments by lateral diffusion. With a similar approach applied to GABAergic synapses, Thomas et al. [67] exploited the irreversible and activity-dependent block of mutated $GABA_A$ receptors by MTSES to demonstrate the mobility of functional $GABA_A$ receptors. This study reported that, at inhibitory synapses of hippocampal pyramidal neurons, the synaptic pool of $GABA_A$ receptors is rapidly replaced independently from receptor intracellular trafficking. The increasing development of photo-reactive pharmacological compounds and modifiers of receptor gating combined to electrophysiological approaches represents nowadays a considerable advantage for precise spatial and temporal control of the activation of receptor subpopulations to ultimately provide real-time description of receptor surface diffusion and intracellular trafficking [1].

2.2 *Single-particle Tracking Techniques*

Although the aforementioned techniques are important to estimate the average value of receptor diffusion at population level, a major breakthrough in the study of receptor lateral mobility is represented by the advent of the single-particle tracking (SPT) techniques that allow the direct visualization of receptor diffusion at the single-molecule level. The possibility to study the mobility of individual molecules, indeed, dramatically increased the accuracy of the quantitative description of receptor diffusion. In this way, it was possible to observe that receptor mobility greatly differs among specific receptor subpopulations and among various membrane domains, a heterogeneity that is crucial for the functional role of receptor diffusion in synaptic transmission [37, 57]. The first pioneering studies that visualized and analyzed the trajectories of individual neurotransmitter receptors diffusing on the neuronal surface exploited latex beads coupled to the receptors of interest by means of primary antibodies and visualized by interference contrast microscopy [10, 44, 60]. This approach, first used with glycine, AMPA and mGlu receptors, revealed the fundamental paradigm of surface receptors diffusion that receptors alternate between periods of high and low mobility correlating with the diffusion in the extrasynaptic and synaptic compartments, respectively. Overall, these studies provided the first direct pieces of evidence of receptor fast diffusion at the neuronal surface. Moreover, they highlighted the key concept that receptor Brownian diffusion (induced by thermal agitation) can be "controlled" and "modulated" at the nanoscale level by transient receptor interactions with the scaffold proteins at specific domains of the neuronal plasma membrane. Although the SPT based on the use of latex beads represented a revolution in the study of receptor lateral mobility, the large size of these beacons (500 nm) represented a major limit for a reliable tracking of receptor diffusion at synapses, where the distance between pre- and postsynaptic elements (synaptic cleft) is only 20 nm. For this purpose, a considerable effort has been made to develop SPT approaches based on smaller fluorescent probes as reporters of receptor mobility. Among them, for instance, small organic fluorescent dyes ($\sim$1 nm) (such as cyanin dyes and Attodyes) are expected to minimally interfere with receptor diffusion [23, 55, 66]. However, these nanoprobes, recognized by the characteristic one-step photobleaching, require strong laser illumination to maximize photon emission and undergo rapid photo-bleaching (<10 s), thus limiting the visualization of the receptor diffusion to short periods [73]. Alternatively, receptor mobility can be probed by using quantum dots (QDs) as reporters. QDs are nanometer-sized semiconductors covered with a ZnS shell and a biocompatible organic coating, giving final fluorescent nanoprobes of 15–25 nm in diameter. Although the QD-based SPT approach can be performed on a limited number of target molecules in each experiment and requires adequate software algorithms for the reconstruction of trajectories (due to the blinking emission of QDs), to date, these semiconductors represent the best available trade-off between size and photo-stability. QDs, indeed, are much smaller than latex beads and, different from organic dyes, they exhibit very low photo-bleaching. The strong photo-stability of QDs allows long-lasting receptor tracking, an essential requirement for the study of receptors diffusion during processes that imply the observation of the same receptor up to 30–40 min such as during long-term synaptic plasticity [70, 71]. Besides their photo-stability, the advantages of QDs for the tracking of receptors are manifold. For instance, QDs possess a very high quantum yield, thus showing an excellent signal-to-noise ratio when illuminated by standard fluorescence lamps. This characteristic is instrumental to achieve considerably high point accuracy in the localization of individual QD-receptor complexes over time. In contrast, as mentioned above, the detection of organic dyes needs strong laser illumination, thus unavoidably determining photo-damage and limiting the observation to brief periods. Importantly, QDs show unique advantageous spectral properties with very broad adsorption and narrow, symmetric emission spectra (the latter depending on the QD characteristics/size). These features make QDs particularly suitable for complex experimental conditions requiring multicolor imaging. An additional advantage of QDs is that they can be easily coupled to biomolecules of interest. Indeed, commercially available QDs can form covalent (or non-covalent) binding to target molecules though surface treatments and functionalization with different reactive groups such as IgGs, biotin, or streptavidin. Functionalization with species-specific IgGs or Fab allows QD binding to primary antibodies targeting the protein to be tracked (Fig. 1). In recent years, streptavidin-functionalized QDs have been extensively used in SPT experiments due to the strong affinity of the biotin–streptavidin interaction that persists during prolonged observation periods. Alternatively, depending on the target molecule and on its engineering with selective tags, functionalization of QDs can include the corresponding binding partner of the tag to achieve a direct and exclusive interaction of QDs with the target molecule. Successful experiments have been performed for instance with scFv, Halotag, polyhistidine, CrAsH, and biotinylated acceptor peptide [19, 29, 32, 36, 63].

As a result of the QD properties and versatility, the QD-based SPT technique has been intensely used over the last decade, providing most of the current knowledge about the receptor diffusion properties. Technological advances allowed further decreasing the QDs size to 10–12 nm [30] to ensure better access to the highly packed synaptic structure. Additional approaches to tag, label, and track surface proteins at single-molecule level have been developed by introducing the α-bungarotoxin binding site (BBS) in the extracellular portion of surface receptors and exploiting the high affinity binding to fluorescent bungarotoxin [23, 24, 59]. In addition to latex beads, SPT techniques based on non-fluorescent probes have been also attempted. For instance, gold particles of ~5 nm have been used to track AMPA receptor in live neurons [38]. In principle, these probes may perform significantly better than QDs, since they are considerably smaller and do not blink or undergo photo-bleaching. Unfortunately, different from latex beads and fluorescent reporters, they can only be detected by photo-thermal imaging, a technique that requires complex experimental setup [7].

3 Role of Receptor Diffusion in Long-Term Synaptic Plasticity

The selective and long-lasting tuning of the synaptic strength induced by external stimuli is believed to underlie important brain functions such as learning and memory. Changes in the postsynaptic signaling have been shown to depend upon both pre- and postsynaptic mechanisms. At the postsynaptic side, the amplitude and

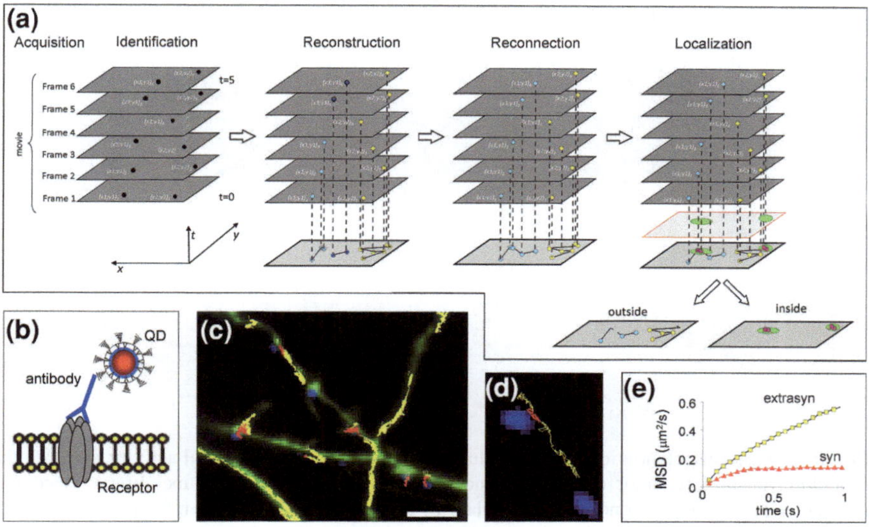

Fig. 1 Experimental approach of the single-particle tracking technique. **a** Schematic representation of the SPT experimental procedure. After the acquisition of sequential images of QD-receptor complexes diffusing in a *x–y* plane, object recognition has to be performed to identify QD and assign them (*x*, *y*) coordinates in each plane. Subsequently, QD-receptor trajectories are reconstructed across adjacent planes with a Vogel's algorithm and then reconnected across QD blinking with a method based on QD maximal allowable displacement during a maximal allowable duration of the dark period. Finally, reconstructed trajectories are distinguished in portions inside and outside compartments of interest (e.g., synapses, EZs, etc.) according to the colocalization of the (*x*, *y*) coordinates of the QD on each plane with those of the compartment. **b** Diagram of surface receptor labeling with a quantum dot (*QD*) through a specific antibody directed against an extracellular epitope of the receptor of interest. **c** Visualization of surface $GABA_A$ receptor diffusion in synaptic and extrasynaptic areas (*red* and *yellow* trajectories, respectively) on the dendrite of a cultured hippocampal neuron (*green*). *Blue spots* represent GABAergic synapses visualized by live labeling of vGAT-Oyster550. Scale bar 10 μm. **d** Up close visualization of the extrasynaptic (*yellow*) and synaptic (*red*) reconstructed trajectory of an individual $GABA_A$ receptor-QD complex, distinguished according to the juxtaposition to the inhibitory presynaptic terminal labeled with vGAT-Oyster550 (*blue*). **e** Mean square displacement curve (*MSD*) of the receptor trajectory reported in (**d**) at synaptic (*red*) and extrasynaptic (*yellow*) areas. The steady state reached by the *red curve* indicates that receptors are confined in synaptic areas, while the linear MSD curve describing the receptor mobility in the extrasynaptic space indicates free Brownian diffusion outside synapses

the kinetics of synaptic currents crucially depend on number of receptors present at the postsynaptic density (PSD) and on the receptor gating properties [41]. Hence, the forms of synaptic potentiation and depression relying on long-lasting changes in the postsynaptic architecture/function are referred to as "postsynaptic plasticity." As demonstrated in the first studies describing the diffusion of receptors at single-molecule level [10, 44], the lateral mobility of receptors at synapses is strongly influenced by the interactions with scaffold proteins that hinder receptor diffusion by acting as "diffusion traps" [15]. Therefore, the modulation of receptor lateral

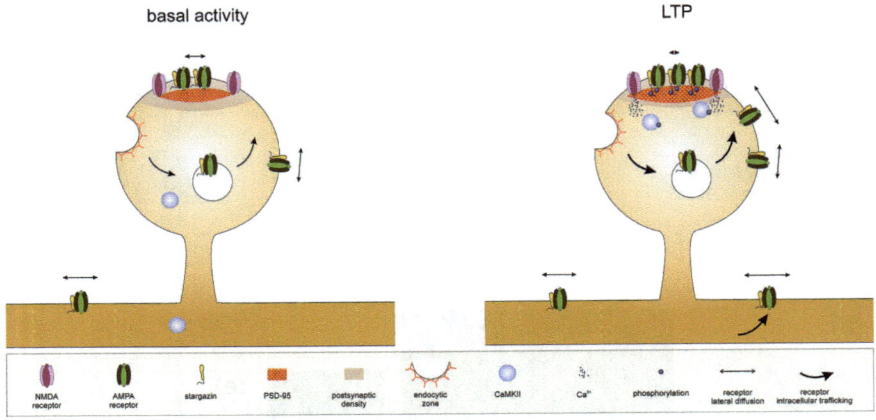

Fig. 2 Long-term potentiation of glutamatergic synapses. *Left* during basal activity, intracellular trafficking ensures receptor delivery to and removal from the neuronal membrane. Receptors internalized at endocytic zones through a clathrin-mediated process can then be recycled. Surface receptors laterally diffuse in the plane of the membrane and exchange between extrasynaptic areas, where they are highly mobile, and synaptic compartments, where they are transiently stabilized by the interaction with the PSD. *Right* during long-term potentiation, AMPA receptor endocytic trafficking is enhanced with promoted exocytosis, leading to a larger number of receptors at the neuronal surface. This pool of activity-dependent exocytosed receptors exhibit higher lateral mobility than preexisting surface receptors. Moreover, upon NMDA receptor activation, $Ca2^{+}$/calmodulin activates and recruits CaMKII to synapses. CaMKII-dependent phosphorylation of the TARP stargazing and of AMPA receptors reinforces the interaction of the AMPAR-stargazing complex to PSD-95, leading to enhanced accumulation and immobilization of AMPA receptors at synapses

diffusion via changes of receptor–scaffold interactions is a fundamental mechanism for setting the number of receptors expressed at postsynaptic sites. At glutamatergic synapses, indeed, several studies have highlighted that LTP is largely dictated by lateral diffusion-mediated dendritic redistribution of AMPA receptors that are likely stabilized at synapses through the increased interaction with scaffold proteins at the glutamatergic PSD [40, 48, 49, 53] (Fig. 2). However, it has not been established yet whether after plasticity induction, the increased AMPA receptor anchoring at synapses occurs either through higher availability of "docking sites" and/or increased receptor affinity scaffold molecules. Bats et al. [6] demonstrated that the binding between stargazing (one of the transmembrane AMPA receptor regulatory proteins, TARPs) and PSD-95 (the main component of glutamatergic density) is crucial for the immobilization of AMPA receptors at synapses. More recently, the demonstration that the interaction between stargazing and PSD95 if favored by the stargazing phosphorylation by CaMKII [48] suggested that, during LTP, the stabilization of AMPAR-stargazing onto preexisting "PSD95 slots" is promoted by CaMKII activity [49]. In addition to the interaction with scaffold proteins, AMPA receptor stabilization at synapses can be regulated also by other processes both in basal activity and during synaptic plasticity. Petrini et al. [53] demonstrated that the presence of endocytic zones (EZs) adjacent to glutamatergic synapses allows a "local receptor recycling" that maintains a pool of mobile surface receptors at synapses. Importantly, this pool of mobile receptors is crucial for receptors increase at synapses during synaptic potentiation. Furthermore, by reversibly trapping AMPA receptors, EZs act as diffusional barriers, limiting the dispersion of receptors from glutamatergic synapses (Fig. 2). The role of EZs in modulating receptor lateral diffusion in basal activity and during glutamatergic plasticity highlights the concept that the redistribution of surface receptors crucially depends on the multiscale dynamic interplay among membrane nanodomains that trap surface receptors and between surface and intracellular compartments with specific hierarchy and affinity [15, 50, 71].

At GABAergic synapses, the role of diffusion on the changes of synaptic strength has been mainly analyzed during inhibitory LTD (iLTD). Sustained neuronal activity, indeed, has been demonstrated to decrease inhibitory synaptic currents due to reduced $GABA_A$ receptor and gephyrin clustering [4]. In the same study, this observation was associated with increased mobility and decreased confinement of $GABA_A$ receptor at synapses. Similarly, [46] reported that activation of glutamatergic synapses (with consequent Ca^{2+} entry through NMDA receptors) led to mobilization and dispersal of $GABA_A$ receptors from GABAergic synapses. Interestingly, these two studies emphasize the role of the phosphatase calcineurin in this form of inhibitory synaptic depotentiation. In particular, Muir et al. [46] found that the lower interaction of $GABA_A$ receptors with the GABAergic PSD and the consequent higher receptor lateral diffusion at inhibitory synapses is due to the dephosphorylation of the residue serine 327 on the $\gamma 2$ subunit of $GABA_A$ receptors, a residue already reported to interfere with $GABA_A$ receptor stability at synapses [72]. It is interesting to point out that increased neuronal activity and the resulting rise of intracellular [Ca^{2+}] immobilize synaptic AMPA receptors [10], while increasing the diffusion of $GABA_A$ receptors at synapses [4, 46]. This opposite effect of neuronal activity on the regulation of AMPA and $GABA_A$R lateral diffusion at synapses may play an important functional role in the coordination of the activity of excitatory and inhibitory systems. Moreover, the evidence that stimuli-inducing potentiation at glutamatergic synapses also triggers concomitant depression at inhibitory synapses [39], along with the aforementioned opposite modulation of synaptic AMPA and $GABA_A$ receptors, suggests that local calcium increase may represent a shared checkpoint to simultaneously orchestrate LTP at glutamatergic synapses and LTD at GABAergic synapses, thus leading to a strong overall unbalance toward excitation.

It is worth mentioning that, in addition to modifications of the number of receptors expressed at the PSD, also changes in the receptor gating properties as well as in the receptor subunit composition represent possible mechanisms underlying postsynaptic plasticity. Indeed, both glutamate and $GABA_A$ receptors can be expressed in various subtypes with specific gating properties ([69, 13]). Consequently, postsynaptic currents mediated by different receptors isoforms will exhibit distinct amplitude, kinetics and calcium permeability, like, for instance, the GluA2 vs GluA1 containing AMPA receptors or $GABA_A$ receptors containing different α1-6 subunits and/or the γ versus δ subunits ([69, 13]). Therefore, significant alterations of the postsynaptic response can occur even in conditions of unchanged number of synaptic receptors through the switch between receptor subtypes. In this context, the stabilization and trapping of diverse receptor subtypes at synapses can be further influenced by the affinity of receptor isoforms for the PSD proteins, following a specific activity-dependent hierarchy.

4 Implications of Receptor Diffusion in Fast Synaptic Signaling

Besides its role in the onset and maintenance of long-term plasticity through the regulation of receptor number at synapses, fast receptor exchange between synaptic and extrasynaptic zones operates a fine tuning of basal synaptic transmission. At conventional central synapses, during repetitive synaptic activation, the postsynaptic response typically decreases in a frequency-dependent manner. At the presynaptic level, indeed, the fatigue of the release machinery depresses the neurotransmitter release, while at the postsynaptic side the accumulation of receptor desensitization may limit receptor activation [74]. After neurotransmitter release, indeed, postsynaptic receptors readily open producing a postsynaptic response. However, following their activation, receptors enter non-conductive (desensitized) state(s) that can persist for hundreds of milliseconds. In this situation, a second event of neurotransmitter release (occurring in tens of millisecond) will generate a smaller response due to the fact that some receptors are non-responsive (Fig. 3). Assuming fast receptor lateral diffusion, after the first release event, desensitized receptors at the synapses would be replaced by extrasynaptic naïve receptors, leading to attenuation of the synaptic response depression [27]. Hence, surface receptor diffusion can modulate the availability of "ready-to-be-activated" receptors at synapses in the millisecond time range. In the current of view of synaptic transmission, the amplitude and duration of synaptic current are determined by the concentration and the release dynamics of neurotransmitter in the synaptic cleft [16]. In addition, lateral diffusion represents a further level of complexity in the modulation of postsynaptic signals at millisecond timescale. Indeed, the level of the steady-state current will be influenced not only by the equilibrium between desensitized and non-desensitized states but also from the rate of exchange between postsynaptic and extrasynaptic receptors. This phenomenon becomes even more pronounced during sustained synaptic activity. In keeping with this, it is interesting to note that the lifetime of AMPA receptors in the desensitized state is compatible with the time needed for the receptor to significantly exchange between synaptic and extrasynaptic space. As a consequence, receptor lateral mobility and receptor gating processes (that determine the receptor intrinsic functional properties) likely cooperate in order to actuate the precise tuning of the postsynaptic current as a function of the frequency of the synaptic

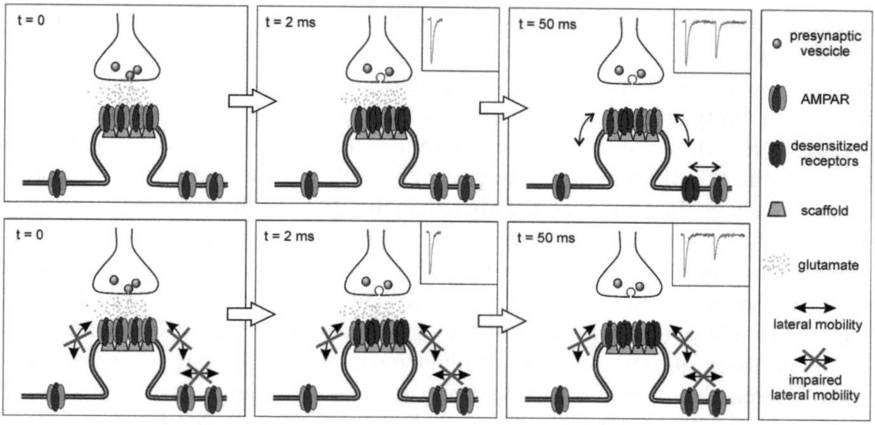

Fig. 3 Functional role of lateral diffusion in the fast modulation of synaptic response. *Upper panels* diagram of fast AMPA receptor exchange at synapses during synaptic transmission. Following glutamate release (*left*), some AMPA receptors undergo desensitization (*middle*) while glutamatergic currents are elicited (*middle inset*). After 50 ms (*right*), lateral mobility allows desensitized receptors to be partially replaced by naïve extrasynaptic receptors, thus limiting the depression of synaptic currents due to the accumulation of desensitized receptors (*right inset*) [27]. *Lower panels* when receptor lateral diffusion is impaired (*left*), the receptors desensitized by a presynaptic release (*middle*) persist at the synapse. As a consequence, a second presynaptic event will elicit a more depressed postsynaptic current (*right inset*) due to the persistence of non-activatable desensitized receptors at synapses (*right*)

activity. In this concern, it could be also hypothesized that receptors showing several kinetically distinct desensitized states (e.g., $GABA_A$ receptors) would "modulate" the synaptic current amplitude over a wide range of receptor diffusion rates and repetitive synaptic activation frequencies [34, 35, 45]. In order to better understand the role of receptor diffusion in the fast modulation of synaptic transmission, it will be important to assess the relation between receptor mobility and receptor gating: does the activated/inactivated receptor state interfere with the ability of the receptor to laterally diffuse in the membrane by altered phospholipid and/or protein interactions and/or conformational states? The assessment of the rules by which mobility and gating are "integrated" will be a fundamental step to understand the role of diffusion in synaptic signaling and neuronal computation.

5 Future Perspectives

One of the clearest demonstrations that receptor lateral diffusion is a highly regulated phenomenon that absolves to precise functional and physiological roles in synaptic transmission is the differential modulation of AMPA and $GABA_A$

receptors mobility at glutamatergic and GABAergic synapses in response to sustained neuronal activity. Indeed, the same change in neuronal activity that immobilizes synaptic AMPA receptors at excitatory synapses also increases $GABA_A$ receptor mobility at inhibitory synapses [4, 53]. Interestingly, this opposite regulation of the mobility of excitatory and inhibitory receptors is associated to glutamatergic LTP and GABAergic iLTD, respectively. Although the role of Ca^{2+}-dependent intracellular signaling has been highlighted in this process, the exact mechanisms of such opposite modulation are still unclear. Further investigations to clarify the mechanisms underlying the coordinated mobility of receptors at both glutamatergic and GABAergic synapses will be crucial to understand the rules of dendritic E/I balance. In addition, it will be a major intellectual challenge to understand how lateral diffusion at synapses can simultaneously operate long-term variations of synaptic strength and short-term fine tuning of synaptic signaling, two processes that coexist at synapses but whose duration differs by several orders of magnitude. Similarly, it will be also important to establish the hierarchy of "diffusive interactions" between diverse membrane nanodomains in relation to their distance and/or spatial organization.

In order to answer these fundamental questions, it will be important to refine our conceptual and technological approach to the study of the multiscale dynamic interaction between proteins. Recent advances in the super-resolution microscopy techniques allowed "counting" molecules at postsynaptic density, thus establishing the precise ratio between receptors and scaffolding proteins [26, 64]. This methodology has to be extended to the manifold players taking places in the transient stabilization and anchoring of receptors at postsynaptic densities [31]. In addition, it will be also important to consider the diffusion and the interactions of the anchoring proteins in relation to their posttranslational modifications. Ultimately, these processes have to be clarified in conditions of synaptic plasticity in which the synapse stability is perturbed to reach a novel level of stability.

The major technological advance that allowed the fine characterization of receptor mobility at the neuronal surface has been the possibility to analyze the behavior of individual receptors. Importantly, the SPT technique revealed that the diffusive properties of receptors show a substantial variability, ranging from immobility to relatively high mobility (up to 1 $\mu m^2/s$). This huge variability suggests that the different lateral diffusion observed among specific receptor fractions are at the base of synapse formation, functioning, and plasticity. However, although the detection of receptor diffusion can be now visualized at super-resolution level [20, 43, 62], the techniques that allow controlling the activity of individual receptors (or very small groups of receptors) are not available yet. The best spatial resolution for the activation/inactivation of ligand-gated receptors, indeed, is diffraction-limited ($\sim$200 nm) being obtained with optical techniques such as laser photo-release of caged compounds (neurotransmitter, antagonists, etc.). Moreover, when using these techniques, the diffusion front generated by the photo-release makes the effective area invaded by the neurotransmitter be even larger than 250–300 nm, thus making impossible to reach the single-protein resolution or to selectively stimulate specific membrane nanodomains. It is also questionable whether diffraction-limited

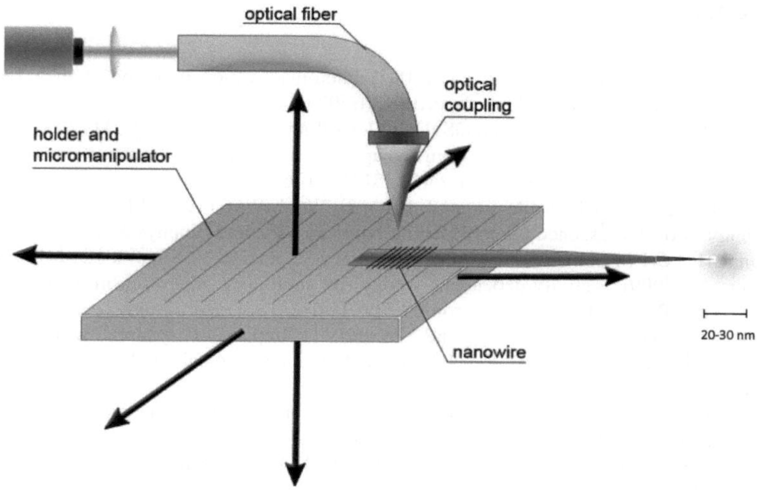

Fig. 4 The spatial confinement of light in the near field. This illumination device is composed by a single nanowire excited from one end by a highly focused laser beam that, through a "backfire coupling," generates a highly localized optical field at the other end of the nanowire. This focused light acts as optical exciting source of light-sensitive receptors expressed at synapses

neurotransmitter uncaging really limits the neurotransmitter activation to individual synapses that range from 100 to 300 nm. In order to overcome these technical limitations, we are developing at IIT a new illumination device capable of focusing light in subdiffraction-limited spots. The methodology for constructing a highly focused beam of light is based on the increment of the localized electric field occurring when a laser beam interacts with a metallic surface with a sharp nanostructure. This phenomenon is at the basis of plasmon polariton technology [54]. The methodological innovation consists in the combination and utilization of nanofabrication techniques to develop structures with spatial control at the nanoscale, such as piezo manipulators or AFM scanning probes. The spatial confinement of light in the near field is comparable to the radius of curvature of tapered nanowires, generating a highly localized beam of light in the order of 20–30 nm (Fig. 4) [18, 22]. By exploiting these tapered nanoprobes, it will be possible, for the first time, to restrict the illumination to individual synapses and/or nanodomains of synapses. In order to be relevant for the investigation of the receptor function at the nanoscale level, this tool has to be used in combination with adequate light-sensitive effectors. For instance, the light-gated ionotropic glutamate receptor (LiGluK2) represents a convenient optogenetic tool that can be effectively switched to either the open/desensitized or closed state by illuminating with 380 nm or >460 nm light, respectively [65]. By studying the diffusion of light-gated ionotropic receptors such as LiGluK2, indeed, it will be possible to explore the receptor diffusion in "controlled conformational states" and to test the functional effects of their transitions between synaptic nanodomains.

References

1. Adesnik H, Nicoll RA, England PM (2005) Photoinactivation of native AMPA receptors reveals their real-time trafficking. Neuron 48(6):977–985
2. Ashby MC, De La Rue SA, Ralph GS, Uney J, Collingridge GL, Henley JM (2004) Removal of AMPA receptors (AMPARs) from synapses is preceded by transient endocytosis of extrasynaptic AMPARs. J Neurosci 24(22):5172–5176
3. Axelrod D, Koppel DE, Schlessinger J, Elson E, Webb WW (1976) Mobility measurement by analysis of fluorescence photobleaching recovery kinetics. Biophys J 16(9):1055–1069
4. Bannai H, Lévi S, Schweizer C, Inoue T, Launey T, Racine V, Sibarita JB, Mikoshiba K, Triller A (2009) Activity-dependent tuning of inhibitory neurotransmission based on $GABA_AR$ diffusion dynamics. Neuron 62(5):670–682
5. Barberis A, Petrini EM, Mozrzymas JW (2011) Impact of synaptic neurotransmitter concentration time course on the kinetics and pharmacological modulation of inhibitory synaptic currents. Front Cell Neurosci 5:6
6. Bats C, Groc L, Choquet D (2007) The interaction between Stargazin and PSD-95 regulates AMPA receptor surface trafficking. Neuron 53(5):719–734
7. Berciaud S, Cognet L, Blab GA, Lounis B (2004) Photothermal heterodyne imaging of individual nonfluorescent nanoclusters and nanocrystals. Phys Rev Lett 93(25):257402
8. Blanpied TA, Scott DB, Ehlers MD (2002) Dynamics and regulation of clathrin coats at specialized endocytic zones of dendrites and spines. Neuron 36(3):435–449
9. Bogdanov Y, Michels G, Armstrong-Gold C, Haydon PG, Lindstrom J, Pangalos M, Moss SJ (2006) Synaptic $GABA_A$ receptors are directly recruited from their extrasynaptic counterparts. EMBO J 25:4381–4389
10. Borgdorff AJ, Choquet D (2002) Regulation of AMPA receptor lateral movements. Nature 417(6889):649–653
11. Bredt DS, Nicoll RA (2003) AMPA receptor trafficking at excitatory synapses. Neuron 40(2):361–379
12. Carroll RC, Beattie EC, von Zastrow M, Malenka RC (2001) Role of AMPA receptor endocytosis in synaptic plasticity. Nat Rev Neurosci 2(5):315–324
13. Cherubini E, Conti F (2001) Generating diversity at GABAergic synapses. Trends Neurosci 24(3):155–162
14. Choquet D, Triller A (2003) The role of receptor diffusion in the organization of the postsynaptic membrane. Nat Rev Neurosci 4(4):251–265
15. Choquet D, Triller A (2013) The dynamic synapse. Neuron 80(3):691–703. doi:10.1016/j.neuron.2013.10.013
16. Clements JD (1996) Transmitter time course in the synaptic cleft: its role in central synaptic function. Trends Neurosci 19:163–171
17. Clements JD, Lester RA, Tong G, Jahr CE, Westbrook GL (1992) The time course of glutamate in the synaptic cleft. Science 258:1498–1501
18. De Angelis F, Liberale C, Coluccio ML, Cojoc G, Di Fabrizio E (2011) Emerging fabrication techniques for 3D nano-structuring in plasmonics and single molecule studies. Nanoscale 3(7):2689–2696
19. Genin E, Carion O, Mahler B, Dubertret B, Arhel N, Charneau P, Doris E, Mioskowski C (2008) CrAsH-quantum dot nanohybrids for smart targeting of proteins. J Am Chem Soc 130(27):8596–8597. doi: 10.1021/ja802987q (Epub 13 Jun 2008)
20. Giannone G, Hosy E, Levet F, Constals A, Schulze K, Sobolevsky AI, Rosconi MP, Gouaux E, Tampé R, Choquet D, Cognet L (2010) Dynamic superresolution imaging of endogenous proteins on living cells at ultra-high density. Biophys J 99(4):1303–1310. doi:10.1016/j.bpj.2010.06.005
21. Giannone G, Hosy E, Sibarita JB, Choquet D, Cognet L (2013) High-content super-resolution imaging of live cell by uPAINT. Methods Mol Biol 950:95–110

22. Giugni A, Torre B, Toma A, Francardi M, Malerba M, Alabastri A, Proietti Zaccaria R, Stockman MI, Di Fabrizio E (2013) Hot-electron nanoscopy using adiabatic compression of surface plasmons. Nat Nanotechnol 8(11):845–852
23. Groc L, Lafourcade M, Heine M, Renner M, Racine V, Sibarita JB, Lounis B, Choquet D, Cognet L (2007) Surface trafficking of neurotransmitter receptor: comparison between single-molecule/quantum dot strategies. J Neurosci 27(46):12433–12437
24. Hannan S, Wilkins ME, Thomas P, Smart TG (2013) Tracking cell surface mobility of GPCRs using α-bungarotoxin-linked fluorophores. Methods Enzymol 521:109–129
25. Harata NC, Aravanis AM, Tsien RW (2006) Kiss-and-run and full-collapse fusion as modes of exo-endocytosis in neurosecretion. J Neurochem 97(6):1546–1570
26. Hastie P, Ulbrich MH, Wang HL, Arant RJ, Lau AG, Zhang Z, Isacoff EY, Chen L (2013) AMPA receptor/TARP stoichiometry visualized by single-molecule subunit counting. Proc Natl Acad Sci USA 110(13):5163–5168. doi:10.1073/pnas.1218765110 (Epub 11 Mar 2013)
27. Heine M, Groc L, Frischknecht R, Beique JC, Lounis B, Rumbaugh G, Huganir RL, Cognet L, Choquet D (2008) Surface mobility of postsynaptic AMPARs tunes synaptic transmission. Science 320:201–205
28. Holcman D, Triller A (2006) Modeling synaptic dynamics driven by receptor lateral diffusion. Biophys J 91(7):2405–2415
29. Howarth M, Takao K, Hayashi Y, Ting AY (2005) Targeting quantum dots to surface proteins in living cells with biotin ligase. Proc Natl Acad Sci USA 102(21):7583–7588 (Epub 16 May 2005)
30. Howarth M, Liu W, Puthenveetil S, Zheng Y, Marshall LF, Schmidt MM, Wittrup KD, Bawendi MG, Ting AY (2008) Monovalent, reduced-size quantum dots for imaging receptors on living cells. Nat Methods 5(5):397–399
31. Hoze N, Nair D, Hosy E, Sieben C, Manley S, Herrmann A, Sibarita JB, Choquet D, Holcman D (2012) Heterogeneity of AMPA receptor trafficking and molecular interactions revealed by superresolution analysis of live cell imaging. Proc Natl Acad Sci USA 109(42):17052–17057. doi: 10.1073/pnas.1204589109 (Epub 3 Oct 2012)
32. Iyer G, Michalet X, Chang YP, Pinaud FF, Matyas SE, Payne G, Weiss S (2008) High affinity scFv-hapten pair as a tool for quantum dot labeling and tracking of single proteins in live cells. Nano Lett 8(12):4618–4623. doi:10.1021/nl8032284
33. Jacob TC, Bogdanov YD, Magnus C, Saliba RS, Kittler JT, Haydon PG, Moss SJ (2005) Gephyrin regulates the cell surface dynamics of synaptic $GABA_A$ receptors. J Neurosci 25:10469–10478
34. Jones MV, Sahara Y, Dzubay JA, Westbrook GL (1995) Desensitized states prolong $GABA_A$ channel responses to brief agonist pulses. Neuron 15(1):181–191
35. Jones MV, Westbrook GL (1998) Defining affinity with the $GABA_A$ receptor. J Neurosci 18(21):8590–8604
36. Kim J, Park HY, Kim J, Ryu J, Kwon do Y, Grailhe R, Song R (2008) Ni-nitrilotriacetic acid-modified quantum dots as a site-specific labeling agent of histidine-tagged proteins in live cells. Chem Commun (Camb) (16):1910–1912. doi: 10.1039/b719434j (Epub 19 Feb 2008)
37. Kusumi A, Nakada C, Ritchie K, Murase K, Suzuki K, Murakoshi H, Kasai RS, Kondo J, Fujiwara T (2005) Paradigm shift of the plasma membrane concept from the two-dimensional continuum fluid to the partitioned fluid: high-speed single-molecule tracking of membrane molecules. Annu Rev Biophys Biomol Struct 34:351–378
38. Lasne D, Blab GA, Berciaud S, Heine M, Groc L, Choquet D, Cognet L, Lounis B (2006) Single nanoparticle photothermal tracking (SNaPT) of 5-nm gold beads in live cells. Biophys J 91(12):4598–4604
39. Lu YM, Mansuy IM, Kandel ER, Roder J (2000) Calcineurin-mediated LTD of GABAergic inhibition underlies the increased excitability of CA1 neurons associated with LTP. Neuron 26(1):197–205
40. Makino H, Malinow R (2009) AMPA receptor incorporation into synapses during LTP: the role of lateral movement and exocytosis. Neuron 64(3):381–390

41. Malinow R, Malenka RC (2002) AMPA receptor trafficking and synaptic plasticity. Annu Rev Neurosci 25:103–126 (Epub 4 Mar 2002)
42. Manley S, Gillette JM, Patterson GH, Shroff H, Hess HF, Betzig E (2008) Lippincott-Schwartz High-density mapping of single-molecule trajectories with photoactivated localization microscopy. J. Nat Methods. 5(2):155–157
43. Manley S, Gillette JM, Lippincott-Schwartz J (2010) Single-particle tracking photoactivated localization microscopy for mapping single-molecule dynamics. Methods Enzymol 475:109–120. doi:10.1016/S0076-6879(10)75005-9
44. Meier J, Vannier C, Sergé A, Triller A, Choquet D (2001) Fast and reversible trapping of surface glycine receptors by gephyrin. Nat Neurosci 4(3):253–260
45. Mott DD, Rojas A, Fisher JL, Dingledine RJ, Benveniste M (2010) Subunit-specific desensitization of heteromeric kainate receptors. J Physiol 588(Pt 4):683–700
46. Muir J, Arancibia-Carcamo IL, MacAskill AF, Smith KR, Griffin LD, Kittler JT (2010) NMDA receptors regulate $GABA_A$ receptor lateral mobility and clustering at inhibitory synapses through serine 327 on the $\gamma 2$ subunit. Proc Natl Acad Sci USA 107(38):16679–16684
47. Nair D, Hosy E, Petersen JD, Constals A, Giannone G, Choquet D, Sibarita JB (2013) Super-resolution imaging reveals that AMPA receptors inside synapses are dynamically organized in nanodomains regulated by PSD95. J Neurosci 33:13204–13224
48. Opazo P, Labrecque S, Tigaret CM, Frouin A, Wiseman PW, De Koninck P, Choquet D (2010) CaMKII triggers the diffusional trapping of surface AMPARs through phosphorylation of stargazin. Neuron 67(2):239–252
49. Opazo P, Sainlos M, Choquet D (2012) Regulation of AMPA receptor surface diffusion by PSD-95 slots. Curr Opin Neurobiol 22(3):453–460
50. Owen DM, Williamson D, Rentero C, Gaus K (2009) Quantitative microscopy: protein dynamics and membrane organisation. Traffic. 10(8):962–971
51. Perestenko PV, Henley JM (2003) Characterization of the intracellular transport of GluR1 and GluR2 alpha-amino-3-hydroxy-5-methyl-4-isoxazole propionic acid receptor subunits in hippocampal neurons. J Biol Chem 278(44):43525–43532 Epub 2003 Aug 8
52. Petralia RS, Wang YX, Wenthold RJ (2003) Internalization at glutamatergic synapses during development. Eur J Neurosci 18(12):3207–3217
53. Petrini EM, Lu J, Cognet L, Lounis B, Ehlers MD, Choquet D (2009) Endocytic trafficking and recycling maintain a pool of mobile surface AMPA receptors required for synaptic potentiation. Neuron 63(1):92–105
54. Raether H (1988) Surface plasmons. Springer, New York
55. Renner M, Choquet D, Triller A (2009) Control of the postsynaptic membrane viscosity. J Neurosci 29(9):2926–2937. doi:10.1523/JNEUROSCI.4445-08.2009
56. Saliba RS, Kretschmannova K, Moss SJ (2012) Activity-dependent phosphorylation of GABAA receptors regulates receptor insertion and tonic current. EMBO J 31(13):2937–2951. doi: 10.1038/emboj.2012.109
57. Saxton MJ, Jacobson K (1997) Single-particle tracking: applications to membrane dynamics. Annu Rev Biophys Biomol Struct 26:373–399
58. Schwille P, Haupts U, Maiti S, Webb WW (1999) Molecular dynamics in living cells observed by fluorescence correlation spectroscopy with one- and two-photon excitation. Biophys J 77(4):2251–2265
59. Sekine-Aizawa Y, Huganir RL (2004) Imaging of receptor trafficking by using alpha-bungarotoxin-binding-site-tagged receptors. Proc Natl Acad Sci USA 101(49):17114–17119 (Epub 24 Nov 2004)
60. Sergé A, Fourgeaud L, Hémar A, Choquet D (2002) Receptor activation and homer differentially control the lateral mobility of metabotropic glutamate receptor 5 in the neuronal membrane. J Neurosci 22(10):3910–3920
61. Shi SH, Hayashi Y, Petralia RS, Zaman SH, Wenthold RJ, Svoboda K, Malinow R (1999) Rapid spine delivery and redistribution of AMPA receptors after synaptic NMDA receptor activation. Science 284(5421):1811–1816

62. Shrivastava AN, Rodriguez PC, Triller A, Renner M (2013) Dynamic micro-organization of P2X7 receptors revealed by PALM based single particle tracking. Front Cell Neurosci 26(7):232. doi:10.3389/fncel.2013.00232
63. So MK, Yao H, Rao J (2008) HaloTag protein-mediated specific labeling of living cells with quantum dots. Biochem Biophys Res Commun 374(3):419–423
64. Specht CG, Izeddin I, Rodriguez PC, El Beheiry M, Rostaing P, Darzacq X, Dahan M, Triller A (2013) Quantitative nanoscopy of inhibitory synapses: counting gephyrin molecules and receptor binding sites. Neuron 79(2):308–321. doi:10.1016/j.neuron.2013.05.013
65. Szobota S, Gorostiza P, Del Bene F, Wyart C, Fortin DL, Kolstad KD, Tulyathan O, Volgraf M, Numano R, Aaron HL, Scott EK, Kramer RH, Flannery J, Baier H, Trauner D, Isacoff EY (2007) Remote control of neuronal activity with a light-gated glutamate receptor. Neuron 54(4):535–545
66. Tardin C, Cognet L, Bats C, Lounis B, Choquet D (2003) Direct imaging of lateral movements of AMPA receptors inside synapses. EMBO J 22(18):4656–4665
67. Thomas P, Mortensen M, Hosie AM, Smart TG (2005) Dynamic mobility of functional $GABA_A$ receptors at inhibitory synapses. Nat Neurosci 8:889–897
68. Tovar KR, Westbrook GL (2002) Mobile NMDA receptors at hippocampal synapses. Neuron 34(2):255–264
69. Traynelis SF, Wollmuth LP, McBain CJ, Menniti FS, Vance KM, Ogden KK, Hansen KB, Yuan H, Myers SJ, Dingledine R (2010) Glutamate receptor ion channels: structure, regulation, and function. Pharmacol Rev 62(3):405–496. doi:10.1124/pr.109.002451
70. Triller A, Choquet D (2005) Surface trafficking of receptors between synaptic and extrasynaptic membranes: and yet they do move! Trends Neurosci 28(3):133–139
71. Triller A, Choquet D (2008) New concepts in synaptic biology derived from single-molecule imaging. Neuron 59(3):359–374
72. Wang Q, Liu L, Pei L, Ju W, Ahmadian G, Lu J, Wang Y, Liu F, Wang YT (2003) Control of synaptic strength, a novel function of Akt. Neuron 38(6):915–928
73. Weiss S (1999) Fluorescence spectroscopy of single biomolecules. Science 283(5408): 1676–1683
74. Zucker RS, Regehr WG (2002) Short-term synaptic plasticity. Annu Rev Physiol 64:355–405

MIX
Papier aus verantwortungsvollen Quellen
Paper from responsible sources
FSC® C105338

If you have any concerns about our products,
you can contact us on
ProductSafety@springernature.com

In case Publisher is established outside the EU,
the EU authorized representative is:
Springer Nature Customer Service Center GmbH
Europaplatz 3, 69115 Heidelberg, Germany

Printed by Libri Plureos GmbH
in Hamburg, Germany